Lecture Notes in Mathematics 1614

Editors:
A. Dold, Heidelberg
F. Takens, Groningen

Springer
Berlin
Heidelberg
New York
Barcelona
Budapest
Hong Kong
London
Milan
Paris
Santa Clara
Singapore
Tokyo

Alexander Koshelev

Regularity Problem for Quasilinear Elliptic and Parabolic Systems

Springer

Author

Alexander Koshelev
Faculty of Mathematics and Mechanics
St. Petersburg University
Bibliotechnaja 2
198904 St. Petersburg, Russia

Cataloging-in-Publication Data.

Die Deutsche Bibliothek - CIP-Einheitsaufnahme

Koŝelev, Aleksandr I.:
Regularity problem for quasilinear elliptic and parabolic systems / Alexander Koshelev. - Berlin ; Heidelberg ; New York ; Barcelona ; Budapest ; Hong Kong ; London ; Milan ; Paris ; Tokyo : Springer, 1995
(Lecture notes in mathematics ; 1614)
ISBN 3-540-60251-8
NE: GT

Mathematics Subject Classification (1991): 35D10, 35K55

ISBN 3-540-60251-8 Springer-Verlag Berlin Heidelberg New York

Typesetting: Camera-ready $\TeX$ output by the author
SPIN: 10479536 46/3142-543210 - Printed on acid-free paper

Contents

In memory of my mother Rebecca.

Preface

This book was written while the author was visiting professor at Heidelberg University, supported by SFB 123 in 1990-1992 and at Stuttgart University Mathematisches Institut A supported by the Deutsche Forschungsgemeinschaft (DFG).

It contains material based on the lectures given to graduate and PhD students at these Universities and the author's most current works concerning regularity of weak solutions for quasilinear elliptic and parabolic systems. Some of the author's earlier results are also presented. The author took this opportunity to include his most recent results concerning the regularity of solutions for some special systems e.g. the Navier-Stokes system. Most of the results are based on coercivity estimates with explicit, sometimes sharp, constants.

I would like give my thanks to Dr. S. Chelkak for all of his help in preparing this book. I also would like to express my gratitude to Prof. W. Jäger and Prof. W. Wendland who made it possible to write and publish this book. The author will never forget the friendly support and fruitful discussions with Prof. S. Hildenbrandt which were a great help. I am thankful to Prof. F. Tomi who encouraged me to present this book to Springer-Verlag for publication. I would like to express my appreciation to Mr. R. Show for taking the time to review my limited English.

St.Petersburg – Stuttgart, 1995 A.Koshelev

Introduction

The smoothness of solutions for quasilinear systems is one of the most important problems in modern mathematical physics. It is impossible to overestimate the significance of these questions not only for theoretical purposes, but for different applications as well. It is clear that the problem of regularity is a part of a more general problem concerning the existence and uniqueness of solutions for systems of partial differential equations.

In this book we concentrate on the second order elliptic and parabolic systems. The last chapter is devoted to the Navier - Stokes system.

The problem of regularity of solutions for elliptic systems was originally formulated by Hilbert as one of his famous problems which at first, was concerned only with the analyticity of solutions for a single general elliptic analytic differential equation of second order with two variables

$$F(x, y, u, u_x, u_y, u_{xx}, u_{xy}, u_{yy}) = 0. \tag{0.0.1}$$

The problem was to prove that any twice continuously differentiable solution must be analytic. If we formulate the same statement in a modern way, it sounds as follows: the ellipticity and analyticity of equation (0.0.1) guarantees that a weak (twice continuously differentiable) solution must be a strong (analytical) one.

The problem was solved by S.Bernstein [1] for at least quasilinear equations. His method was based on the following main principles: 1) maximum principle, which was known earlier for analytic functions, 2) method of continuation of parameter; later this method was generalized by Leray and Schauder [1] and formulated as a homotopy principle, 3) method of apriori estimates. The apriori estimates belong to the so-called class of coercivity.

We can illustrate these estimates for the Poisson equation in a bounded domain Ω

$$\Delta u = f \tag{0.0.2}$$

with the condition

$$u \mid_{\partial\Omega} = 0. \tag{0.0.3}$$

Let X be a space of functions which are defined on Ω and let the derivatives D^2u of these functions also be defined. Suppose that f belongs to X. The estimate

$$\|D^2u\|_X \leq C_X\|f\|_X + C\|u\|_X \tag{0.0.4}$$

where C_X and C are positive constants which do not depend on f and u we shall call a coercivity estimate. S.Bernstein was the first to show that inequality (0.0.4) holds true for $X = \mathcal{L}_2$ and $C_{\mathcal{L}_2} = 1$. Of course, it was supposed by S.Bernstein that f is at least continuous and the second derivatives of the solution of the problem (0.0.2),(0.0.3) are also continuous.

As we shall see later, C_X has a crucial influence on the regularity of solutions for the systems which we consider in this book.

As we have mentioned earlier, S.Bernstein obtained his results for the two-dimensional case. It took about fifty years to find the solution of the Hilbert problem considered for one second-order elliptic equation with an arbitrary number of independent variables.

In 1930, Petrovsky [1] proved that the smooth solution of an analytic elliptic system is also analytic. The paper by Oleinik [1] gives a survey of results relating to the smoothness of solutions for boundary value problems of elliptic equations and systems. The results of Schauder [1], Leray, Schauder [1], De Giorgi [1], Moser [1], Nirenberg [1], Campanato [1] and many other mathematicians are of great importance with regard to this problem. The most complete results concerning this problem can be found in the book by Ladyzhenskaya, Ural'tseva [1].
The situation with parabolic equations of second order is even more complex. Here the breakthrough result was obtained by Nash [1]. Important results and surveys can be found in the monographs of Ladyzhenskaya, Solonnikov,Ural'tseva [1] and Krylov [1]. It is not the purpose of this book to give a description of complete results obtained in the problem of regularity for second order elliptic and parabolic equations. The reader can get a more or less complete picture from the monographs mentioned.
At the present time, the problem of regularity can be formulated as follows. Let a $2l$ order system in m-dimensional space have a weak solution belonging to some Sobolev space W. The question is: what should the additional restrictions on the coefficients of the systems, boundary surface and boundary conditions be in order for the weak solution to become regular, i.e. Hölder continuous, differentiable, etc?
In recent years, a significant number of papers have been devoted to the study of smoothness of solutions for the above mentioned systems.
In 1930-1940, Morrey [1]-[3] obtained the most complete results for the second-order elliptic systems ($l=1$ and $m=2$), concerning the analyticity and differentiability of weak solutions. In Giaquinta [2] regularity for the solutions of some general variational problems, so-called minimizers, is proved.
In his paper [1], Frehse showed that if $u \in W_s^{(l)}$ and $ls=m$, then the weak solution of the problem is bounded. Under the same or slightly different conditions bounding l, s and m, Widman [1]-[2] proved the solution is Hölder continuous. A somewhat more general result was obtained by Solonnikov [1].
In 1968, Almgren [1] proved for second-order systems that if the data of the boundary value problem are smooth, the weak solution can lose its smoothness only on a set with the Hausdorff $(m-1)$-dimensional measure equal to zero. Later these results were extended by Morrey [4] to arbitrary ordered systems. Significant results in this direction obtained by Giusti [1], Giaquinta [1],[2] and other mathematicians formed the so-called theory of partial regularity of solutions.The partial regularity approach for the Navier-Stokes systems was implemented by Caffarelli, Kohn and Nirenberg [1].
The situation for parabolic systems was more uncertain. For example, Campanato [2] proved in the two-dimensional case that under certain natural conditions the weak bounded solution of some class of parabolic systems will be Hölder-continuous, both in space and time. An additional asumption concerning weak solution boundedness in $\mathcal{L}_\infty(Q)$ makes this result different from the analogous theorem proved by Morrey for elliptic systems in the two-dimensional case.
However, Giaquinta and Giusti [1] proved earlier that the weak solution for the parabolic system is regular with the possible exception of a singular closed set. They studied its Hausdorff measure.
Recently in an article by Nečas and Sverak [1] the regularity of solutions was proved for parabolic systems of small dimensions ($m \leq 4$) where the coefficients depended only on the gradient of solutions. The Hölder continuity of the first derivatives was also

obtained for the two-dimensional case with respect to space variables. The situation for systems is considerably different from the case of a single second-order quasilinear elliptic or parabolic equation with natural smoothness conditions on the coefficients and the domain in which the boundary value problem is solved.
In 1968, it was established in a series of examples for the case when the dimension m of the space in which the problem is solved is sufficiently large ($m \geq 3$), that there exists a nonsmooth solution for the smooth boundary value problem (Mazja [1], Giusti and Miranda [1]). This fact holds true even for a linear elliptic system of divergent form (De Giorgi [2]). For parabolic systems a similar result was obtained by Stara, John, Mali [1].
Hence, there arises the question of singling out the more or less precise class of systems for which the weak solution is at least Hölder continuous. Such an approach was applied by Cordes [1] for a singular second-order linear elliptic non-divergent form equation with bounded coefficients. He proved that if the dispersion of the spectrum for the ellipticity matrix is bounded by some explicit constant, then the solution will be Hölder continuous.
We shall also consider systems for which the spectrum of the so-called ellipticity or parabolicity matrix satisfies stronger conditions than positiveness and boundedness. The dispersion of the spectrum should be connected with the rate of asymmetry of the matrix mentioned above. For this class of systems, we prove the weak solution is regular (Hölder continuous, differentiable with Hölder continuous derivatives, ...). We have shown that the conditions which single out this class are sharp for elliptic systems and are at least unavoidable for parabolic systems.
It is well known that the maximum principle is not valid for general elliptic and parabolic systems. Hence, this most fruitful and strong analytical method, which has been applied to scalar second order equations, cannot be used here.
The results obtained in this book are based on two main ideas:
1) The universal iterative method which converges to the solution not only in weak ("energetic") norms, but also in strong ones. 2)Coercivity estimates of the type (0.0.4) in singular weighted Sobolev spaces with explicit, sometimes precise constants C_X.
The sharpness of some results indicates this approach may in some sense provide an optimal method for investigation. However, the author was not able to prove the conditions and estimates obtained lead to precise results in parabolic systems and the Navier-Stokes system. On the other hand, the estimates which are presented in this book allow us to find explicit constants for the norms of some singular integral operators and to prove the Liouville theorem for both elliptic and parabolic systems.
Below, we outline in brief the contents of each chapter.
This monograph consists of six chapters. Chapter 1 is devoted to principal definitions and some results concerning the existence of solutions and convergence of the universal iterative methods in "energetic" spaces.
We consider a bounded domain Ω in the m-dimensional Euclidean space $R^m (m \geq 2)$, whose boundary is a sufficiently smooth closed surface Γ. Inside Ω a system of equations with respect to the vector function $u = (u^{(1)}, \ldots, u^{(N)})$

$$L(u) \equiv \sum_{0 \leq |\beta| \leq l} (-1)^{|\beta|} D^\beta a_\beta(x; D^{\tilde{\beta}} u) = 0 \quad (0 \leq |\tilde{\beta}| \leq l), \tag{0.0.5}$$

is given. Here $\beta = (\beta_1, \ldots, \beta_m)$ and $\tilde{\beta} = (\widetilde{\beta_1} \ldots, \widetilde{\beta}_m)$ are multi-indices,

$D^\beta = D_1^{\beta_1} \dots D_m^{\beta_m}$, D_i is the operator of differentiation with respect to x_i, D^0 is the identity operator and $|\beta| = \beta_1 + \dots + \beta_m$.
With regard to N-dimensional coefficients $a_\beta(x, p_{\tilde{\beta}})$, we assume that certain conditions of smoothness are satisfied. We assume that for any collection of N-dimensional real vectors $\xi_\beta = (\xi_\beta^{(1)}, \dots, \xi_\beta^{(N)})$, any $x \in \overline{\Omega}$, and $p_{\tilde{\beta}}$, the inequalities

$$(0.0.6) \qquad \sum_{i,k=1}^{N} \sum_{0 \le |\beta|, |\tilde{\beta}| \le l} \frac{\partial a_\beta^{(i)}}{\partial p_{\tilde{\beta}}^{(k)}} \xi_\beta^{(i)} \xi_{\tilde{\beta}}^{(k)} \ge \mu_0 (1 + |p|^2)^{\frac{s-2}{2}} \sum_{0 \le |\beta| \le l} |\xi_\beta|^2,$$

$$\|\frac{\partial a_\beta}{\partial p_{\tilde{\beta}}}\| \le \nu_0 (1 + |p|^2)^{\frac{s}{2}}$$

with $s > 1$, $\mu_0, \nu_0 = \text{const} > 0$ and $|p|^2 = \sum_{i=1}^{N} \sum_{0 \le |\beta| \le l} |p_\beta^{(i)}|^2$ hold true. This means the system is strong elliptic. In addition to the latter conditions, certain conditions relating to the behaviour of functions a_β, $\frac{\partial a_\beta}{\partial x}$ are necessary when $p \to \infty$. They provide the existence and uniquieness of the weak solution for the system (0.0.5) with the following boundary conditions

$$(0.0.7) \qquad (u - \varphi)|_\Gamma = \frac{\partial (u - \varphi)}{\partial \nu}|_\Gamma = \dots = \frac{\partial^{l-1} (u - \varphi)}{\partial \nu^{l-1}}|_\Gamma = 0,$$

where ν is the normal to Γ and φ is a trace of some function from $W_p^{(l)}(\Omega)$. The existence of a weak solution of problem (0.0.5), (0.0.7) for any dimension m was first proved in a number of papers (Vishik [1] , Browder [1], Minty [1] ...).
Among these papers, the most important for us is the paper by Vishik in which the existence of a weak solution, belonging to the space of $W_p^{(l+1)}(\Omega')$ (with Ω' essentially contained in Ω) is proved.
Taking into account the numerical applications, the existence of the weak solution can be obtained by the application of the following iterative process

$$(0.0.8) \qquad \sum_{k=0}^{l} (-1)^k \Delta^k u_{n+1} = \sum_{k=0}^{l} (-1)^k \Delta^k u_n - \varepsilon L(u_n) \; (\varepsilon = \text{const} > 0),$$

where the iterations satisfy conditions (0.0.7). This process was proposed by the author (Koshelev [3], [4]).
This process converges for $s = 2$ in the energetic norm under simple natural restrictions beginning with an arbitrary initial iteration $u_0 \in W_2^{(l)}(\Omega)$. Therefore, we call process (0.0.8) a universal one.
For $l = 1, 1 < s \le 2$, when the system (0.0.5) can degenerate, we also consider an iterative process with penalty

$$(0.0.9) \qquad \Delta u_{n+1,\delta} = \Delta u_{n,\delta} - \varepsilon L_\delta(u_{n,\delta}), \quad \delta = const > 0,$$

where $L_\delta = L + \delta\Delta$. In the author's paper (Koshelev [4]), it is proved that there exists a subsequence δ_n such that if $n \to +\infty$ then u_{n,δ_n} tends to the solution of (0.0.5),(0.0.7) in the energetic norm. Further, this allows us to obtain the regularity of solutions for the problem under consideration from the boundedness of strong norms for u_{n,δ_n}.
In chapter 1 we also discuss the parabolic system

$$(0.0.10) \qquad \partial_t u - L(u) = 0$$

in a cylinder $Q = (0,T) \times \Omega$ with boundary conditions (0.0.10) and initial conditions $u|_{t=0} = 0$. (The coefficients of $L(u)$ can also depend on t). Under general assumptions, an iterative method

$$\varepsilon \partial_t u_{n+1} - \Delta u_{n+1} = - \Delta u_n + \varepsilon L(u_n) \tag{0.0.11}$$

with the above-mentioned boundary and initial conditions converges. We provide the proof of convergence of this method in the energetic norm. Initially, it was done by Chistyakov[1].

The second chapter is devoted mainly to Hölder continuity of weak solutions for nondegenerate second-order elliptic systems with bounded nonlinearities in divergence form where A is the matrix of the left-hand side quadratic form in (0.0.5), i.e. the matrix

$$A = \left\{ \frac{\partial a_k^{(i)}}{\partial p_l^{(j)}} \right\} \; (i,j = 1, ..., N; k,l = 0, ..., m). \tag{0.0.12}$$

We assume A^+ and A^- are respectively the symmetric and skew-symmetric parts of A. Denote the eigenvalues of A^+ by λ_i and the infimum and supremum of λ_i respectively by λ and Λ. We also denote the upper boundary of the eigenvalues of the matrix $C = A^+A^- - A^-A^+ - (A^-)^2$ by σ and suppose that $\lambda > 0$ and $\Lambda < \infty$. Let

$$K^2 = \begin{cases} \sigma(\lambda^2 + \sigma)^{-1}, & \sigma \geq \frac{\lambda(\Lambda-\lambda)}{2}, \\ \frac{(\Lambda-\lambda)^2+4\sigma}{(\Lambda+\lambda)^2}, & \sigma \leq, \frac{\lambda(\Lambda-\lambda)}{2}, \end{cases} \tag{0.0.13}$$

$\alpha = 2 - m - 2\gamma (0 < \gamma < 1)$ and

$$1 - \frac{\alpha(\alpha + m - 2)}{2(m-1)} > 0. \tag{0.0.14}$$

In chapter II, we prove that if some natural conditions concerning the coefficients $a_k(x,p)$, domain Ω and boundary conditions are satisfied, and

$$K \sqrt{1 - \frac{\alpha(m-2)}{m-1}} \left[1 - \frac{\alpha(\alpha + m - 2)}{2(m-1)} \right]^{-1} < 1 \tag{0.0.15}$$

is true, then the weak solution of (0.0.5), (0.0.7) ($l = 1$) is Hölder continuous with exponent

$$\gamma = \frac{2 - m - \alpha}{2}. \tag{0.0.16}$$

This was proved by the author in [11] and [15]. The above mentioned result follows from the coercivity estimate (0.0.4) for $X = \mathcal{L}_{2,\alpha}(B_\delta)$ where $\mathcal{L}_{2,\alpha}(B_\delta)$ is the space of square summable functions with the weight $|x - x_0|^\alpha$ and B_δ is a ball in R^m with the center x_0 and radius δ. More precisely, the solution of the problem

$$\Delta u = div f, \quad u|_{\partial B_\delta} = 0 \tag{0.0.17}$$

satisfies the inequality

$$(0.0.18)\quad \int_{B'} |\nabla u|^2 |x-x_0|^\alpha dx \le \left[1-\frac{\alpha(m-2)}{m-1}+\eta\right]\left[1-\frac{\alpha(\alpha+m-2)}{2(m-1)}\right]^{-2} \times$$
$$\times \int_{B_\delta} |f|^2|x-x_0|^\alpha dx + C\int_{B_\delta} |f|^2 dx$$

where η is an arbitrary small positive constant (C is as usualy an unessential nonnegative constant). From this inequality follows the analogous estimate for the singular operator

$$(0.0.19)\quad J(f)=\frac{\partial}{\partial x_k}\frac{1}{(m-2)|S|}\int_{R^m} f^{(k)}(y)|y-x|^{2-m}dy,$$

where $|S|$ is the surface of a unit sphere. More precisely

$$(0.0.20)\quad J(f)\le \left[1-\frac{\alpha(m-2)}{m-1}+\eta\right]\left[1-\frac{\alpha(\alpha+m-2)}{2(m-1)}\right]^{-2} \times$$
$$\times \int_{B_\delta} |f|^2|x-x_0|^\alpha dx + C\int_{B_\delta} |f|^2 dx.$$

This and analogous estimates can be obtained using the Stein's result [1], but his method does not provide us with an explicit constant in either (0.0.18) or (0.0.20). It should also be mentioned that the estimate (0.0.15) is sharp for a small γ. In fact, for a symmetric A ($A^- = 0$) we have

$$K=\frac{\Lambda-\lambda}{\Lambda+\lambda}$$

and the condition (0.0.15) gives

$$(0.0.21)\quad \frac{\Lambda-\lambda}{\Lambda+\lambda}\sqrt{1+\frac{(m-2)^2}{m-1}}<1.$$

The example given in 2.5 shows this condition to be sharp, which means that if (0.0.21) is false, then there exists such a system for which the weak solution is discontinuous. Chapter III is devoted to the applications of the results obtained in chapters I and II. Here we consider some elasto-plastic problems for media with hardening. It is proved that the method of elastic solutions converges both in energetic and strong norms if the anisotropy of the material is small enough. This is guaranteed by an inequality of type (0.0.15). It is shown that if this condition is false, there exists a system, whose solution has a finite energy and a discontinious displacement. Analytically, the result is based on the so-called Korn inequality for weighted spaces

$$(0.0.22)\quad \sum_{i,k=1}^{m}\int_{\Omega}\left[D_k u^{(i)}+D_i u^{(k)}\right]^2 r^\alpha dx \ge \min\left\{2\frac{(\alpha+m)^2}{(\alpha+m)^2-4\alpha},\frac{2m+\alpha}{m}\right\}\times$$
$$\times\sum_{k=1}^{m}\int_{\Omega}|\nabla u^{(k)}|^2 r^\alpha dx - C\int_{\Omega}\Big(\sum_{k=1}^{m}|\nabla u^{(k)}|^2+|u|^2\Big)dx.$$

The main results of this Chapter are published in the author's monograph [15] and in [13],[14]. Chapter III also contains an exact form of the Liouville theorem, which was proved by the author in [15] (Chapter 4).
In chapter IV, we consider additional regularity properties of a weak solution for second-order elliptic systems, for example, Hölder continuity of the first derivatives up to the boundary of the domain.
The results , which can be found in 4.5 and 4.6, are based on the explicit constants for coercivity estimates of the type (0.0.4) with $C = 0$ in $\mathcal{L}_{2,\alpha}$, $\alpha = 2 - m - \gamma$ $(0 < \gamma < 1)$. For $m \geq 2$ and $B_R(x_0)$ with boundary condition $u|_{\partial B_R} = 0$ the estimate is as follows:

$$\int_{B_R} |D^2u|^2|x - x_0|^\alpha dx \leq C_\alpha \int_{B_R} |\triangle u|^2|x - x_0|^\alpha dx \tag{0.0.23}$$

and C_α is given by formula (4.2.35).
In spite of the explicit form of C_α , which is given in 4.2, it is impossible to apply (0.0.23) to particular cases. Therefore, we have the following additional inequalities

$$\int_{B_R} |D'^2u|^2 r^\alpha \zeta dx \leq (1 + M_\gamma^2 + \eta) \int_{B_R} |\triangle u|^2 r^\alpha \zeta dx + \tag{0.0.24}$$

$$+C\left\{\left(\int_{B_R} |D'^2u|^2 r^\alpha \zeta dx\right)^{\frac{m}{m+2\gamma}} \left(\int_{B_R} |Du|^2 dx\right)^{\frac{2\gamma}{m+2\gamma}} + \int_{B_R} (|Du|^2 + |u|^2) dx\right\},$$

where

$$M_\gamma^2 = \frac{(m - 2 + 2\gamma)\{(1 + \gamma)^2 + [2 - (1 - \gamma)^2]m\}}{(m + 1 + \gamma)^2(1 - \gamma)^2} \tag{0.0.25}$$

and ζ is a smooth cut-off function.
The estimate (0.0.24) is based on the multiplicative inequality

$$|u(0)|^2 \leq C \left(\int_{B_R} |D'u|^2 r^\alpha dx\right)^{\frac{m}{m+2\gamma}} \left(\int_{B_R} |u|^2 dx\right)^{\frac{2\gamma}{m+2\gamma}}$$

and can be applied to obtain the regularity of weak solutions. In section 4.6 of chapter IV, we obtain the following statement: let the eigenvalues λ_i of the symmetrix matrix

$$A = \left\{\frac{\partial a_i}{\partial p_j}\right\} \quad (i, j = 0, ..., m)$$

satisfy the following inequalities

$$\frac{\lambda}{1 + |p|^s} \leq \lambda_j \leq \frac{\Lambda}{1 + |p|^s}$$

with $\Lambda, \lambda = \text{const} > 0$ and $0 \leq s \leq 1$; if the relation

$$\frac{\left(1 + \frac{m-2}{m+1}\right)[1 + (m - 2)(m - 1)]}{\left(1 + \frac{m-2}{m+1}\right)[1 + (m - 2)(m - 1)] - 1} \frac{\lambda}{\Lambda} > 1$$

holds, then under some conditions of smallness, the weak solution of system (0.0.5) with condition (0.0.3) satisfies the Hölder condition. These results were published by the author in [18].

Chapter V is devoted to second-order parabolic systems which we consider in cylinder $Q = (0,T) \times \Omega$ with a finite $T > 0$. The results which are proved there are based primarily on two lemmas for a function $w(t,x)$ satisfying the parabolic equation

$$\varepsilon \partial_t w + \Delta w = f \tag{0.0.26}$$

in $Q_R = (0,T) \times B_R$ with boundary conditions

$$w|_{\partial B_R} = w|_{t=T} = 0. \tag{0.0.27}$$

<u>Lemma 5.2.1</u> Let β' be an arbitrary number, $\beta + m - 4 > 0$ and $0 \leq \beta < m$. Then the weak solution of the problem (0.0.26), (0.0.27) satisfies the following estimate

$$\int_{Q_R} |\Delta w|^2 r^\beta \zeta dx dt \leq \frac{m}{m-\beta} \int_{Q_R} |f|^2 r^\beta \zeta dx dt + C(R) \int_{Q_R} |D^2 w|^2 r^{\beta'} dx dt. \tag{0.0.28}$$

<u>Lemma 5.2.2</u> If $0 \leq \beta < m$ and $\beta + m - 4 > 0$, then for the weak solution of problem (0.0.26), (0.0.27) the inequality

$$\int_{Q_R} |\Delta w|^2 r^\beta dx dt \leq \frac{m}{m-\beta} \int_{Q_R} |f|^2 r^\beta dx dt$$

holds.

In chapter V, we prove the Hölder continuity of the weak solution for both t and x for the parabolic system (0.0.10) under additional assumptions concerning the differentiability of the system's coefficients with respect to x. It is also assumed that the boundary conditions

$$u|_{\partial\Omega} = u|_{t=0} = 0 \tag{0.0.29}$$

and the inequality analogous to (0.0.15) for $m \geq 3$

$$\frac{m}{2(1-\gamma)} \left[1 - \frac{(2-m-2\gamma)(m-1)}{(1-\gamma)^2} \right] M_\gamma K < 1 \tag{0.0.30}$$

are satisfied.

If we take a small γ, then for $m \geq 3$ the inequality (0.0.30) has the form

$$\frac{m}{2} [1 + (m-2)(m-1)] \left(1 + \frac{m-2}{m+1} \right) K < 1 \tag{0.0.31}$$

So for $m \geq 3$, the condition (0.0.31) guarantees the Hölder continuity of weak solutions both in x and in t for the problem (0.0.10), (0.0.29). It is also proved here for $m = 2$, the condition (0.0.31) takes the form

$$\sqrt{2} K < 1. \tag{0.0.32}$$

This inequality was obtained with the help of S. Chelkak. The last relations were proved by the author in [19].

As we see in contrast to the elliptic case, under conditions (0.0.31) and (0.0.32) parabolicity does not guarantee the Hölder continuity of a weak solution in the parabolic case.

In chapter V, some coercivity inequalities are proved. We shall mention here only one inequality of this type.

Theorem 5.3.2 Suppose that $u \in L_2\{0,T;W^{(2)}_{2,\alpha}(B_R)\}$ satisfies only the second of the conditions (0.0.29), ($u=0$ when $t=0$). Then the estimate

$$(0.0.33)\quad \int_{Q_R} |D'^2u|^2 r^\alpha \zeta dx dt \le \left(1+M_\gamma^2+\eta\right) A^2_{\alpha,m} \int_{Q_R} |\varepsilon \partial_t u - \Delta u|^2 r^\alpha \zeta dx dt +$$

$$+C\left\{\left(\int_{Q_R} |D'^2u|^2 r^\alpha \zeta dx dt\right)^{\frac{m}{m+2\gamma}} \left(\int_{Q_R} |Du|^2 dx dt\right)^{\frac{2\gamma}{m+2\gamma}} + \int_{Q_R} |Du|^2 dx dt\right\}$$

holds, where C does not depend on $\varepsilon > 0$, $\alpha = 2-m-2\gamma$ $(0<\gamma<1)$ and

$$(0.0.34)\qquad A^2_{\alpha,m} = \begin{cases} 1-\frac{\alpha}{2+\alpha}-\frac{2+\alpha}{(1+\alpha/4)^2\alpha}, & m=2, \\ \frac{m}{(m+\alpha)}, & m>2. \end{cases}$$

In chapter V, we provide applications for problems related to the 'blow-up' problem for some coupled systems. The Liouville theorem for parabolic systems is also proved in this chapter.

Chapter VI, the last chapter in the book, is devoted to Stokes and the Navier-Stokes system in a bounded domain Ω. In this chapter, we consider mainly the problem of the existence of strong solutions for the nonstationary Navier-Stokes system. From the results of Ladyzhenskaya [1] and Solonnikov [3] it follows that for small Reynolds numbers and some smoothness assumptions concerning the boundary of the domain and the massive forces, there exists for the first boundary problem a continuous regular (for example, Hölder continuous, etc.) solution. In these results the constants, which estimate the solution pointwise, have an implicit form. With the help of the coercivity inequalities with explicit constants, we obtain some explicit estimates of the strong solution for finite time.

The first two sections contain coercivity estimates for both stationary and nonstationary Stokes systems. We shall now give two examples of the inequalities which are proved in chapter VI . Some of these results were obtained with the help of A.Wagner and published in the paper of Chelkak and Koshelev[1].

We begin by considering the stationary Stokes system

$$(0.0.35)\qquad \begin{cases} \Delta u + \nabla p = f, \\ \operatorname{div} u = 0 \end{cases}$$

with condition (0.0.3). Suppose $\int_\Omega p dx = 0$ and x_0 is an arbitrary point of Ω with $\operatorname{dist}(x_0, \partial\Omega) > 0$ (R_0 =const and $R < R_0$). If $u \in L_2\{\delta, T; W^{(1)}_{2,\alpha}(B_R)\}$ then the following estimates for weak solution u, p

$$(0.0.36)\quad \int_{B_R(x_0)} |\nabla p|^2|x-x_0|^\alpha dx \le \left[1+\frac{(m-2)^2}{m-1}+O(\gamma)\right] \int_{B_R(x_0)} |f|^2|x-x_0|^\alpha dx + +C\int_{B_R(x_0)} |p|^2 dx$$

and

$$(0.0.37)\quad \int_{B_R(x_0)} |D'^2u|^2|x-x_0|^\alpha \zeta dx \le \left\{1+\left[1+\frac{(m-2)^2}{m-1}\right]^{1/2}\right\}^2 \times$$

$$\times\left[1+\frac{(m-2)}{m-1}+O(\gamma)\right] \int_{B_R(x_0)} |f|^2|x-x_0|^\alpha dx +$$

$$+C\left[\left(\int_\Omega |D'^2u|^2|x-x_0|^\alpha\zeta dx\right)^{\frac{m}{m+2\gamma}}\left(\int_\Omega |D'u|^2dx\right)^{\frac{2\gamma}{m+2\gamma}}+\int_\Omega (|D'u|^2+|f|^2)dx\right]$$

are true and C is independent of x_0 (theorem 6.1.1).
Further on, we consider the nonstationary Stokes system

$$(0.0.38)\quad \begin{cases} \partial_t u - \nu\,\Delta\, u + \nabla p = f \\ \mathrm{div} u = 0 \end{cases}$$

with conditions (0.0.29). We prove if $f \in L_2\{0,T; L_{2,\alpha}(\Omega)\}$ with $\alpha = 2-m-2\gamma$ $(0<\gamma<1)$ satisfying (2.3.20), then the solution of the system (0.0.36) with the boundary condition (0.0.27) satisfies the estimates

$$(0.0.39)\quad \int_{Q_R} |\nabla p|^2 r^\alpha \zeta dxdt \le (N_\gamma^2+\eta)\int_{Q_R} |f|^2 r^\alpha\zeta dxdt +$$

$$+C\left[\left(\int_{Q_R} |\nabla p|^2 r^\alpha\zeta dxdt\right)^{\frac{m}{m+2\gamma}}\left(\int_{Q_R} |f|^2 dxdt\right)^{\frac{2\gamma}{m+2\gamma}}+\int_{Q_R} |f|^2 dxdt\right]$$

and

$$(0.0.40)\quad \int_{Q_R} |D'^2u|^2 r^\alpha\zeta dxdt \le$$

$$\le \frac{2A^2_{\alpha,m}}{\nu^2}(1+M_\gamma^2+\eta)(1+N_\gamma)^2\int_{Q_R} |f|^2 r^\alpha\zeta dxdt +$$

$$+C\left(\int_{Q_R} |D'^2u|^2 r^\alpha\zeta dxdt\right)^{\frac{m}{m+2\gamma}}\left(\int_{Q_R} |Du|^2 dxdt\right)^{\frac{2\gamma}{m+2\gamma}} +$$

$$+\left[\left(\int_{Q_R} |\nabla p|^2 r^\alpha\zeta dxdt\right)^{\frac{m}{m+2\gamma}}\left(\int_{Q_R} |f|^2 dxdt\right)^{\frac{2\gamma}{m+2\gamma}}+\int_{Q_R} |f|^2 dxdt\right],$$

where

$$(0.0.41) \qquad N_\gamma^2 = \max\{[1 - \frac{\alpha(m-2)}{m-1}][1 - \frac{\alpha(\alpha+m-2)}{2(m-1)}]^{-2}, 2m^2(1+M_\gamma^2)\}$$

and $1 - 2^{-1}(\alpha+m-2)\alpha(m-1)^{-1} > 0$. The third and the fourth sections are devoted to the Navier-Stokes system

$$\partial_t u - \nu \Delta u + u^{(k)} D_k u + \nabla p = f(x,t),$$
$$\text{div} u = 0$$

with homogeneous boundary conditions (0.0.29) in Q_R. It is proved that if $\alpha = 2 - m - 2\gamma$ $(0 < \gamma < 1)$ and the Reynolds number

$$\nu^{-1}(\sup_{x_0} \int_Q |f|^2 |x - x_0|^\alpha dx dt)^{1/2}$$

is sufficiently small, then there exists a Hölder continuous solution in both t and x of the problem, satisfying the estimate

$$(0.0.42) \qquad \sup_{x_0 \in \Omega} \left[\int_{Q_R} (|\partial_t u|^2 + |D'^2 u|^2) r^\alpha dx dt \right]^{1/2} \leq$$

$$\leq \frac{2}{\nu^2} A_{\alpha,m} \left((1 + M_\gamma^2 + \eta)(1 + N_\gamma) \right)^{1/2} \sup_{x_0 \in \Omega} \left[\int_{Q_R} |f|^2 r^\alpha dx dt + o(\frac{1}{\nu^2}) \right],$$

where $A_{\alpha,m}$, M_γ and N_γ are defined by (0.0.24), (0.0.34) and (0.0.41). It follows from (0.0.42) that $\max|u(x,t)|$ is finite.

List of Notation

C: unessential nonnegative constant.
η : sufficiently small positive constant.
R^m: m-dimensional Euclidean space.
$R^m_+(R^m_-) = R^m \cap (x_m > 0)((x_m < 0))$.
$x(x_1, \ldots, x_m)$-vector in R^m with components x_i.
xy-scalar product in R^m.
$|x|$: length of x.
$B_\delta(x_0)$ ball in R^m with center x_0 and radius δ.
$B = B_1(0)$.
$B^+_\delta(x_0)(B^-_\delta(x_0)) = B_\delta(x_0) \cap (x_m > 0)((x_m < 0))$.
$B_{\delta_1,\delta_2}(x_0) = B_{\delta_1}(x_0) \backslash B_{\delta_2}(x_0)(\delta_2 < \delta_1)$.
$B^+_{\delta_1,\delta_2}(x_0)(B^-_{\delta_1,\delta_2}(x_0)) = B_{\delta_1,\delta_2} \cap (x_m > 0)((x_m < 0))$.
S: unit sphere in R^m.
Ω: bounded domain in R^m.
$\Omega_\delta(x_0) = \Omega \cap B_\delta(x_0)$.
$u(x) = \{u^{(1)}(x), \ldots, u^{(N)}(x)\}$: vector-function with N scalar functional components $u^{(j)}(x)$ defined on Ω.
$D_i = \frac{\partial}{\partial x_i}(i = 1, \ldots, m)$.
$D^\alpha = \prod_{k=1}^m D_k^{\alpha_k}(\alpha = (\alpha_1, \ldots, \alpha_m)$: multi-index.
$D_0 = D^0 = I$: unit operator.
$D^\ell u D^\ell v = \sum_{0 \le |\alpha| \le \ell} D^\alpha u D^\alpha v(|\alpha| = \sum_{i=1} \alpha_i)$.

$$|D^\ell u|^2 = D^\ell u D^\ell u.$$

$$DuDv = D^1 u D^1 v.$$

$$D'^\ell u D'^\ell v = \sum_{|\alpha| = \ell} D^\alpha u D^\alpha v.$$

$$|D'^\ell u|^2 = D'^\ell u D'^\ell u.$$

$$D'uD'v = D'^1 u D'^1 v.$$

$W_p^{(\ell)}(\Omega)$: Sobolev space of vector functions, defined on Ω with all weak p-summable derivatives up to the order ℓ; the norm is defined by the equality

$$\|u\|_{W_p^{(\ell)}(\Omega)} = (\int_\Omega |D^\ell u|^p dx)^{\frac{1}{p}}.$$

Analogously

$$\|u\|_{W_p^{(\ell)}(\Omega)} = (\int_\Omega |D'^\ell u|^p dx)^{\frac{1}{p}}$$

and

$$\|u\|'_{W_p^{(\ell)}(\Omega; x_0)} = (\int_\Omega |D'^\ell u|^p dx)^{\frac{1}{p}}$$

(with sufficient homogeneous conditions).
$H_{p,\ell,\alpha}(\Omega) \subset W^{(\ell)}_{p,\alpha}(\Omega; x_0)$ for $\forall x_0 \in \Omega$ with finite norm

$$\|u\|_{H_{p,\ell,\alpha}(\Omega)} = \sup_{x_0 \in \Omega} \|u\|_{W^{(\ell)}_{p,\alpha}(\Omega; x_0)}$$

or

$$\|u\|'_{H_{p,\ell,\alpha}(\Omega)} = \sup_{x_0 \in \Omega} \|u\|_{W^{(\ell)\prime}_{p,\alpha}(\Omega, x_0)}.$$

$$H_{2,\ell,\alpha} = H_{\ell,\alpha}.$$

$$H_{2,1,\alpha} = H_{\alpha}.$$

$C^{k,\alpha}(\Omega)$: space of functions defined on Ω with all derivatives of order k satisfying Hölder condition with the exponent α.

$$C^{0,\alpha}(\Omega) \equiv C^{\alpha}(\Omega)$$

$$C^{0,0}(\Omega) \equiv C(\Omega).$$

$Q = (0, T) \times \Omega$: cylinder with $\forall T = const > 0$.

$$(t, x) \in Q.$$

$$Q_\delta(x_0) = Q \cap ((0, T) \times B_\delta(x_0)).$$

$u(t; x)$: functions defined on Q.
$\partial_t u = \dot{u}$: derivative with respect to t.
$W^{k,\ell}_p(Q)$: Sobolev space of functions, defined on Q, possessing all derivatives up to the order k with respect to t and up to the order ℓ with respect to x, which are p-summable; the norm is defined by the equality

$$\|u\|_{W^{k,\ell}_p(Q)} = \left(\int_Q (|\partial^k_t u|^p + |D^\ell u|^p)dx\right)^{\frac{1}{p}}.$$

Analogously to $W^{(\ell)}_{p,\alpha}(\Omega; x_0)$ and $H_{\ell,p,\alpha}(\Omega)$ are defined the spaces $W^{k,\ell}_p(Q; x_0)$ with the norm

$$\|u\|_{W^{k,\ell}_{p,\alpha}(Q;x_0)} = \left(\int_Q (|\partial^k_t u|^p + |D^\ell u|^p)|x - x_0|^\alpha dx\right)^{\frac{1}{p}}$$

and

$$\|u\|_{H^{k,\ell}_{p,\alpha}(Q)} = sup_{x_0 \in \Omega} \|u\|_{W^{k,\ell}_{p,\alpha}(Q;x_0)}.$$

A: matrix of ellipticity (parabolicity).
A^+: symmetric part of A.
A^-: skew-symmetric part of A.
matrix $C = A^+A^- - A^-A^+ - (A^-)^2$.
$\{\lambda_i\}$: eigenvalues of A^+.

$\{\sigma_i\}$: eigenvalues of C.
$\lambda = inf \lambda_i$, $\Lambda = sup \lambda_i$.
$\sigma = sup \sigma_i$.

$$K = \begin{cases} \sigma(\sigma + \lambda^2)^{-1} & \text{for } \sigma \geq \lambda(\Lambda - \lambda)/2 \\ [(\Lambda - \lambda)^2 + 4\sigma](\Lambda + \lambda)^{-2} & \text{for } \sigma \leq \lambda(\Lambda - \lambda)/2 \end{cases} .$$

Chapter 1

Weak solutions and the universal iterative process

1.1 Quasilinear elliptic and parabolic systems and systems with bounded nonlinearities

Let Ω be a finite domain in $R^m (m \geq 2)$ with a sufficiently smooth boundary $\partial\Omega$. It means that the boundary $\partial\Omega$ can be locally represented by a continuous, Hölder continuous, k-times differentiable functions, etc. On these domains we shall consider a class of N-dimensional vector functions and agree that these functions belong to some functional space if all its components belong to this space. Let $u(x) = \left\{u^{(1)}(x), \ldots, u^{(N)}(x)\right\}$ be such a function. Consider a system

(1.1.1) $$L(u) \equiv \sum_{0\leq|\beta|\leq l} (-1)^{|\beta|} D^\beta a_\beta(x; D^{\tilde\beta} u) = 0 \quad (0 \leq |\tilde\beta| \leq l),$$

where $\beta = (\beta_1, \ldots, \beta_m)$ and $\tilde\beta = (\tilde\beta_1, \ldots, \tilde\beta_m)$ are multiindices, $D^\beta = D_1^{\beta_1} \ldots D_m^{\beta_m}$, $D_i (i = 1, \ldots, m)$ is the operator of differentiation with respect to x_i , D^0 is the identity operator and $|\beta| = \beta_1 + \ldots + \beta_m$.
Concerning the N-dimensional coefficients $a_\beta(x, p_{\tilde\beta}) = \left\{a_\beta^{(1)}(x, p_{\tilde\beta}), \ldots, a_\beta^{(N)}(x, p_{\tilde\beta})\right\}$ we assume they satisfy the following conditions: **1)** Caratheodory condition, i.e. they are measurable with respect to x for all finite $p_{\tilde\beta}$ and continuously differentiable with respect to $p_{\tilde\beta}$ for almost all $x \in \Omega$; **2)** For any collection of N-dimensional real vectors $\xi_\beta = \left\{\xi_\beta^{(1)}, \ldots, \xi_\beta^{(N)}\right\}$, any $x \in \overline\Omega$ and $p_{\tilde\beta}$, the inequalities

(1.1.2) $$\nu_0 \left(1 + |p|^2\right)^{(s-2)/2} \sum_{0\leq|\beta|\leq l} |\xi_\beta|^2 \geq \sum_{i,k=1}^N \sum_{0\leq|\beta|,|\tilde\beta|\leq l} \frac{\partial a_\beta^{(i)}}{\partial p_{\tilde\beta}^{(k)}} \xi_\beta^{(i)} \xi_{\tilde\beta}^{(k)} \geq$$
$$\geq \mu_0 \left(1 + |p|^2\right)^{(s-2)/2} \sum_{0\leq|\beta|\leq l} |\xi_\beta|^2,$$

(1.1.3) $$\left|\frac{\partial a_\beta^{(i)}}{\partial p_{\tilde\beta}^{(k)}}\right| \leq \nu_0 \left(1 + |p|^2\right)^{(s-2)/2}$$

hold, where $1 < s \leq 2$ and $\mu_0, \nu_0 = \text{const} > 0$ and $|p|^2 = \sum_{i=1}^{N} \sum_{0\leq|\beta|\leq l} |p_\beta^{(1)}|^2$.

3) for some $q > 1$ and any $u \in W_q^{(l)}(\Omega')$ with the coefficients $a_\beta\left(x; D^{\tilde{\beta}}u\right) \in \mathcal{L}_{q/(s-1)}(\Omega')$, where Ω' is a subdomain of Ω (in particular the whole of Ω). Denote the matrix of the form (1.1.2) by

$$A = \left\{ \frac{\partial a_\beta^{(i)}}{\partial p_{\tilde{\beta}}^{(k)}} \right\}. \tag{1.1.4}$$

Denote the size of this matrix by $M \times M$. Let $A = A^+ + A^-$, where A^+ and A^- are symmetric and skew-symmetric parts of A, respectively. Let λ_i be the eigenvalues of A^+.

From inequalities (1.1.2) it follows that all λ_i satisfy

$$\mu_0 \left(1 + |p|^2\right)^{\frac{s-2}{2}} \leq \lambda_i \leq \nu_0 \left(1 + |p|^2\right)^{\frac{s-2}{2}},$$

where μ_0, ν_0 are the same constants as in (1.1.2). So for $s < 2$ and $|p| \to +\infty$, the eigenvalues of A^+ tend to zero. In this case system (1.1.1) is called a degenerating one. When $s = 2$, system (1.1.1) will be a nondegenerating one and all λ_i are estimated from below and from above by positive numbers.

Consider also the matrix

$$C = A^+A^- - A^-A^+ - (A^-)^2. \tag{1.1.5}$$

Assume the upper bound for the eigenvalues σ_i of C by σ. Let $\lambda = \inf \lambda_i$ and $\Lambda = \sup \lambda_i$. Here the infimum and supremum are taken over all $i = 1, \ldots, M$, and $(x,p) \in \Omega \times R^M$. Due to the boundedness of A elements the value σ is also bounded. So, system (1.1.1) will be a system with bounded nonlinearities and from assumptions follows that $\lambda > 0$ and $\Lambda < +\infty$.

Lemma 1.1.1. *The inequality*

$$\|I - \varepsilon A\| \leq K_\varepsilon \tag{1.1.6}$$

holds, where

$$K_\varepsilon^2 = \sup_{i,x,p}(1 - \varepsilon\lambda_i)^2 + \varepsilon^2\sigma. \tag{1.1.7}$$

Proof. Let $\chi \in R^M$. Consider $|(I - \varepsilon A)\chi|$. Setting $A = A^+ + A^-$, we get

$$|(I - \varepsilon A)\chi|^2 = \left[I - \varepsilon(A^+ + A^-)\right]\chi \cdot \left[I - \varepsilon(A^+ + A^-)\right]\chi =$$
$$= |(I - \varepsilon A^+)\chi|^2 - \varepsilon(I - \varepsilon A^+)\chi \cdot A^-\chi - \varepsilon A^-\chi \cdot (I - \varepsilon A^+)\chi + \varepsilon^2 A^-\chi \cdot A^-\chi.$$

As $(A^-)^* = -A^-$, we have $A^-\chi \cdot \chi = 0$. Hence,

$$|(I - \varepsilon A)\chi|^2 = \left|(I - \varepsilon A^+)\chi\right|^2 + \varepsilon^2(A^+A^- - A^-A^+)\chi \cdot \chi + \varepsilon^2|A^-\chi|^2.$$

Taking into account that $|A^-\chi|^2 = -(A^-)^2\chi \cdot \chi$, we come to the identity

$$|(I - \varepsilon A)\chi|^2 = |(I - \varepsilon A^+)\chi|^2 + \varepsilon^2 C\chi \cdot \chi.$$

Therefore,

$$|(I-\varepsilon A)\chi|^2 = \sum_{i=1}^{M}(1-\varepsilon\lambda_i)^2\xi_i^2 + \varepsilon^2\sum_{i=1}^{M}\sigma_i\zeta_i^2,$$

where ξ and ζ are two vectors connected with χ by an orthogonal transformation. As $|\xi| = |\zeta| = |\chi|$,

$$|(I-\varepsilon A)\chi|^2 \le \left[\sup_{i,x,p}(1-\varepsilon\lambda_i)^2 + \varepsilon^2\sigma\right]|\chi|^2.$$

□

We shall now show how to choose ε such that K_ε is minimal.

Lemma 1.1.2. *If* $\sigma \ge \frac{\lambda(\Lambda-\lambda)}{2}$, *then*

(1.1.8)
$$K^2 = \inf_{\varepsilon>0} K_\varepsilon^2 = \frac{\sigma}{\sigma+\lambda^2},$$

and the value K *is attained when* $\varepsilon = (\sigma+\lambda^2)^{-1}\lambda$. *If* $\sigma \le \frac{\lambda(\Lambda-\lambda)}{2}$, *then*

(1.1.9)
$$K^2 = \inf_{\varepsilon>0} K_\varepsilon^2 = \frac{(\Lambda-\lambda)^2+4\sigma}{(\Lambda+\lambda)^2}$$

and the value K *is attained, when* $\varepsilon = 2(\Lambda+\lambda)^{-1}$.

Proof. We have mentioned that $\lambda > 0$ and $\Lambda < +\infty$. It is easy to see that the inequality

$$\sigma + \Lambda^2 > 0$$

holds.
In fact,

$$K_\varepsilon^2 = \max\left\{1-2\lambda\varepsilon+(\sigma+\lambda^2)\varepsilon^2, 1-2\Lambda\varepsilon+(\sigma+\Lambda^2)\varepsilon^2\right\} \ge 0.$$

If we suppose that $\sigma+\Lambda^2 \le 0$, for a certain ε, then K_ε^2 becomes negative.
Let $f(\varepsilon) = 1-2\lambda\varepsilon+(\sigma+\lambda^2)\varepsilon^2$ and $F(\varepsilon) = 1-2\Lambda\varepsilon+(\sigma+\Lambda^2)\varepsilon^2$. Hence, $f'(0) = -2\lambda$, $F'(0) = -2\Lambda$ and $f'(0) \ge F'(0)$. Suppose $\varepsilon^* > 0$ is the root of the equation $f(\varepsilon) = F(\varepsilon)$. Consequently, for $0 \le \varepsilon \le \varepsilon^*$, we have $f(\varepsilon) \ge F(\varepsilon)$ and for $\varepsilon \ge \varepsilon^*$ we get $f(\varepsilon) \le F(\varepsilon)$. As $\varepsilon^* = 2(\Lambda+\lambda)^{-1}$, we have

$$K_\varepsilon^2 = \begin{cases} 1-2\lambda\varepsilon+(\sigma+\lambda^2)\varepsilon^2, & 0 \le \varepsilon \le \varepsilon^*, \\ 1-2\Lambda\varepsilon+(\sigma+\Lambda^2)\varepsilon^2, & \varepsilon \ge \varepsilon^*. \end{cases}$$

Moreover,

$$K_{\varepsilon^*}^2 = \frac{(\Lambda-\lambda)^2+4\sigma}{(\Lambda+\lambda)^2} \ge 0.$$

Consider the following two cases: 1) $\sigma > -\lambda^2$; 2) $\sigma \le -\lambda^2$. Let us first assume $\sigma > -\lambda^2$. Then the graph of the function $f(\varepsilon)$ is convex downwards. Let ε^{**} be the point of $\min f(\varepsilon)$. If $\varepsilon^{**} < \varepsilon$, then $\inf K_\varepsilon^2$ is attained at $\varepsilon = \varepsilon^{**}$. In this case

$$\inf K_\varepsilon^2 = K_{\varepsilon^{**}}^2 = \frac{\sigma}{\sigma+\lambda^2}$$

and it is clear that $\sigma \geq 0$. If $\varepsilon^* < \varepsilon^{**}$, then $\inf K_\varepsilon^2$ is reached for $\varepsilon = \varepsilon^*$. Hence,

$$\inf K_\varepsilon^2 = \frac{(\Lambda - \lambda)^2 + 4\sigma}{(\Lambda + \lambda)^2} \geq 0.$$

Additionaly, it follows from this inequality that $(\Lambda - \lambda)^2 + 4\sigma \geq 0$. It is easy to see that the equality (1.1.8) holds when $\sigma \geq 2^{-1}\lambda(\Lambda - \lambda)$, and (1.1.9) when $\sigma \leq 2^{-1}\lambda(\Lambda - \lambda)$. Therefore the lemma is proved for the case, where $\sigma + \lambda^2 > 0$. In the second case, when $\sigma + \lambda^2 \leq 0$, the graph of the function $f(\varepsilon)$ is a straight line, or is convex upward. In this case $\inf K_\varepsilon^2 = K_{\varepsilon^*}^2$ and can be calculated with the help of (1.1.9). □

Remark 1.1.1. It can be seen from (1.1.8) and (1.1.9) that

$$K < 1. \tag{1.1.10}$$

Let $Q = (0, T) \times \Omega$ be a cylinder in which a parabolic system

$$\partial_t u + L(u) = 0 \tag{1.1.11}$$

will be considered. In this case we assume the Caratheodory condition is satisfied in Q, the inequalities (1.1.2),(1.1.3) hold uniformly with respect to t.
In this book we shall mainly consider second order parabolic systems, which can be written in the following form

$$\partial_t u - L(u) \equiv \partial_t u - \left[\sum_{i=1}^{m} D_i a_i(t, x, Du) - a_0(t, x, Du)\right] = 0 \tag{1.1.12}$$

Sometimes we denote by $\dot{u}$ the derivative $\partial_t u$ with respect to t . Consider along with system (1.1.1) the boundary conditions of the first boundary value problem

$$u\Big|_{\partial\Omega} = \varphi_0, \quad \frac{\partial u}{\partial \nu}\Big|_{\partial\Omega} = \varphi_1, \ldots, \frac{\partial^{l-1} u}{\partial \nu^{l-1}}\Big|_{\partial\Omega} = \varphi_{l-1}. \tag{1.1.13}$$

where ν is an outward normal to $\partial\Omega$.
If $\varphi_j \in W_q^{(l-j-1/q)}(\partial\Omega)$, $j = 0, 1, \ldots, l-1$, then there exists such a function $\varphi \in W_q^{(l)}(\Omega)$, that with its normal derivatives up to the $(l-1)$ order coincides with the right-hand side terms of (1.1.13) on $\partial\Omega$. Then conditions (1.1.13) can be transformed into homogenous equalities

$$u\Big|_{\partial\Omega} = \frac{\partial u}{\partial \nu}\Big|_{\partial\Omega} = \ldots = \frac{\partial^{l-1} u}{\partial \nu^{l-1}}\Big|_{\partial\Omega} = 0 \tag{1.1.14}$$

and conditions 1)–3) will still hold. Let C_0^∞ be the linear space of infinitely differentiable functions with a compact support. Denote the closure of C_0^∞ with supp $u \subset \Omega$ in the norm of $W_p^{(l)}(\Omega)$ by $\overset{\circ}{W}{}_p^{(l)}(\Omega)$. For systems (1.1.11) and (1.1.12) we shall add homogeneous boundary conditions (1.1.14) with $l-1$ and initial condition

$$u\Big|_{t=0} = 0. \tag{1.1.15}$$

The nonhomogeneous condition can be transfered to (1.1.15) in the same way as for (1.1.13).

1.2 Universal iterative process and weak solutions; nondegenerated systems with bounded nonlinearities

In this section we shall first introduce some notations.
Let $u(x)$ and $v(x)$ be vector-functions with N components l-times differentiable in R^m. Denote

(1.2.1) $$D^l u \cdot D^l v = \sum_{0 \le |\beta| \le l} D^\beta u \cdot D^\beta v,$$

where $D^\beta u \cdot D^\beta v$ is a scalar product in R^N and $\beta = (\beta_1, \ldots, \beta_m)$ are multiindices,

(1.2.2) $$D'^l u D'^l v = \sum_{|\beta|=l} D^\beta u D^\beta v,$$

(1.2.3) $$\left|D^l u\right|^2 = D^l u \cdot D^l u = \sum_{k=0}^{l} \sum_{|\beta|=k} D^\beta u \cdot D^\beta u,$$

and

(1.2.4) $$\left|D'^l u\right|^2 = D'^l u \cdot D'^l u = \sum_{|\beta|=l} D^\beta u \cdot D^\beta u.$$

We shall also write

(1.2.5) $$D^1 u \cdot D^1 v = Du \cdot Dv \text{ and}$$

(1.2.6) $$D'^1 u \cdot D'^1 v = D'u \cdot D'v \quad \text{(usually we shall omit the dots).}$$

Generally the weak solution $u \in \overset{\circ}{W}{}_s^{(l)}(\Omega)$ of problem (1.1.1), (1.1.13) satisfies for $\forall v \in \overset{\circ}{W}{}_s^{(l)}(\Omega)$ the integral identity

(1.2.7) $$\int_\Omega a_\beta(x; D^l u) D^\beta v dx = 0,$$

where $a_\beta D^\beta v$, for fixed β, is a scalar product in R^N of a_β and $D^\beta v$ and instead of $\tilde{\beta}$ we shall write l. We shall call $W_s^{(l)}(\Omega)$ an energetic space for problem (1.1.1), (1.1.13). Under these general assumptions the existence of a weak solution for problem (1.1.1), (1.1.13), was proved by Vishik[1],Browder [1] Minty[1], Nečas[1] and Lions [1], etc. ... . So, the existence of a weak solution is now a well established fact. Taking into account numerical problems, we shall also prove the existence of the problem under consideration by means of the universial iterative process.
Consider the following iterative process

(1.2.8) $$\int_\Omega D^l u_{n+1} D^l v dx = \int_\Omega D^l u_n D^l v dx - \varepsilon \int_\Omega a_\beta(x; D^l u_n) D^\beta v dx,$$

where $u_0 \in W_2^{(l)}(\Omega), \forall v \in \overset{\circ}{W}{}_2^{(l)}(\Omega)$, $\varepsilon > 0$ and, for $n \ge 0$, the function u_n satisfies boundary conditions (1.1.13), This process was introduced for a single second order elliptic equation by the author (Koshelev [3]). The convergence of the method was first proved in an energetic norm. For general elliptic systems, this was proved in [4] by the author. It is sometimes useful to apply a simpler process than (1.2.8)

$$\int_\Omega D'^l u_{n+1} D'^l v\,dx = \int_\Omega D'^l u_n D'^l v\,dx - \varepsilon \int_\Omega a_\beta(x; D^l u_n) D^\beta v\,dx. \tag{1.2.9}$$

Process (1.2.9) is generally applied in the case, when $a_\beta \equiv 0$ for $|\beta| < l$ and a_β depends only on x the l-th order derivatives of u. If the functions u_n, u_{n+1} and a_β are sufficiently smooth, processes (1.2.8) and (1.2.9) are equivalent to the systems of differential equations of the order $2l$. In particular, if u_n and u_{n+1} are $2l$-times, a_β are $|\beta|$-times ($0 \le |\beta| \le l$) continuously differentiable, then integrating by parts, respectively l and $|\beta|$-times, so that the differentiation transfers to the first factors, we come to the following systems

$$\sum_{j=0}^{l} (-1)^j \Delta^j u_{n+1} = \sum_{j=0}^{l} (-1)^j \Delta^j u_n - \varepsilon L(u_n) \tag{1.2.8'}$$

and

$$(-1)^l \Delta^l u_{n+1} = (-1)^l \Delta^l u_n - \varepsilon L(u_n). \tag{1.2.9'}$$

Naturally, boundary conditions (1.1.13) are satisfied.

Lemma 1.2.1. *Suppose that 1)–3) hold and suppose that the following conditions are satisfied:*

1) *the boundary* $\partial\Omega \in C^{2l-1}$;

2) *the coefficients* $a_\beta(x; p_{\tilde\beta})$ *are* $|\beta|$*-times continuously differentiable with respect to all variables;*

3) *if* $u \in W_q^{(2l)}(\Omega)$ *with some* $q > \frac{m}{l}$;

4) *the right-hand side terms in (1.1.13) are traces of some function from* $W_q^{(2l)}(\Omega)$ *and respectively its derivatives with regard to* ν.

If $u_0 \in W_q^{(2l)}(\Omega)$, *then all the iterations* u_n *exist and belong to* $W_q^{(2l)}(\Omega)$.

Proof. The proof is evident and follows immediately from the embedding theorems. □

Consider now the case, when conditions 1) – 3) are satisfied with $s = 2$ (bounded nondegenerated systems).

Theorem 1.2.1. *Let conditions 1)–3) be satisfied with* $q = 2$ *and by virtue of this the inequality*

$$\left| \frac{\partial a_\beta^{(i)}}{\partial p_{\tilde\beta}^{(k)}} \right| < \nu_0 \tag{1.2.10}$$

holds.

Then there exists the weak solution $u \in W_2^{(l)}\left(\overset{\circ}{W}{}_2^{(l)}\right)$ *of problems* (1.1.1), (1.1.13) *((1.1.14)), and processes* (1.2.8), (1.2.9) *converge* in* $W_2^{(l)}(\Omega)$, *with* $\forall u_0 \in W_2^{(l)}\left(\overset{\circ}{W}{}_2^{(l)}\right)$ *and appropriately chosen* $\varepsilon > 0$.

Proof. If $u_n \in W_2^{(l)}(\Omega)$, then from conditions 1)–3) of section 1.1 it follows that the right-hand side terms of (1.2.8) and (1.2.9) are linear functionals on $v \in W_2^{(l)}$. Therefore these terms can be represented under Riesz theorem by the integrals $\int\limits_\Omega D^l f_n D^l v dx$ and $\int\limits_\Omega D^{\prime\prime} f_n D^{\prime\prime} v dx$, respectively. It follows then that $u_{n+1} = f_n$ and all u_n, $n = 0, 1, \dots n$, belong to $W_2^{(l)}\left(\overset{\circ}{W}{}_2^{(l)}\right)$.

Let us now show that under appropriately chosen $\varepsilon > 0$ processes (1.2.8) and (1.2.9) converge. We do it now for example for (1.2.8). Subtracting two successive identities (1.2.8), we get

$$\int\limits_\Omega D^l(u_{n+1} - u_n) D^l v dx =$$
$$= \int\limits_\Omega \left\{D^\beta(u_n - u_{n-1}) - \varepsilon\left[a_\beta(x; D^l u_n) - a_\beta(x; D^l u_{n-1})\right]\right\} D^\beta v dx.$$

Apply the Hölder inequality to the right-hand side. Then

$$\left|\int\limits_\Omega D^l(u_{n+1} - u_n) D^l v dx\right|^2 \leq$$
$$\leq \sum_{0 \leq |\beta| \leq l} \int\limits_\Omega \left|D^\beta(u_n - u_{n-1}) - \varepsilon\left[a_\beta(x; D^l u_n) - a_\beta(x; D^l u_{n-1})\right]\right|^2 dx \int\limits_\Omega \left|D^l v\right|^2 dx.$$

Taking $v = u_{n+1} - u_n$, we have

(1.2.11)
$$\int\limits_\Omega \left|D^l(u_{n+1} - u_n)\right|^2 dx \leq \int\limits_\Omega \left\{\left|D^l(u_n - u_{n-1})\right|^2 - \right.$$
$$-2\varepsilon\left[a_\beta(x; D^l u_n) - a_\beta(x; D^l u_{n-1})\right] D^\beta(u_n - u_{n-1}) +$$
$$\left. + \varepsilon^2 \sum_{0 \leq |\beta| \leq l} \left|a_\beta(x; D^l u_n) - a_\beta(x; D^l u_{n-1}\right|^2\right\} dx.$$

It is obvious that after using the finite difference formula and (1.1.2), the inequality

$$\int\limits_\Omega \left[a_\beta(x; D^l u_n) - a_\beta(x; D^l u_{n-1})\right] D^\beta(u_n - u_{n-1}) dx \geq \mu_0 \int\limits_\Omega \left|D^l(u_n - u_{n-1})\right|^2 dx$$

is true. To show this, consider the expression $\Omega(u) = a_\beta(x; D^l u) D^\beta v$, where $v \in W_2^{(l)}(\Omega)$. Under the formula of finite differences, we have

$$\Omega(u_n) - \Omega(u_{n-1}) = \frac{\overline{\partial a_\beta}}{\partial p_{\widetilde{\beta}}^{(k)}} D^{\widetilde{\beta}}\left(u_n^{(k)} - u_{n-1}^{(k)}\right) D^\beta v,$$

---

*In the case of process (1.2.9), a_β depends only on x and $D^{\widetilde{\beta}} u$, where $|\widetilde{\beta}| = l$. So, inequalities (1.1.2) must be satisfied only for $\xi_\beta = 0$, where $|\beta| < l$.

where the bar over the derivative indicates that its arguments lie between u_{n-1} and u_n. Substituting v by $u_n - u_{n-1}$, we get

$$\left[a_\beta(x; D^l u_n) - a_\beta(x; D^l u_{n-1})\right] D^\beta(u_n - u_{n-1}) =$$
$$= \overline{\frac{\partial a_\beta}{\partial p^{(k)}_{\tilde\beta}}} D^{\tilde\beta}\left(u_n^{(k)} - u_{n-1}^{(k)}\right) D^\beta(u_n - u_{n-1}).$$

After integrating and applying (1.1.2) we get the required inequality. Likewise, we arrive at the inequality

$$\sum_{0\le|\beta|\le l} \int_\Omega \left|a_\beta(x; D^l u_n) - a_\beta(x; D^l u_{n-1})\right|^2 dx \le$$
$$\le C \int_\Omega \left|D^l(u_n - u_{n-1})\right|^2 dx.$$

Then from (1.2.11) we get

$$\int_\Omega \left|D^l(u_{n+1} - u_n)\right|^2 dx \le \left(1 - 2\mu_0\varepsilon + C\varepsilon^2\right) \int_\Omega \left|D^l(u_n - u_{n-1})\right|^2 dx.$$

If we take $\varepsilon > 0$ sufficiently small, then the inequality $1 - 2\mu_0\varepsilon + C\varepsilon^2 < 1$ holds. □

It is evident from the proof that processes (1.2.8) and (1.2.9) converge at the rate of the geometric progression with the common ratio, depending on ε.
In the following theorems we determine the values of ε under which the optimal convergency holds.

Theorem 1.2.2. *Let inequalities* (1.1.2) *for* $s = 2$ *be satisfied. Universal iterative process* (1.2.8) *for problem* (1.1.1), (1.1.14) *converges in* $W_2^{(l)}(\Omega)$ *to the weak solution, starting from* $\forall u_0 \in \overset{\circ}{W}{}_2^{(l)}(\Omega)$ *for*

(1.2.12) $$\varepsilon = \frac{\lambda}{\sigma + \lambda^2},$$

when $\sigma \ge \frac{1}{2}\lambda(\Lambda - \lambda)$ *and*

(1.2.13) $$\varepsilon = \frac{2}{\Lambda + \lambda},$$

when $\sigma \le \frac{1}{2}\lambda(\Lambda - \lambda)$.
The process converges at the rate of the geometric progression with the common ratio K, *where* K *is defined by equalities* (1.1.8) *and* (1.1.9). *Process* (1.2.8) *converges in* $W_2^{(l)}(\Omega)$ *(not necessarily in an optimal way) for* $K_\varepsilon < 1$.

Proof. Consider once more the difference between two successive equalities (1.2.8). Then

$$\int_\Omega D^l(u_{n+1} - u_n) D^l v dx =$$
$$= \int_\Omega \left\{D^\beta(u_n - u_{n-1}) - \varepsilon\left[a_\beta(x; D^l u_n) - a_\beta(x; D^l u_{n-1})\right]\right\} D^\beta v dx.$$

Using the finite difference theorem, we obtain

$$\int_\Omega D^l(u_{n+1}-u_n)D^l v dx = \tag{1.2.14}$$

$$= \int_\Omega \left[D^\beta(u_n-u_{n-1}) - \frac{\overline{\partial a_\beta}}{\partial p^{(k)}_{\tilde\beta}} D^{\tilde\beta}\left(u_n^{(k)}-u_{n-1}^{(k)}\right)\right] D^\beta v dx.$$

Take the first projection of the vector $u_n \in R^N$ and construct a column of all its derivatives of an order not higher than l. Extend this column by all derivatives of the second projection etc., up to the N-th projection and its derivatives. So, we get a vector which we denote by $U_n \in R^M$. A similar operation will be performed with the vectors u_n, $n = 0, 1, \dots$, and v. Taking into account the form of the matrix A, we have

$$\int_\Omega D^l(u_{n+1}-u_n)D^l v dx = \int_\Omega (I-\varepsilon\overline{A})(U_n-U_{n-1})V dx, \tag{1.2.15}$$

where the bar over the matrix A denotes that the elements of A are calculated, when its arguments lie between $D^{\tilde\beta}u_{n-1}$ and $D^{\tilde\beta}u_n$. From this follows the inequality

$$\int_\Omega D^l(u_{n+1}-u_n)D^l v dx \le \int_\Omega \|I-\varepsilon\overline{A}\|\ |U_n-U_{n-1}||V|dx,$$

where $|\cdot|$ denotes the length of the vector in R^M and $\|\cdot\|$ the norm of the matrix $R^M \to R^M$. Taking $\sup\|I-\varepsilon\overline{A}\|$ outside the sign of integral, we come to the estimate

$$\int_\Omega D^l(u_{n+1}-u_n)D^l v dx \le \sup_{x,p}\|I-\varepsilon\overline{A}\| \int_\Omega |U_n-U_{n-1}||V|dx.$$

Further we take $V = U_{n+1} - U_n$ and notice that

$$|U_{n+1}-U_n|^2 = |D^l(u_{n+1}-u_n|^2.$$

After applying the Hölder inequality and cancelling $\left[\int_\Omega |D^l(u_{n+1}-u_n)|^2dx\right]^{1/2}$ we obtain

$$\left[\int_\Omega |D^l(u_{n+1}-u_n)|^2dx\right]^{1/2} \le \sup_{x,p}\|I-\varepsilon\overline{A}\| \left[\int_\Omega |D^l(u_n-u_{n-1})|^2dx\right]^{1/2}.$$

Applying (1.1.6), we get that the iterative process under consideration converges when $K_\varepsilon < 1$. Take ε, as it is prescribed by lemma 1.1.2 and the proof of the theorem is completed. □

Remark 1.2.1. *If the matrix A is symmetric, then $A^- = 0$ and $\sigma = 0$. Hence*

$$K = \frac{\Lambda-\lambda}{\Lambda+\lambda}. \tag{1.2.16}$$

It is obvious that for any $n \geq 0$ the inequalities

$$\|u_n - u\| \leq K^n \|u_0 - u\| \tag{1.2.17}$$

hold, where u is a weak solution of problem (1.1.1), (1.1.14) and $\|\cdot\|$ denotes the energetic norm, i.e. the norm in $W_2^{(l)}$.
Consider now the case, when a_β depends only on x and $D^{\tilde{\beta}}u$, where $|\tilde{\beta}| = l$ and $a_\beta = 0$, for $|\beta| < l$. Let A' be matrix

$$A' = \left\{ \frac{\partial a_\beta^{(i)}}{\partial p_{\tilde{\beta}}^{(k)}} \right\}, \quad |\beta| = |\tilde{\beta}| = l, \tag{1.2.18}$$

with dimension $M' \times M'$. Let $\lambda'_i, i = 1, \ldots, M'$, be the eigenvalues of $A'^+, \lambda' = \inf_{i,x,p} \lambda'_i$, $\Lambda' = \sup_{i,x,p} \lambda'_i$. The matrix C' is denoted by a formula, similar to (1.1.5) and $\sigma' = \sup_{i,x,p} \sigma'_i$, where $\sigma'_i, i = 1, \ldots, M'$ are the eigenvalues of C'. Let the coefficients a_β of the system (1.1.1) satisfy conditions 1)–3) for $s = 2$. Denote by K'_ε and K', respectively, the values K_ε and K for matrix A'. The following theorem is true.

Theorem 1.2.3. *Let a_β depend only on x, the derivatives of the order l, and $a_\beta = 0$ for $|\beta| < l$. If conditions 1)–3) with $s = 2$ for $\xi_\beta = 0$, when $|\beta| < l$, hold, then the iterative process* (1.2.9) *for problem* (1.1.1), (1.1.14) *converges in $W_2^{(l)}$ to a weak solution, starting with $\forall u_0 \in \overset{\circ}{W}{}_2^{(l)}(\Omega)$, when*

$$\varepsilon = \frac{\lambda'}{\sigma' + \lambda'^2}, \tag{1.2.19}$$

for $\sigma' \geq \frac{1}{2}\lambda'(\Lambda' - \lambda')$, and

$$\varepsilon = \frac{2}{\Lambda' + \lambda'}, \tag{1.2.20}$$

for $\sigma' \leq \frac{1}{2}\lambda'(\Lambda' - \lambda')$. The process converges at the rate of the geometric progression with the common ratio K'. The process also converges for $K'_\varepsilon < 1$.

The proof does not differ from that of the previous theorem. Note that in the case of the symmetric matrix we have

$$K' = \frac{\Lambda' - \lambda'}{\Lambda' + \lambda'} < 1. \tag{1.2.21}$$

In the last three theorems $\partial\Omega$ should satisfy only the general conditions for the possibility to integrate by parts.
Now we consider the parabolic system (1.1.12) with conditions (1.1.13) and (1.1.15). By $W_q^{k,l}$ we shall denote the Sobolev space of functions posessing all weak derivatives of order k with respect to t and of order l with respect to x. The norm in these spaces will be defined by the formula

$$\|u\|_q^{k,l} = \left(\int_Q \left[|\partial_t^k u|^q + |D^l u|^q \right] dx dt \right)^{1/q}. \tag{1.2.22}$$

In the case of $k = l = 0$ we shall write $\mathcal{L}_q(Q)$ instead of $W_q^{0,0}$.
For problems (1.1.11), (1.1.14) and (1.1.15) we consider the iterational process

$$\varepsilon\partial_t u_{n+1} + \sum_{j=0}^{l}(-1)^j\Delta^j u_{n+1} = \sum_{j=0}^{l}(-1)^j\Delta^j u_n - \varepsilon L(u_n) \tag{1.2.23}$$

with some $\varepsilon = \text{const} > 0$ and all $u_n(t,x)$ satisfying conditions (1.1.14) and (1.1.15).
In the weak form process (1.2.23) can be written in the form

$$\varepsilon\int_Q \partial_t u_{n+1} v dx dt + \int_Q D^l u_{n+1} D^l v dx dt = \int_Q D^l u_n D^l v dx dt -$$

$$-\varepsilon \sum_{0\le|\beta|\le l} \int_Q a_\beta(t;x,D^l u_n) D^\beta v dx dt \tag{1.2.24}$$

where $v(t,x)$ is sufficiently smooth. The following result was proved by Chistjakov [1].

Theorem 1.2.4. *Let $s=2$, $\partial\Omega \subset C^{2l-1,\gamma}$ $(\gamma > 0)$, conditions 1) and 2) of section 1.1 are satisfied and $L(u) \in \mathcal{L}_2(Q)$ for each $u \in \mathcal{L}_2\left\{(0,T); W_2^{(2l)}(\Omega)\right\}$. Suppose also that $u_0 \in \mathcal{L}_2\left\{(0,T); W_2^{(2l)}(\Omega)\right\}$. Then the iterative process* (1.2.24) *converges in $W_2^{0,1}(Q)$ at a geometric progression rate to the weak solution of* (1.1.11), (1.1.14), (1.1.15) *under conditions on ε given by theorem 1.2.2. with the common rate K.*

Proof. As $u_0 \in \mathcal{L}_2\left\{(0,T), W_2^{(2l)}(\Omega)\right\}$ then the right-hand side of (1.2.23) will belong to $\mathcal{L}_2(Q)$.
It follows from estimates in $W_2^{1,2}(Q)$ (see Ladyženskaya, Solonnikov and Ural'tseva [1]) that $u_1 \in W_2^{1,2l}(Q)$ and therefore all iterations will belong to the same space. Then we can close the identity (1.2.24) in $W_2^{1,l}(Q)$ with respect to v and replace v with the iteration u_{n+1}. The right-hand side of (1.2.24) can be estimated exactly as in theorem 1.2.2. Subtracting two successive equalities (1.2.24) we come to the following relation

$$\varepsilon\int_Q (u_{n+1}-u_n)\partial_t(u_{n+1}-u_n) dx dt + \int_Q \left|D^l(u_{n+1}-u_n)\right|^2 dx dt \le$$

$$\le \sup_{x,t,p}\|I-\varepsilon A\|^2 \int_Q \left|D^l(u_n-u_{n-1})\right|^2 dx dt.$$

Take into account the first term on the left-hand side can be integrated with respect to t giving a positive term we shall have

$$\int_Q |D^l(u_{n+1}-u_n)|^2 dx dt \le K_\varepsilon^2 \int_Q |D^l(u_n-u_{n-1})|^2 dx dt$$

□

Remark 1.2.2. *The existence of weak solution for problem (1.1.11), (1.1.14), (1.1.15) can be proved under less restrictive conditions.*

1.3 Degenerating elliptic systems with bounded nonlinearities

In this section we consider the existence of solution for problem (1.1.1), (1.1.13) under the condition that inequalities (1.1.2) hold when $s \leq 2$. We begin with two lemmas.

Lemma 1.3.1. *If the functions a_β satisfy conditions 1)-3) in section 1.1 with $s \leq 2$, then, for $\forall u, z \in W_2^{(l)}$, we get*

$$\left[a_\beta(x; D^l u) - a_\beta(x; D^l z)\right] D^\beta(u-z) \geq \tag{1.3.1}$$
$$\geq C \sum_{0\leq|\beta|\leq l} \left|a_\beta(x; D^l u) - a_\beta(x; D^l z)\right|^2,$$

where C is a positive constant.

Proof. Consider the expression

$$\frac{\partial a_\beta}{\partial p_{\tilde{\beta}}} D^{\tilde{\beta}} u v_\beta = \frac{\partial a_\beta^{(i)}}{\partial p_{\tilde{\beta}}^{(k)}} D^{\tilde{\beta}} u^{(k)} v_\beta^{(i)},$$

where $v_\beta(x)$, $0 \leq |\beta| \leq l$, is an arbitrary system of smooth functions and the summation runs over the repeated indices.
Let

$$A(u) = \left[D^\beta u - \delta a_\beta(x; D^l u)\right] v_\beta,$$

where δ is a positive constant to be chosen later. As $a_\beta(x; p_{\tilde{\beta}})$ are continuously differentiable with respect to $p_{\tilde{\beta}}$ for almost all $x \in \Omega$, we have

$$A(u) - A(z) = \left[D^\beta(u-z) - \delta \overline{\frac{\partial a_\beta}{\partial p_{\tilde{\beta}}}} D^{\tilde{\beta}}(u-z)\right] v_\beta.$$

In the above expression the bar over the derivatives denotes their arguments are $\overline{p}_{\tilde{\beta}} = D^{\tilde{\beta}} u + \theta D^{\tilde{\beta}}(z-u)$, $0 < \theta < 1$. The values of $|p|$, corresponding to $\overline{p}_{\tilde{\beta}}$, we denote by $|\overline{p}|$. Applying the Hölder inequality, we get

$$|A(u) - A(z)| \leq$$
$$\leq \left[\left|D^l(u-z)\right|^2 - 2\delta \overline{\frac{\partial a_\beta}{\partial p_{\tilde{\beta}}}} D^{\tilde{\beta}}(u-z) D^\beta(u-z) + \right.$$
$$\left. + \delta^2 \sum_{0\leq|\beta|\leq l} \left|\overline{\frac{\partial a_\beta}{\partial p_{\tilde{\beta}}}} D^{\tilde{\beta}}(u-z)\right|^2\right]^{1/2} \left(\sum_{0\leq|\beta|\leq l} |v_\beta|^2\right)^{1/2}.$$

Applying once more the Hölder inequality to the last term of the first factor, we obtain

$$|A(u) - A(z)| \leq$$
$$\leq \left[\left|D^l(u-z)\right|^2 - 2\delta \overline{\frac{\partial a_\beta}{\partial p_{\tilde{\beta}}}} D^{\tilde{\beta}}(u-z) D^\beta(u-z) + \right.$$
$$\left. + \delta^2 \sum_{0\leq|\beta|,|\tilde{\beta}|\leq l} \left|\overline{\frac{\partial a_\beta}{\partial p_{\tilde{\beta}}}}\right|^2 \left|D^l(u-z)\right|^2\right]^{1/2} \left(\sum_{0\leq|\beta|\leq l} |v_\beta|^2\right)^{1/2}.$$

From (1.1.2) it follows that

$$\overline{\frac{\partial a_\beta}{\partial p_{\widetilde{\beta}}}} D^{\widetilde{\beta}}(u-z) D^\beta(u-z) \geq \frac{\mu_0}{\left(1+|\overline{p}|^2\right)^{(2-s)/2}} \left|D^l(u-z)\right|^2$$

and

$$\sum_{0\leq|\beta|,|\widetilde{\beta}|\leq l} \left|\frac{\partial a_\beta}{\partial p_{\widetilde{\beta}}}\right|^2 \leq \frac{C}{(1+|\overline{p}|^2)^{2-s}}.$$

As

$$(1+|\overline{p}|^2)^{2-s} \geq (1+|\overline{p}|^2)^{(2-s)/2},$$

we have

$$\sum_{0\leq|\beta|,|\widetilde{\beta}|\leq l} \left|\frac{\partial a_\beta}{\partial p_{\widetilde{\beta}}}\right|^2 \leq \frac{C}{(1+|\overline{p}|^2)^{(2-s)/2}}.$$

Then

$$|A(u)-A(z)| \leq$$

$$\leq \left[1 - \frac{2\mu_0\delta}{(1+|\overline{p}|^2)^{(2-s)/2}} + \frac{\delta^2 C}{(1+|\overline{p}|^2)^{(2-s)/2}}\right]^{1/2} \left|D^l(u-z)\right| \left(\sum_{0\leq|\beta|\leq l} |v_\beta|^2\right)^{1/2}.$$

Take $v_\beta = D^\beta(u-z) - \delta\left[a_\beta(x;D^l u) - a_\beta(x;D^l z)\right]$ in A. Hence,

$$A(u)-A(z) = \left\{D^\beta(u-z) - \delta\left[a_\beta(x;D^l u) - a_\beta(x;D^l z)\right]\right\} \times$$
$$\times \left\{D^\beta(u-z) - \delta\left[a_\beta(x;D^l u) - a_\beta(x;D^l z)\right]\right\}$$

and

$$A(u)-A(z) = \left|D^l(u-z)\right|^2 - 2\delta\left[a_\beta(x;D^l u) - a_\beta(x;D^l z)\right] D^\beta(u-z) + \tag{1.3.2}$$
$$+\delta^2 \sum_{0\leq|\beta|\leq l} \left|a_\beta(x;D^l u) - a_\beta(x;D^l z)\right|^2 = \sum_{0\leq|\beta|\leq l} |v_\beta|^2.$$

From this equality it follows that

$$\sum_{0\leq|\beta|\leq l} |v_\beta|^2 \leq$$

$$\leq \left[1 - \frac{2\mu_0\delta}{(1+|p|^2)^{(2-s)/2}} + \frac{\delta^2 C}{(1+|p|^2)^{(2-s)/2}}\right]^{1/2} \left|D^l(u-z)\right| \left(\sum_{0\leq|\beta|\leq l} |v_\beta|^2\right)^{1/2}.$$

Cancelling the common factors and then raising to the second power, we obtain the inequality

$$\sum_{0\leq|\beta|\leq l} |v_\beta|^2 \leq \left[1 - \frac{2\mu_0\delta}{(1+|p|^2)^{(2-s)/2}} + \frac{\delta^2 C}{(1+|p|^2)^{(2-s)/2}}\right] \left|D^l(u-z)\right|^2.$$

Choose $\delta > 0$, but sufficiently small, so that the value of the bracket is less than the value of the unit. Then, raising the right-hand side term, we get

$$\sum_{0\le|\beta|\le l} |v_\beta|^2 \le \left|D^l(u-z)\right|^2.$$

Using (1.3.2) and cancelling the term $\left|D^l(u-z)\right|^2$ on both sides, we come to the required estimate. □

Lemma 1.3.2. *If the conditions of the previous lemma are satisfied, then for* $\forall u, z \in W_s^{(l)}$, $1 < s \le 2$, *the inequality*

$$\int_\Omega \left[a_\beta(x; D^l u) - a_\beta(x; D^l z)\right] D^\beta(u-z)dx \ge \tag{1.3.3}$$

$$\ge C\frac{\|u-z\|^2_{W_s^{(l)}}}{1+\|u-z\|^{2-s}_{W_s^{(l)}} + \|u\|^{2-s}_{W_s^{(l)}}}$$

holds.

Proof. Similar to the proof of lemma 1.3.1, we can get the equality

$$B(u,z) \equiv \int_\Omega \left[a_\beta(x; D^l u) - a_\beta(x; D^l z)\right] D^\beta(u-z)dx =$$

$$= \int_\Omega \frac{\overline{\partial a_\beta}}{\partial p_{\widetilde{\beta}}} D^{\widetilde{\beta}}(u-z)D^\beta(u-z)dx,$$

where the arguments of the derivatives $\overline{\partial a_\beta}/\partial p_{\widetilde{\beta}}$ are $\overline{p}_{\widetilde{\beta}} = D^{\widetilde{\beta}}u + \theta(D^{\widetilde{\beta}}\ z - D^{\widetilde{\beta}}u)$, $0 < \theta < 1$.
Taking into account the right-hand side of the first inequality (1.1.2), we come to

$$B(u,z) \ge \mu_0 \int_\Omega \frac{\left|D^l(u-z)\right|^2 dx}{\left[1+\sum\limits_{0\le|\beta|\le l} |D^\beta u + \theta D^\beta(z-u)|^2\right]^{(2-s)/2}}.$$

By means of an elementary inequality $(a+b)^2 \le 2(a^2+b^2)$, we get

$$B(u,z) \ge \frac{\mu_0}{2^{(2-s)/2}} \int_\Omega \frac{\left|D^l(u-z)\right|^2 dx}{\left[1+|D^l u| + |D^l(u-z)|^2\right]^{(2-s)/2}}. \tag{1.3.4}$$

Consider the integral

$$\int_\Omega \left|D^l(u-z)\right|^s dx =$$

$$= \int_\Omega \frac{\left|D^l(u-z)\right|^s}{\left[1+|D^l u|^2 + |D^l(u-z)|^2\right]^{\frac{(2-s)s}{4}}} \left[1+\left|D^l u\right|^2 + \left|D^l(u-z)\right|^2\right]^{\frac{(2-s)s}{4}} dx.$$

Applying the Hölder inequality with exponents $2/s$ and $2/(2-s)$ to the right-hand side, we have

$$\int\limits_{\Omega} \left|D^l(u-z)\right|^s dx \leq$$

$$\leq \left\{ \int\limits_{\Omega} \frac{\left|D^l(u-z)\right|^2 dx}{\left[1+|D^l u|^2+|D^l(u-z)|^2\right]^{(2-s)/2}} \right\}^{s/2} \left\{ \int\limits_{\Omega} \left[1+\left|D^l u\right|^2+\left|D^l(u-z)\right|^2\right]^{s/2} dx \right\}^{(2-s)/2}$$

From the above it follows that

$$\int\limits_{\Omega} \frac{\left|D^l(u-z)\right|^2 dx}{\left[1+|D^l u|^2+|D^l(u-z)|^2\right]^{(2-s)/2}} \geq \frac{\|u-z\|^2_{W_s^{(l)}}}{\left\{ \int\limits_{\Omega} \left[1+|D^l u|^2+|D^l(u-z)|^2\right]^{s/2} dx \right\}^{(2-s)/s}} \geq$$

$$\geq C \frac{\|u-z\|^2_{W_s^{(l)}}}{1+\|u-z\|^{2-s}_{W_s^{(l)}}+\|u\|^{2-s}_{W_s^{(l)}}}.$$

The last inequality and estimate (1.3.4) lead to (1.3.3). □

Let us prove now the following statement.

Theorem 1.3.1. *Let conditions* 1) – 3) *with* $1 < s \leq 2$ *and* $q = 2$ *be satisfied and the weak solution* u *of problem* (1.1.1), (1.1.14) *belongs to* $\overset{\circ}{W}{}_2^{(l)}$. *If the initial iteration* u_0 *belongs to* $\overset{\circ}{W}{}_2^{(l)}$, *then iterative process* (1.2.8) *converges in the norm of* $W_s^{(l)}$ *to solution of problem* (1.1.1), (1.1.14).

Proof. It is easy to see that all iterations of process (1.2.8) belong to $\overset{\circ}{W}{}_2^{(l)}$. Under the assumption of the theorem the initial iteration u_0 belongs to $\overset{\circ}{W}{}_2^{(l)}$. In this case the iteration u_1 can be established from the system

$$\sum_{j=0}^{l}(-1)^j \Delta^j u_1 = \sum_{j=0}^{l}(-1)^j \Delta^j u_0 - \varepsilon L(u_0)$$

as a weak solution satisfying boundary conditions (1.1.14). As $a_\beta(x; D^l u_0) \in \in \mathcal{L}_{2/(2-s)} \subset \mathcal{L}_2$, then $u_1 \in \overset{\circ}{W}{}_2^{(l)}(\Omega)$. Subtracting from the left and right-hand sides in equality (1.2.8) the expression

$$\int\limits_{\Omega} D^l u D^l v dx$$

and using relation (1.2.7), we get

$$\int\limits_{\Omega} D^l(u_{n+1}-u) D^l v dx = \tag{1.3.5}$$

$$= \int\limits_{\Omega} \left\{ D^\beta(u_n - u) - \varepsilon \left[a_\beta(x; D^l u_n) - a_\beta(x; D^l u)\right] \right\} D^\beta v dx.$$

Denote the right-hand side by I. Then after using the Hölder inequality, we have

$$|I| \leq \left\{ \|u_n - u\|^2_{W_2^{(l)}} - 2\varepsilon \int_\Omega \left[a_\beta(x; D^l u_n) - a_\beta(x; D^l u)\right] D^\beta(u_n - u)dx + \right.$$

$$\left. +\varepsilon^2 \sum_{0\leq|\beta|\leq l} \int_\Omega \left|a_\beta(x; D^l u_n) - a_\beta(x; D^l u)\right|^2 dx \right\}^{1/2} \left[\int_\Omega |D^l v|^2 dx\right]^{1/2}.$$

Next from (1.3.1) it follows that

$$|I| \leq \left\{ \|u_n - u\|^2_{W_2^{(l)}} - \right. \tag{1.3.6}$$

$$\left. -\varepsilon(2 - \varepsilon C^{-1}) \int_\Omega \left[a_\beta(x; D^l u_n) - a_\beta(x; D^l u)\right] D^\beta(u_n - u)dx \right\}^{1/2} \|v\|_{W_2^{(l)}},$$

where C is the constant from (1.3.1).
Let $\varepsilon < 2C^{-1}$ and $\varepsilon(2 - \varepsilon C) = \varrho > 0$. Take $v = u_{n+1} - u$. Hence with the help of (1.3.6) from (1.3.5) after cancellation, we have

$$\|u_{n+1} - u\|^2_{W_2^{(l)}} \leq \|u_n - u\|^2_{W_2^{(l)}} -$$

$$-\varrho \int_\Omega \left[a_\beta(x; D^l u_n) - a_\beta(x; D^l u)\right] D^\beta(u_n - u)dx.$$

Using (1.3.3), we get

$$\|u_{n+1} - u\|^2_{W_2^{(l)}} \leq \|u_n - u\|^2_{W_2^{(l)}} - \varrho C \frac{\|u_n - u\|^2_{W_s^{(l)}}}{1 + \|u_n - u\|^{2-s}_{W_s^{(l)}} + \|u\|^{2-s}_{W_s^{(l)}}}. \tag{1.3.7}$$

Therefore, the series

$$\sum_{n=0}^{\infty} \frac{\|u_n - u\|^2_{W_s^{(l)}}}{1 + \|u_n - u\|^{2-s}_{W_s^{(l)}} + \|u\|^{2-s}_{W_s^{(l)}}}$$

converges and

$$\lim_{n\to+\infty} \frac{\|u_n - u\|^2_{W_s^{(l)}}}{1 + \|u_n - u\|^{2-s}_{W_s^{(l)}} + \|u\|^{2-s}_{W_s^{(l)}}} = 0.$$

Inequality (1.3.7) leads to the boundedness of $\|u_n - u\|_{W_2^{(l)}}$. Then under embedding theorem $\|u_n - u\|_{W_s^{(l)}}$ is also bounded and $\lim\limits_{n\to+\infty} \|u_n - u\|_{W_s^{(l)}} = 0$. □

1.4 Regularization of the universal iterative process

In this section we shall give another proof of the solvability of the problem under consideration in the energetic space $W_s^{(l)}$, when $1 < s \leq 2$. The main point of the proof is the convergence of some regularized iterative process in $W_s^{(l)}$. Let

$$L_\delta(u) \equiv (-1)^{|\beta|} D^\beta a_{\beta,\delta}(x; D^l u) \quad (\delta > 0) \tag{1.4.1}$$

be a regularized elliptic operator with

(1.4.2) $$a_{\beta,\delta} = \delta D^{\beta}u + a_{\beta}(x; D^{l}u).$$

Consider the system

(1.4.3) $$L_{\delta}(u) = 0$$

with boundary conditions (1.1.14).

To solve problem (1.4.3), (1.1.14) we shall apply the following iterative process

(1.4.4) $$\sum_{j=0}^{l}(-1)^{j}\Delta^{j}u_{n+1} = \sum_{j=0}^{l}(-1)^{j}\Delta^{j}u_{n} - \varepsilon L_{\delta}(u_{n}),$$

where $a_{\beta,\delta}$ are defined by (1.4.2). The equality (1.4.4) can be written in an integral form

(1.4.5) $$\int_{\Omega} D^{l}u_{n+1}D^{l}v\,dx = \int_{\Omega} D^{l}u_{n}D^{l}v\,dx - \varepsilon\int_{\Omega} a_{\beta,\delta}(x; D^{l}u_{n})D^{\beta}v\,dx.$$

It is evident, when $\delta > 0$, system (1.4.3) is a system with bounded nonlinearites. This follows from the fact that for it the system of inequalities (1.1.2) holds with $s = 2$. In fact, the equality

$$\frac{\partial a^{(i)}_{\beta,\delta}}{\partial p^{(k)}_{\tilde{\beta}}} = \delta\delta_{\beta\tilde{\beta}}\delta_{ik} + \frac{\partial a^{(i)}_{\beta}}{\partial p^{(k)}_{\tilde{\beta}}}$$

follows from (1.4.2). Here $\delta_{\beta\tilde{\beta}} = 1$, when the indices coincide and $\delta_{\beta\tilde{\beta}} = 0$, when $\beta \neq \tilde{\beta}$. Then

$$\frac{\partial a^{(i)}_{\beta,\delta}}{\partial p^{(k)}_{\tilde{\beta}}}\xi^{(i)}_{\beta}\xi^{(k)}_{\tilde{\beta}} = \delta\sum_{0\le|\beta|\le l}|\xi_{\beta}|^{2} + \frac{\partial a^{(i)}_{\beta}}{\partial p^{(k)}_{\beta}}\xi^{(i)}_{\beta}\xi^{(k)}_{\tilde{\beta}} \ge$$

$$\ge \left[\delta + \frac{\mu_{0}}{(1+|p|^{2})^{(2-s)/2}}\right]\sum_{0\le|\beta|\le l}|\xi_{\beta}|^{2} \ge \delta\sum_{0\le|\beta|\le l}|\xi_{\beta}|^{2}.$$

So a part (1.1.2) for system (1.4.3) is checked. To get the remaining part and to obtain the second inequality, we perform the following

$$\left|\frac{\partial a^{(i)}_{\beta,\delta}}{\partial p^{(k)}_{\tilde{\beta}}}\right| \le \delta + \frac{\nu_{0}}{(1+|p|^{2})^{(2-s)/2}} \le \delta + \nu_{0}.$$

This proves the validity of the remaining inequalities (1.1.2). Under theorem 1.2.1, problem (1.4.3), (1.1.14) has a unique solution $u \in \overset{\circ}{W}{}^{(l)}_{2}(\Omega)$ and universal iterative process (1.2.8) converges to u in $\overset{\circ}{W}{}^{(l)}_{2}(\Omega)$. Denote by u_{δ} the solution of problem (1.4.3), (1.1.14). Then the following result is valid.

Theorem 1.4.1. *If* $1 < s \le 2$, *conditions* 1)–2) *for system* (1.1.1) *are satisfied and* 3) *holds for* $q = 2(s-1)$. *Then problem* (1.1.1), (1.1.14) *has a unique solution* u *in* $\overset{\circ}{W}{}^{(l)}_{s}(\Omega)$ *and in this space* $u_{\delta_k} \to u$, *where* $\{\delta_k\}$ *is some sequence tending to zero.*

Proof. Let us first show that the inequality

$$(1.4.6) \qquad \sum_{0\le|\beta|\le l}\int_\Omega |a_\beta|^{\frac{s}{s-1}}dx \le C_1\|u\|^s_{W_s^{(l)}} + C_2$$

holds.
In fact from conditions 1)–3) of 1.1 we have

$$\sum_{0\le|\beta|\le l}\int_\Omega |a_\beta|^{\frac{s}{s-1}}dx \le$$
$$\le C\sum_{0\le|\beta|\le l}\left[\int_\Omega \left|a_\beta(x,D^l u) - a_\beta(x,0)\right|^{\frac{s}{s-1}}dx + \int_\Omega |a_\beta(x,0)|^{\frac{s}{s-1}}dx\right] \le$$
$$\le C_1\int_\Omega \frac{|D^l u|^{\frac{s}{s-1}}dx}{(1+|D^l u|^2)^{\frac{(2-s)s}{2(s-1)}}} + C_2 \le C_1\int_\Omega |D^l u|^s dx + C_2.$$

Let us show now that under the conditions of the theorem the inequality

$$(1.4.7) \qquad \int_\Omega a_\beta(x;D^l u)D^\beta u dx \ge \mu_1\int_\Omega |D^l u|^s dx - C$$

holds ($\mu_1 = \text{const} > 0$).
Introduce the expression

$$(1.4.8) \qquad A(u,v) = \int_\Omega a_\beta(x;D^l u)v_\beta dx.$$

Then

$$A(u,v) - A(0,v) = \int_\Omega \left[a_\beta(x;D^l u) - a_\beta(x;0)\right] v_\beta dx.$$

Under the conditions of the theorem

$$A(u,v) - A(0,v) = \int_\Omega \overline{\frac{\partial a_\beta}{\partial p_{\tilde\beta}}} D^{\tilde\beta} u v_\beta dx,$$

where the arguments in $\overline{\frac{\partial a_\beta}{\partial p_{\tilde\beta}}}$ are some intermediate values $\vartheta D^{\tilde\beta}u$, $0 < \vartheta < 1$. Take $v_\beta = D^\beta u$ in the last equality. Transfering $A(0,v)$ to the other side of the equality and taking into account (1.4.8) and (1.1.2) we get

$$\int_\Omega a_\beta(x;D^l u)D^\beta u dx \ge \mu_0\int_\Omega \frac{|D^l u|^2}{(1+|\overline{p}|^2)^{(2-s)/2}}dx - \int_\Omega a_\beta(x;0)D^\beta u dx.$$

From $0 < \vartheta < 1$ it follows that $1+|\overline{p}|^2 \le 1+|p|^2 = 1+|D^l u|^2$. Under assumption 3) of section 1.1 the function $a_\beta(x,0) \in \mathcal{L}_{s/(s-1)}$. The elementary inequality

$$|ab| \le \eta|a|^s + C|b|^{s/(s-1)},$$

where η is (as usual in this book) an arbitrary positive constant gives

$$\int_\Omega a_\beta(x; D^l u) D^\beta u dx \geq \mu_0 \int_\Omega \frac{|D^l u|^2}{(1+|D^l u|^2)^{(2-s)/2}} dx - \eta \int_\Omega |D^l u|^s dx - C =$$

$$= \mu_0 \int_\Omega (1+|D^l u|^2)^{s/2} dx - \mu_0 \int_\Omega \frac{dx}{(1+|p|^2)^{(2-s)/2}} - \eta \int_\Omega |D^l u|^s dx - C.$$

Obviously, if we take $|p| = 0$ in the middle term on the right-hand side we get inequality (1.4.7).
As u_δ is a weak solution of problem (1.4.3),(1.1.14), we come to an integral identity

$$\int_\Omega a_{\beta,\delta}(x; D^l u_\delta) D^\beta v dx = 0, \tag{1.4.9}$$

which is true for $\forall v \in \overset{\circ}{W}{}_2^{(l)}(\Omega)$. It follows from $s \leq 2$ and $u_\delta \in \overset{\circ}{W}{}_2^{(l)}(\Omega)$ that it is possible to take $v = u_\delta$. Then, using (1.4.2) we have

$$\delta \int_\Omega |D^l u_\delta|^2 dx + \int_\Omega a_\beta(x; D^l u_\delta) D^\beta u_\delta dx = 0, \tag{1.4.10}$$

and (1.4.7) gives the following estimate

$$\|u_\delta\|_{W_s^{(l)}(\Omega)} \leq C, \tag{1.4.11}$$

where C does not depend on δ. Applying (1.4.6), we arrive at the inequality

$$\|a_\beta(x; D^l u_\delta)\|_{W_{s'}^{(l)}(\Omega)} \leq C, \quad s' = \frac{s}{s-1}, \tag{1.4.12}$$

where C also does not depend on δ. Inequalities (1.4.11), (1.4.12) along with weak compactness of the ball in $W_s^{(l)}$ give the following weak limits*

$$\lim_{\delta\to 0} u_\delta = u^*(x) \quad \text{and} \quad \lim_{\delta\to 0} a_\beta(x; D^l u_\delta) = a_\beta^*(x).$$

Passing to the limit in (1.4.9), we get

$$\int_\Omega a_\beta^*(x) D^\beta v dx = 0. \tag{1.4.13}$$

Let us now show that the equality

$$\int_\Omega a_\beta(x; D^l u^*) D^\beta v dx = 0$$

holds. This will complete the proof of the theorem. For this purpose, we shall prove that $u_\delta \to u^*$ in $W_s^{(l)}(\Omega)$, when $\delta \to 0$. Take any sufficiently small $\eta > 0$ and a smooth function w such that the inequality

$$\|u^* - w\|_{W_s^{(l)}} < \eta \tag{1.4.14}$$

*We denote the subsequence of δ, for which u^* and a^* exist by the same symbol δ.

holds.
Consider the integral

$$J_\delta = \int_\Omega \left[a_{\beta,\delta}(x; D^l u_\delta) - a_{\beta,\delta}(x; D^l w)\right] D^\beta(u_\delta - w)dx, \tag{1.4.15}$$

where $a_{\beta,\delta}$ are defined by (1.4.2).
Integral (1.4.15) can be represented as the difference between the following two integrals

$$J_1 = \int_\Omega a_{\beta,\delta}(x; D^l u_\delta) D^\beta(u_\delta - w)dx \tag{1.4.16}$$

and

$$J_2 = \int_\Omega a_{\beta,\delta}(x; D^l w) D^\beta(u_\delta - w)dx. \tag{1.4.17}$$

Taking $v = u_\delta - w$ in (1.4.9), we get

$$J_1 = 0. \tag{1.4.18}$$

Consider integral (1.4.17). This integral can be introduced in the form

$$\begin{aligned} J_2 &= \delta \int_\Omega D^\beta w D^\beta(u_\delta - w)dx + \int_\Omega a_\beta(x; D^l w) D^\beta(u_\delta - w)dx = \\ &= \delta \int_\Omega D^\beta w D^\beta(u_\delta - u^*)dx + \delta \int_\Omega D^\beta w D^\beta(u^* - w)dx + \\ &+ \int_\Omega a_\beta(x; D^l w) D^\beta(u_\delta - u^*)dx + \int_\Omega a_\beta(x; D^l w) D^\beta(u^* - w)dx. \end{aligned}$$

Denote successively the integrals on the right-hand side by J_3, J_4, J_5 and J_6. Then we can have

$$J_2 = \delta J_3 + \delta J_4 + J_5 + J_6. \tag{1.4.19}$$

As $u_\delta \to u$ in a weak sense, when $\delta \to 0$, then if $\delta < \delta_1$ ($\delta_1 > 0$ and sufficiently small), we get

$$\delta |J_3| < \eta. \tag{1.4.20}$$

Using the Hölder inequality for J_4, we get

$$|J_4| \leq \|w\|_{W_{s'}^{(l)}} \|u^* - w\|_{W_s^{(l)}}.$$

Take $\delta_2 > 0$ so small, that from $\forall \delta < \delta_2$ it follows that $\delta \|w\|_{W_{s'}^{(l)}} < 1$. Then from the previous inequality and (1.4.14) we come to

$$\delta |J_4| < \eta. \tag{1.4.21}$$

It is obvious that the inequality

$$\|a_\beta(x; D^l w)\|_{\mathcal{L}_{s'}} < C \tag{1.4.22}$$

holds (C does not depend on δ). This follows from inequality (1.4.6) which can be introduced in the form

$$\|a_\beta(x; D^l w)\|^{s'}_{\mathcal{L}_{s'}} \leq C_1 \|w\|^s_{W^{(l)}_s} + C_2.$$

Then

$$\|a_\beta(x; D^l w)\|^{s'}_{\mathcal{L}_{s'}} \leq C_1 \left(\|w - u^*\|_{W^{(l)}_s} + \|u^*\|_{W^{(l)}_s}\right)^s + C_2.$$

Using (1.4.14), we come to (1.4.22).
Let us now estimate the integral J_6. Applying the Hölder inequality to J_6, we find

$$|J_6| \leq \sum_{0 \leq |\beta| \leq l} \|a_\beta(x; D^l w)\|_{\mathcal{L}_{s'}} \|u^* - w\|_{W^{(l)}_s}.$$

Then it follows from (1.4.14) and (1.4.22) that

(1.4.23) $$|J_6| < C\eta.$$

It remains now to give the estimate for J_5. As $u_\delta \to u^*$ weakly, when $\delta \to 0$, then for some $\delta_3 > 0$ and $\forall \delta < \delta_3$, we get

(1.4.24) $$|J_5| < \eta.$$

Taking the min δ_k $(k = 1, 2, 3)$ with the help of (1.4.20), (1.4.21), (1.4.23) and (1.4.24) we arrive at

$$|J_2| < C\eta.$$

Using equalities (1.4.15) – (1.4.18), we get

$$|J_\delta| < C\eta.$$

Then it follows from (1.4.2) and lemma 1.3.2. that

$$\left| \int_\Omega \left[a_\beta(x; D^l u_\delta) - a_\beta(x; D^l w)\right] D^\beta (u_\delta - w) dx \right| < C\eta$$

and inequality (1.3.3) gives

(1.4.25) $$\frac{\|u_\delta - w\|^2_{W^{(l)}_s}}{1 + \|u_\delta - w\|^{2-s}_{W^{(l)}_s} + \|w\|^{2-s}_{W^{(l)}_s}} \leq C\eta.$$

Using (1.4.14), we get

$$\|w\|_{W^{(l)}_s} \leq C.$$

Let us now show that

(1.4.26) $$\|u_\delta - w\|_{W^{(l)}_s} < C\eta^{1/2}.$$

It is sufficient to have (1.4.26) for small η. Denote $t = \|u_\delta - w\|_{W^{(l)}_s}$ and $1 + \|w\|^{2-s}_{W^{(l)}_s} = a^2$. Then (1.4.25) can be introduced in the form

$$\frac{t^2}{a^2 + t^{2-s}} < C\eta \quad (1 < s \leq 2).$$

Let us first have $t \leq a^{2/(2-s)}$. Then from the previous relation it follows that $t < C\eta^{1/2}$. If $t > a^{2/(2-s)}$, we have $t < C\eta^{1/s}$ from the same inequality. Therefore, for sufficiently small η, we get inequality (1.4.26). This inequality shows that $u_\delta \to u^*$ in $W_s^{(l)}$. Hence

$$0 = \int_\Omega a_{\beta,\delta}(x; D^l u_\delta) D^\beta v dx \longrightarrow \int_\Omega a_\beta(x; D^l u^*) D^\beta v dx.$$

□

This theorem was proved at first in the author's paper [4].

Corollary 1.4.1 *Consider the iterative process* (1.4.4) *for system* (1.4.3) *with boundary conditions* (1.1.14), *which can be also written in the weak form. Denote the iterations of* (1.4.1) *by* $u_n^{(\delta)}$. *From theorems* 1.3.1 *and* 1.4.1 *it follows that there exists a subsequence of* $u_{n_k}^{(\delta_k)}$ *which converges to the solution* $u(x)$ *of* (1.4.3), (1.1.14) *in* $W_s^{(l)}(\Omega)$.

In fact, for any $\delta > 0$ the solution of (1.4.3), (1.1.14) belongs to $\overset{\circ}{W}{}_2^{(l)}(\Omega)$. Therefore, $u_n^{(\delta)}$ according to theorem 1.3.1 tends to u_δ in $W_s^{(l)}(\Omega)$. Since some subsequence u_{δ_k} of u_δ converges in $W_s^{(l)}(\Omega)$ to $u(x)$ then the diagonal subsequence $u_{n_k}^{(\delta_k)}$ will also converge to u in $W_s^{(l)}(\Omega)$. □

Chapter 2

Regularity of solutions for non degenerated quasilinear second order elliptic systems of the divergent form with bounded nonlinearities

2.1 Some functional spaces and preliminary results

In the previous chapter we have used the Sobolev spaces $W_s^{(l)}(\Omega)$ and $W_s^{k,l}(Q)$. We have also considered the space $C^{k,\gamma}$. This was a space of k-times continuously differentiable functions, where the derivatives of the order k satisfy the Hölder condition with the exponent γ. The norm in C^γ is

(2.1.1) $$||u||_{k,\gamma} = \sup_{x,y\in\overline{\Omega};|\beta|=\kappa} \frac{|D^\beta u(x) - D^\beta u(y)|}{|x-y|^\gamma} + \sup_{x\in\overline{\Omega}} |u(x)|.$$

When $k = 0$, the norm will be denoted by $|u|_\gamma$ (the norm in C will be denoted by $|u|_0$). We shall often use the Sobolev spaces $W_p^{(l)}(\Omega)$, the spaces of vector functions $u(x)$ with every $u^{(i)}(x)(i = 1,\ldots,N)\mathcal{L}_p$ - integrable ($p \geq 1$) in the considered domain along with all their derivatives of the order l. The norm in $W_p^{(l)}$ is

$$||u||_{W_p^{(l)}(\Omega)} = \left(\int_\Omega |D^l u|^p dx\right)^{1/p}.$$

We shall also consider the spaces $\mathcal{L}_{p,\alpha}(\Omega, x_0)$ of the functions which are p-summable with the weight $|x - x_0|^\alpha$ and the norm

$$||u||_{\mathcal{L}_{p,\alpha}(\Omega,x_0)} = \left(\int_\Omega |u|^p |x-x_0|^\alpha dx\right)^{1/p}.$$

The functional spaces $W^{(l)}_{p,\alpha}(\Omega, x_0)$ with the norm

$$\|u\|_{W^{(l)}_{p,\alpha}(\Omega,x_0)} = \left(\int_\Omega |D^l u|^p |x - x_0|^\alpha \right)^{\frac{1}{p}}$$

are similarly denoted. It is known, that the class of C^∞ functions is dense in $W^{(l)}_{p,\alpha}$. Let $\alpha = 2 - m - 2\gamma, 0 < \gamma < 1$. Consider a space of functions from $W^{(l)}_p(\Omega)$ with a finite norm

$$\|u\|_{l,p,\alpha} = \sup_{x_0 \in \overline{\Omega}} \left(\int_\Omega |D^l u|^p |x - x_0|^\alpha dx \right)^{1/p} . \tag{2.1.2}$$

We denote this space by $H_{l,p,\alpha}$. Instead of $H_{1,p,\alpha}$ we shall use $H_{p,\alpha}$ and, when $p = 2$ simply H_α. So, $W^{(1)}_2$ coincides with H_0 and

$$\|u\|_{1,p,\alpha} = \|u\|_{p,\alpha} = \sup_{x_0 \in \overline{\Omega}} \left(\int_\Omega |Du|^p |x - x_0|^\alpha dx \right)^{1/p} . \tag{2.1.3}$$

In particular,

$$\|u\|_\alpha = \sup_{x_0 \in \overline{\Omega}} \left(\int_\Omega |Du|^2 |x - x_0|^\alpha dx \right)^{1/2} . \tag{2.1.4}$$

When $\alpha = 0$ and $p = 2$ (i.e. $W^{(1)}_2$), the norm of H_0 will be simply denoted by $\|\cdot\|$. In some cases we shall consider certain other expressions for the norms in $H_{l,p,\alpha}$. For example, the norm can be defined by the formula

$$\|u\|'_{l,p,\alpha} = \sup_{x_0 \in \overline{\Omega}} \left[\int_\Omega \left(|D^{l'} u|^p + |u|^p \right) |x - x_0|^\alpha dx \right]^{1/p} . \tag{2.1.2'}$$

We shall denote this space by $H'_{l,p,\alpha}$. Later in the corollary to the theorem 2.1.2 we shall show that the spaces $H_{l,p,\alpha}$ and $H'_{l,p,\alpha}$ with the norms (2.1.2) and (2.1.2′) are equivalent.
The H_α spaces for $m = 2$ were introduced by Nirenberg [1]. With the help of these spaces he proved the smoothness of solutions of the boundary value problems for the second order quasilinear equations with two variables. For the case of arbitrary $m > 2$, Cordes [1] proved some important theorems concerning regular solutions of elliptic equations. Later Krein et al. have considered a general theory of these spaces (see the paper by Gluŝko [1] and Gluŝko and Krein [1]). It is well known that the function u is not necessarily continuous, when $u \in W^{(1)}_2(\Omega)(m \geq 2)$. But when $u \in H_\alpha$, where $\alpha = 2 - m - 2\gamma, 0 < \gamma < 1$, the function is continuous and satisfies the Hölder condition with the exponent γ . These properties of H_α and similar spaces, introduced by Morrey [1]-[4] and Campanato[1], are very important in the considerations, concerning regular solutions of boundary value problems for elliptic equations.

Consider now some properties of the spaces $H_{l,p,\alpha}$. Let $\delta > 0$ and $B_\delta(x_0)$ be a ball with the centre x_0, radius δ and $\Omega_\delta(x_0) = \Omega \cap B_\delta(x_0)$. Let us set $\delta_0 > 0$, and consider the following norm in $H_{l,p,\alpha}$

$$\|u\|''_{l,p,\alpha} = \sup_{\delta \le \delta_0, x_0 \in \overline{\Omega}} \left(\int_{\Omega_\delta(x_0)} |D^l u|^p |x - x_0|^\alpha dx \right)^{1/p}. \tag{2.1.5}$$

We shall show that the norms (2.1.2) and (2.1.5) are equivalent. Because of

$$\sup_{\delta \le \delta_0, x_0 \in \overline{\Omega}} \int_{\Omega_\delta(x_0)} |D^l u|^p |x - x_0|^\alpha dx \le \sup_{x_0 \in \overline{\Omega}} \int_{\Omega} |D^l u|^p |x - x_0|^\alpha dx,$$

we have an inequality

$$\|u\|''_{l,p,\alpha} \le \|u\|_{l,p,\alpha}. \tag{2.1.6}$$

It follows from (2.1.2) that

$$\|u\|^p_{l,p,\alpha} \le \sup_{x_0 \in \overline{\Omega}} \int_{\Omega_\delta(x_0)} |D^l u|^p |x - x_0|^\alpha dx + \sup_{x_0 \in \overline{\Omega}} \int_{\Omega / \Omega_\delta(x_0)} |D^l u|^p |x - x_0|^\alpha dx.$$

With $\alpha < 0$ and $x \in \Omega/\Omega_{\delta_0}(x_0)$, the inequality $|x - x_0|^\alpha \le \delta^\alpha$ holds. Then we have

$$\|u\|^p_{l,p,\alpha} \le \sup_{\delta \le \delta_0, x_0 \in \overline{\Omega}} \int_{\Omega_\delta(x_0)} |D^l u|^p |x - x_0|^\alpha dx + \delta_0^\alpha \int_\Omega |D^l u|^p dx. \tag{2.1.7}$$

Let us cover Ω with a finite set of balls $B_{\delta_0}(x_k)$ and let r_k be the distance between x and x_k. When $x \in B_{\delta_0}(x_k)$ then $1 \le r_k^\alpha \delta_0^{-\alpha}$ and

$$\begin{aligned} \int_\Omega |D^l u|^p dx &\le \sum_k \delta_0^{-\alpha} \int_{\Omega_{\delta_0}(x_k)} |D^l u|^p r_k^\alpha dx \le \\ &\le \delta_0^{-\alpha} C(\delta_0) \sup_{\delta \le \delta_0, x_0 \in \overline{\Omega}} \int_{\Omega_{\delta_0}(x_0)} |D^l u|^p |x - x_0|^\alpha dx. \end{aligned}$$

Therefore we have

$$\|u\|_{l,p,\alpha} \le C \|u\|''_{l,p,\alpha}.$$

This inequality together with (2.1.6) proves that the norms (2.1.2) and (2.1.5) are equivalent.

Later we shall use the Hardy inequality (Hardy, Littelwood, Polya [1]). Let $p > 1, s \ne 1, f \in \mathcal{L}_{p,p-s}(R^1_+, 0)$ $\left(R^1_+ = \{x : x \ge 0\}\right)$ and $F(r) = \int_0^r f(\rho) d\rho$ $(s > 1)$, $F(r) = \int_r^{+\infty} f(\rho) d\rho$ $(s < 1)$. Then the following inequality is true

$$\int_0^{+\infty} r^{-s} |F(r)|^p dr \le \left(\frac{p}{|s-1|} \right)^p \int_0^{+\infty} r^{-s} (r|f|)^p dr. \tag{2.1.8}$$

Suppose $u(r)$ is a scalar function with $u' = 0$ for $r > \delta > 0$, $f(r) = u' \in \mathcal{L}_{p,\alpha+m-1}(R^1_+, 0), m \geq 2, \alpha = p - m - p\gamma, 0 < \gamma < 1$ and $s = p\gamma + 1$. Using the Hardy inequality (2.1.8), we get

$$\int_0^\delta |u(r) - u(0)|^p r^{\alpha+m-3} dr \leq \tag{2.1.9}$$

$$\leq \frac{1}{\gamma^p} \int_0^\delta |u'(r)|^p r^{\alpha+m-1} dr.$$

It follows from the Hardy's inequality that this relation is also true for $\delta = +\infty$. Let us now prove some results concerning embedding $H_{p,\alpha}$ in $C^{0,\gamma}$ spaces. We start with a simple result for $p = 2$.

Lemma 2.1.1. *Let $u(r)$ be a scalar function, defined on $[0, \delta]$ and let $\alpha = 2 - m - 2\gamma, 0 < \gamma < 1$. If $u \in W^{(1)}_{2,\alpha+m-1}([0, \delta], 0)$, then*

$$|u(0)|^2 \leq \eta \int_0^\delta |u'|^2 r^{\alpha+m-1} dr + C \int_0^\delta |u|^2 r^{m-1} dr. \tag{2.1.10}$$

Proof. It is sufficient to prove (2.1.10) for smooth functions. Hence

$$u(0) = u(r) - [u(r) - u(0)],$$

and

$$|u(0)|^2 \leq 2\left[|u(r)|^2 + |u(r) - u(0)|^2\right].$$

Let us multiply this inequality by r^{m-1} and integrate it over a small interval $[0, \varepsilon], \delta > \varepsilon > 0$. Then we have

$$m^{-1}\varepsilon^m |u(0)|^2 \leq 2 \int_0^\varepsilon |u(r)|^2 r^{m-1} dr + 2 \int_0^\varepsilon |u(r) - u(0)|^2 r^{m-1} dr.$$

Consider the last term on the right-hand side. Obviously,

$$\int_0^\varepsilon |u(r) - u(0)|^2 r^{m-1} dr = \int_0^\varepsilon r^{-\alpha+2} |u(r) - u(0)|^2 r^{\alpha+m-3} dr \leq$$

$$\leq \varepsilon^{m+2\gamma} \int_0^\varepsilon |u(r) - u(0)|^2 r^{\alpha+m-3} dr \leq \varepsilon^{m+2\gamma} \int_0^\delta |u(r) - u(0)|^2 r^{\alpha+m-3} dr.$$

Now apply (2.1.9). We now have

$$\varepsilon^m |u(0)|^2 \leq \frac{2m\varepsilon^{m+2\gamma}}{\gamma^2} \int_0^\delta |u'(r)|^2 r^{\alpha+m-1} dr + C \int_0^\delta |u(r)|^2 r^{m-1} dr.$$

If we divide both parts by ε^m and set $\frac{2m\varepsilon^{2\gamma}}{\gamma^2} = \eta$, we get (2.1.10). □

We suppose $\partial\Omega$ satisfies the following two conditions: 1)there exists a positive constant A_Ω that for all $x_0 \in \overline{\Omega}$ the relation

$$|\Omega \cap B_\delta(x_0)| > A_\Omega \delta^m \tag{2.1.11}$$

is satisfied; 2) let $\Omega'_\delta = B_\delta(x_0) \cap \partial\Omega$; then there exists $\delta_0 > 0$ such that for all $x_0 \in \overline{\Omega}$ and $\delta < \delta_0$ the surface Ω'_δ can be represented with either the help of equation $r = F(\theta)$ where F is continious and (r, θ) are the spherical coordinates in R^m with the center x_0, or the surface is a part of a conic surface with vertex x_0. It is obvious then that if $\partial\Omega$ satisfies locally the Lipschitz condition then the conditions 1) and 2) are satisfied.

Theorem 2.1.1. *If $p > 1, m \geq 2, \alpha = p - m - p\gamma, 0 < \gamma < 1$ and $\partial\Omega$ locally satisfies Lipschitz condition, then*

$$H_{p,\alpha} \subset C^{0,\gamma}$$

and

$$\sup_{x_1, x_2 \in \overline{\Omega}} \frac{|u(x_1) - u(x_2)|}{|x_1 - x_2|^\gamma} \leq \frac{2^{1+\frac{m}{p}}}{A_\Omega^{1/p}\gamma} \|u\|_{p,\alpha}. \tag{2.1.12}$$

Proof. It is sufficient to prove this theorem for smooth functions. Take a point $x_0 \in \Omega$ as the centre of a ball $B_\delta(x_0)$, where δ is sufficiently small. Let ρ, θ be spherical coordinates with the origin x_0 and $0 \leq \rho \leq F(\theta)$, where θ lies on the unit sphere S and the equation of $\partial\Omega_\delta(x_0)$ has the form $\rho = F(\theta)$. Then

$$u(x) - u(x_0) = \int_0^\rho \frac{\partial u}{\partial \tau} d\tau,$$

where $\frac{\partial}{\partial\tau}$ is the directional derivative. If we fix θ and extend $\frac{\partial u}{\partial \tau}$ with zero outside the interval $[0, F(\theta)]$, then with the help of (2.1.9) we have

$$\int_0^{F(\theta)} |u(x) - u(x_0)|^p |x - x_0|^{-1-p\gamma} d\rho \leq$$
$$\leq \frac{1}{\gamma^p} \int_0^{F(\theta)} |D'u|^p |x - x_0|^{p-1-p\gamma} d\rho.$$

After integration over S * we get

$$\int_{\Omega_\delta(x_0)} |u(x) - u(x_0)|^p |x - x_0|^{-m-p\gamma} dx \leq \tag{2.1.13}$$
$$\leq \frac{1}{\gamma^p} \int_{\Omega_\delta(x_0)} |D'u|^p |x - x_0|^{p-m-p\gamma} dx.$$

Take two arbitrary points $x_1, x_2 \in \overline{\Omega}$. We assume that the distance between these points is less than δ, where $\delta > 0$ is sufficiently small. Let $\overline{x}$ be the centre of the straight segment between x_1 and x_2 and $y \in B_{\frac{|x_1 - x_2|}{2}}(\overline{x}) \cap \Omega$. Obviously,

$$|u(x_2) - u(x_2)|^p \leq 2^{p-1} \left[|u(x_1) - u(y)|^p + |u(x_2) - u(y)|^p \right].$$

*If $x_0 \in \partial\Omega$ then part of $\partial\Omega_\delta$ can coincide with a cone with the vertex x_0. In this case the integration is extended only over part of S.

As $|x_1 - x_2| \geq |x_1 - y|$ and $|x_1 - x_2| \geq |x_2 - y|$, then

$$\frac{|u(x_1) - u(x_2)|^p}{|x_1 - x_2|^{m+p\gamma}} \leq 2^{p-1}\left[\frac{|u(x_1) - u(y)|^p}{|x_1 - y|^{m+p\gamma}} + \frac{|u(x_2) - u(y)|^p}{|x_2 - y|^{m+p\gamma}}\right].$$

Integrating this inequality over $B_{\frac{|x_1-x_2|}{2}}(\overline{x}) \cap \Omega$ and applying (2.1.11) we get

$$\frac{A_\Omega}{2^m}|u(x_1) - u(x_2)|^p|x_1 - x_2|^{-p\gamma} \leq$$
$$\leq 2^{p-1} \int\limits_{B_{\frac{|x_1-x_2|}{2}}(\overline{x})\cap\Omega} |u(x_1) - u(y)|^p|x_1 - y|^{-m-p\gamma}dy +$$
$$+2^{p-1} \int\limits_{B_{\frac{|x_1-x_2|}{2}}(\overline{x})\cap\Omega} |u(x_2) - u(y)|^p|x_2 - y|^{-m-p\gamma}dy.$$

The right-hand side will be larger if we integrate in the first term over $\Omega_{|x_1-x_2|}(x_1)$ and in the second over $\Omega_{|x_1-x_2|}(x_2)$. Now we come to (2.1.12). □

Remark 2.1.1. *Take* $\Omega = B_R(x_0)$, $p = 2$ *and* $\alpha = 2 - m - 2\gamma (0 < \gamma < 1)$. *Then the left-hand side of inequality* (2.1.13) *can be written in the form*

$$\int\limits_{B_R(x_0)} |u(x) - u(x_0)|^2|x - x_0|^{\alpha-2}dx \leq \frac{1}{\gamma^2} \int\limits_{B_R(x_0)} |D'u|^2|x - x_0|^\alpha dx. \tag{2.1.14}$$

Theorem 2.1.2. *If the condition of the previous theorem are satisfied,then with an arbitrary* $\eta > 0$, *we have*

$$|u(x_0)| \leq \eta \left(\int\limits_{\Omega_\delta(x_0)} |D'u|^p|x - x_0|^\alpha dx\right)^{1/p} + C\|u\|_{\mathcal{L}_p} \tag{2.1.15}$$

and

$$|u|_0 \leq \eta\|u\|_{p,\alpha} + C\|u\|_{\mathcal{L}_p}. \tag{2.1.16}$$

Proof. First we shall prove (2.1.15). Introduce the spherical system of coordinates (ρ, θ) with the pole x_0. Consider the domain $\Omega_\delta(x_0)$ and take an arbitrary point x inside this domain. Hence

$$u(x_0) = u(x) - \int\limits_0^\rho \frac{\partial u}{\partial \tau}d\tau.$$

From this follows

$$|u(x_0)|^p \leq 2^{p-1}\left[|u(x)|^p + \left|\int\limits_0^\rho \frac{\partial u}{\partial \tau}d\tau\right|^p\right].$$

Integrating over $\Omega_\delta(x_0)$, we have

$$\text{mes}\Omega_\delta(x_0)|u(x_0)|^p \leq 2^{p-1}\left[\int\limits_\Omega |u|^p dx + \int\limits_{\Omega_\delta(x_0)} \left|\int\limits_0^\rho \frac{\partial u}{\partial \tau}d\tau\right|^p dx\right].$$

Suppose that the point $(\overline{\rho}, \theta)$ belongs to $\partial\Omega_\delta(x_0)$. From the property 2) of $\partial\Omega$ it follows that

$$\int_{\Omega_\delta(x_0)} \left| \int_0^\rho \frac{\partial u}{\partial \tau} d\tau \right|^p dx \leq \int_{S_\delta(x_0)} dS \int_0^{\overline{\rho}} \left| \int_0^\rho \frac{\partial u}{\partial \tau} d\tau \right|^p \rho^{m-1} d\rho =$$

$$= \int_S dS \int_0^{\overline{\rho}} \rho^{p-\alpha} \rho^{\alpha+m-1-p} \left| \int_0^\rho \frac{\partial u}{\partial \tau} d\tau \right|^p d\rho \leq \delta^{m+p\gamma} \int_{S_\delta(x_0)} dS \int_0^{\overline{\rho}} \rho^{-1-p\gamma} \left| \int_0^\rho \frac{\partial u}{\partial \tau} d\tau \right|^p d\rho,$$

where $S_\delta(x_0)$ is a part of the unit sphere S. Using inequality (2.1.11) we come to

$$\delta^m |u(x_0)|^p \leq C \left[\delta^{m+p\gamma} \int_{\Omega_\delta(x_0)} |D^l u|^p |x - x_0|^\alpha dx + \|u\|^p_{\mathcal{L}_p} \right].$$

Setting $C\delta^{p\gamma} = \eta^p$, we come to (2.1.15). In order to prove (2.1.16) we must take sup over $x_0 \in \overline{\Omega}$ on both sides of the last inequality. □

Corollary 2.1.1 *The norms in $H_{l,p,\alpha}$ and $H'_{l,p,\alpha}$ are equivalent.*

Proof. From (2.1.2) and (2.1.2′) it follows that $\|u\|'_{l,p,\alpha} \leq \|u\|_{l,p,\alpha}$. Let us now show that the norm $\|u\|_{l,p,\alpha}$ can be estimated by a constant multiplied by $\|u\|'_{l,p,\alpha}$. Hence it is sufficient to prove that for $1 \leq s \leq l$ the inequality

$$\|u\|_{l-s,p,\alpha} \leq \eta \|u\|_{l,p,\alpha} + C\|u\|_{\mathcal{L}_p} \tag{2.1.17}$$

holds true. From (2.1.15) it follows that

$$\left|D'^{l-s} u\right| \leq \eta \|u\|_{l-s+1,p,\alpha} + C \left\|D^{l-s} u\right\|_{\mathcal{L}_p} \leq \eta \|u\|_{l,p,\alpha} + C \left\|D^{l-1} u\right\|_{\mathcal{L}_p}.$$

Using the very well known relation

$$\|D^{l-1} u\|_{\mathcal{L}_p} \leq \eta \|D^l u\|_{\mathcal{L}_p} + C\|u\|_{\mathcal{L}_p}$$

and taking into account the fact that $\alpha < 0$, we have

$$\int_\Omega |D^l u|^p dx \leq C \int_\Omega |D^l u|^p |x - x_0|^\alpha dx,$$

$$|D'^{l-s} u| \leq \eta \|u\|_{l,p,\alpha} + C\|u\|_{\mathcal{L}_p}$$

and

$$\left|D'^{l-s} u\right|^p \leq \eta \|u\|^p_{l,p,\alpha} + C\|u\|^p_{\mathcal{L}_p}.$$

Multiplying both sides by $|x - x_0|^\alpha$, integrating over Ω and using the fact that $\int_\Omega |x - x_0|^\alpha dx < +\infty \text{with} \alpha > -m$, we come to the following inequality

$$\int_\Omega \left|D'^{l-s} u\right|^p |x - x_0|^\alpha dx \leq \eta \|u\|^p_{l,p,\alpha} + C\|u\|^p_{\mathcal{L}_p}.$$

Now take the sup over all $x_0 \in \overline{\Omega}$, use the elementary inequality $|a + b|^{1/p} \leq |a|^{1/p} + |b|^{1/p}$; then the proof of (2.1.17) is completed. □

Next we shall prove an inequality

$$(2.1.17') \qquad \int\limits_{\Omega_\delta(x_0)} |u|^2|x-x_0|^\alpha dx \le \eta \int\limits_{\Omega_\delta(x_0)} |D'u|^2|x-x_0|^\alpha dx + C\int\limits_\Omega |u|^2 dx$$

with $\delta > 0$ sufficiently small. Analogous to the proof of theorem 2.1.2 we come to

$$\int\limits_{\Omega_\delta(x_0)} |u|^2|x-x_0|^\alpha dx \le 2\int\limits_S dS \int\limits_0^{\overline{\rho}} |u(x_0)|^2 \rho^{\alpha+m-1} d\rho +$$

$$+2\int\limits_S dS \int\limits_0^{\overline{\rho}} \rho^{\alpha+m-1} \left| \int\limits_0^\rho \frac{\partial u}{\partial \tau} d\tau \right|^2 d\rho.$$

Estimating the right-hand side by the application (2.1.9) and (2.1.10), we get

$$\int\limits_{\Omega_\delta(x_0)} |u|^2|x-x_0|^\alpha dx \le \frac{2}{\alpha+m}\int\limits_S \rho^{-\alpha+m}\left[\eta\int\limits_0^{\overline{\rho}} \left|\frac{\partial u}{\partial \rho}\right|^2 \rho^{\alpha+m-1} d\rho + \right.$$

$$\left. +C_1 \int\limits_0^{\overline{\rho}} |u|^2 \rho^{m-1} d\rho \right] dS + C_2 \int\limits_S \rho^{-2} \int\limits_0^{\overline{\rho}} \left|\frac{\partial u}{\partial \rho}\right|^2 \rho^{\alpha+m-1} d\rho dS.$$

As $\overline{\rho} \le \delta$ and $\left|\frac{\partial u}{\partial \rho}\right|^2 \le |D'u|^2$, we come to

$$\int\limits_{\Omega_\delta(x_0)} |u|^2|x-x_0|^\alpha dx \le \left(\frac{2\delta^{\alpha+m}}{\alpha+m}\eta + C_2\delta^2\right) \int\limits_{\Omega_\delta(x_0)} |D'u|^2|x-x_0|^\alpha dx +$$

$$+\frac{2C_1\delta^{\alpha+m}}{\alpha+m} \int\limits_{\Omega_\delta(x_0)} |u|^2 dx.$$

This completes the proof of (2.1.17').
We shall also consider the space $\widetilde{W}^{(2)}_{2,\tilde{\alpha}}(R^m, x_0)$ which comprises all functions with a finite norm

$$(2.1.18) \qquad \|u\|_{\widetilde{W}^{(2)}_{2,\tilde{\alpha}}(R^m,x_0)} = \left(\sum_{k=0}^{2} \int\limits_{R^m} |D'^k u|^2 |x-x_0|^{\tilde{\alpha}-4+2k} dx\right)^{1/2},$$

where x_0 is a fixed point in R^m. The space $\widetilde{W}^{(2)}_{2,\tilde{\alpha}}(R^m, x_0)$ will be applied in cases where $\tilde{\alpha} \in (m-4, m), m \ge 4$ and $\tilde{\alpha} \in (1,2), m = 3$. It can be shown that the space of function C^∞, which is finite on R^m is dense in $\widetilde{W}^{(2)}_{2,\tilde{\alpha}}(R^m, x_0)$.
The closure in the norm $W^{(l)}_{2,\alpha}$ of the space of smooth functions which vanish in the neighbourhood of $\partial\Omega$ we define as $\overset{\circ}{W}{}^{(l)}_{2,\alpha}$. In the following lemma we establish an inequality analogous to that of Fridrichs-Poincaré.

Lemma 2.1.2. *Let* $\alpha > -m$ *and* $v \in \overset{\circ}{W}{}^{(1)}_{2,\alpha}(B_\rho(x_0), x_0)$*. Then*

$$(2.1.19) \qquad \|v\|_{\mathcal{L}_{2,\alpha}(B_\rho)} \le C\rho\|D'v\|_{\mathcal{L}_{2,\alpha}(B_\rho)},$$

where $\mathcal{L}_{2,\alpha}(B_\rho) = \mathcal{L}_{2,\alpha}(B_\rho(x_0), x_0)$.

Proof. It is sufficient to prove the lemma for a smooth and finite function v. In this case

$$v(x) = -\int_r^\rho \frac{\partial v}{\partial r} dr,$$

where $r = |x - x_0|$. Let S be a unit sphere in R^m. Applying the Hölder inequality, we get

$$\|v\|^2_{\mathcal{L}_{2,\alpha}(B_\rho)} = \int_{B_\rho} |v|^2 r^\alpha dx = \int_S dS \int_0^\rho |v|^2 r^{\alpha+m-1} dr =$$

$$= \int_S dS \int_0^\rho \left| \int_r^\rho \frac{\partial v}{\partial \tau} d\tau \right|^2 r^{\alpha+m-1} dr \leq$$

$$\leq \int_S dS \int_0^\rho \left(\int_r^\rho \tau^{-\alpha-m+1} d\tau \right) \left(\int_r^\rho \left| \frac{\partial v}{\partial \tau} \right|^2 \tau^{\alpha+m-1} d\tau \right) r^{\alpha+m-1} dr \leq$$

$$\leq \int_S dS \int_0^\rho \left| \frac{\partial v}{\partial \tau} \right|^2 \tau^{\alpha+m-1} d\tau \int_0^\rho \int_r^\rho \tau^{-\alpha-m+1} d\tau r^{\alpha+m-1} dr =$$

$$= \frac{\rho^2}{2(\alpha+m)} \int_S dS \int_0^\rho \left| \frac{\partial v}{\partial r} \right|^2 r^{\alpha+m-1} dr = C^2 \rho^2 \int_{B_\rho} \left| \frac{\partial v}{\partial r} \right|^2 r^\alpha dx \leq$$

$$\leq C^2 \rho^2 \int_{B_\rho} |D'v|^2 r^\alpha dx = C^2 \rho^2 \|D'v\|^2_{\mathcal{L}_{2,\alpha}(B_\rho)}.$$

□

Corollary 2.1.2 *If* $v \in \overset{\circ}{W}{}^{(l)}_{2,\alpha}(B_\rho)$, *then we have*

(2.1.20) $$\|v\|_{W^{(l)}_{2,\alpha}(B_\rho)} \leq (1 + C_1\rho^2 + C_2\rho^l) \left\| D^{\prime l} v \right\|_{\mathcal{L}_{2,\alpha}(B_\rho)}.$$

Estimate (2.1.20) is derived from (2.1.19) with the help of the mathematical induction method.

□

Later we shall need inequality (2.1.15) for the special case, when the domain Ω is the ball $B_\delta(x_0)$.

Lemma 2.1.3. *If* $u \in W^{(1)}_{2,\alpha}(B)(B = B_1(x_0))$, *then the inequality*

(2.1.21) $$|u(x_0)|^2 < \eta \int_B |D'u|^2 r^\alpha dx + C_0(\eta) \int_B |u|^2 dx$$

holds with $\alpha = 2 - m - 2\gamma (0 < \gamma < 1)$ *and*

(2.1.22) $$C_0(\eta) = 2m|S|^{-(m/2\gamma+1)} \gamma^{-m/2\gamma} \eta^{-m/2\gamma}.$$

A separate proof is not necessary because repeating the proof of theorem 2.1.2 for $p = 2, \Omega = B_\delta(x_0)$, paying attention to the constants, is sufficient.
The powers of the integrals on the right-hand sides of inequalities (2.1.15), (2.1.21) are equal to one. These inequalities belong to the so-called class of linear inequalities; however in some problems it is important to have the so-called multiplicative inequalities, which we shall obtain in the following lemma.

Lemma 2.1.4. *If $u \in W^{(1)}_{2,\alpha}(B_R)$ and $u = 0$ on ∂B_R, then the inequality*

$$(2.1.23) \qquad |u(x_0)|^2 \le C \left(\int_{B_R} |D'u|^2 r^\alpha dx \right)^{\frac{m}{m+2\gamma}} \left(\int_{B_R} |u|^2 dx \right)^{\frac{2\gamma}{m+2\gamma}}$$

holds.

Proof. Substituting in (2.1.21) the expression (2.1.22) we get

$$|u(x_0)|^2 < \eta \int_{B_R} |D'u|^2 r^\alpha dx + C\eta^{-m/2\gamma} \int_{B_R} |u|^2 dx.$$

Then substituting

$$\eta = \left(\int_{B_R} |D'u|^2 r^\alpha dx \right)^{-2\gamma/(m+2\gamma)} \left(\int_{B_R} |u|^2 dx \right)^{2\gamma/(m+2\gamma)}$$

we come to (2.1.23) (if $|D'u| = 0$ then $u \equiv 0$ and (2.1.23) is trivial). □

Consider now a special case of Ω. Let T_κ be a spherical cone with a plane angle 2κ near the vertex.
We introduce the spherical coordinates $(r, \Theta_1, \dots, \Theta_{m-1})$ in such a way, that the axes of the cone coincides with the axes x_m and $x_m = r\cos\Theta_{m-1}$ $(0 \le \Theta_{m-1} \le \kappa$, $0 \le \Theta_i \le 2\pi (i = 1, \dots, m-2))$. Let $\Omega = T_\kappa \cap B_{\delta_0}(0) (\delta_0 \ge 0)$. Take x with $0 < |x| \le \delta_0$ and consider $\Omega_x = T_\kappa \cap B_{|x|}(0)$. For any smooth $u(x)$, which is determined in Ω we can write

$$\frac{|u(x) - u(0)|^2}{|x|^{m+2\gamma}} \le \frac{2|u(x) - u(y)|^2}{|x|^{m+2\gamma}} + \frac{2|u(y) - u(0)|^2}{|x|^{m+2\gamma}},$$

with $y \in \Omega_x$. Taking into account that

$$|x| \ge H|x - y|,$$

where

$$H = \begin{cases} 1 & 0 \le \kappa \le \pi/3 \\ \frac{1}{2\sin\frac{\kappa}{2}} & \pi/3 \le \kappa \le \pi, \end{cases}$$

we come to

$$\frac{|u(x) - u(0)|^2}{|x|^{m+2\gamma}} \le \frac{2}{H^{m+2\gamma}} \frac{|u(x) - u(y)|^2}{|x - y|^{m+2\gamma}} + 2\frac{|u(y) - u(0)|^2}{|y|^{m+2\gamma}}.$$

Integrating over Ω_x, we have

$$C\frac{|u(x) - u(0)|^2}{|x|^{2\gamma}} \le \frac{2}{H^{m+2\gamma}} \int_{\Omega_x} |u(x) - u(y)|^2 |x - y|^{-m-2\gamma} ds +$$
$$+2 \int_{\Omega_x} |u(y) - u(0)|^2 |y|^{-m-2\gamma} dy.$$

Taking the sphere $B_{|x|}(x)$ and enlarging the first integral by extending the integration over $B_{|x|}(x) \cap \Omega$ we get

(2.1.24)
$$\frac{|u(x)-u(0)|^2}{|x|^{2\gamma}} \leq C\left[\int\limits_{B_{|x|}(x)\cap\Omega} |u(x)-u(y)|^2|x-y|^{-m-2\gamma}dy + \right.$$
$$\left. + \int\limits_{\Omega_x} |u(y)-u(0)|^2|y|^{-m-2\gamma}dy\right].$$

Consider first the integral

$$I_1 = \int\limits_{B_{|x|}(x)\cap\Omega} |u(x)-u(y)|^2|x-y|^{-m-2\gamma}dy.$$

Introducing the spherical coordinates with the pole x and applying Hardy inequality (2.1.9), we come to estimate

$$I_1 \leq C \int\limits_{B_{|x|}(x)\cap\Omega} |\nabla u|^2|x-y|^{2-m-2\gamma}dy.$$

For the integral

$$I_2 = \int\limits_{\Omega_x} |u(y)-u(0)|^2|y|^{-m-2\gamma}dy$$

with the help of the above method we come to the inequality

$$I_2 \leq C\int\limits_{\Omega_x} |\nabla u|^2|y|^{2-m-2\gamma}dy$$

and we obtain

(2.1.25)
$$\frac{|u(x)-u(0)|^2}{|x|^{2\gamma}} \leq$$
$$\leq C\left[\int\limits_{B_{|x|}(x)\cap\Omega} |\nabla u|^2|x-y|^{\alpha}dy + \int\limits_{\Omega_x} |u(y)-u(0)|^2|y|^{\alpha}dy\right].$$

2.2 Estimates for ordinary differential operators.

Let $u_s(r), s = 0,1\ldots,$ be a system of scalar functions, defined on segment $[0,\delta]$, where δ is a positive constant. Consider on the same segment a smooth nonnegative function $\zeta(r)$ which satisfies the following conditions:

(2.2.1)
$$\zeta(r) = \begin{cases} 1 & \text{for } 0 \leq r \leq \delta/2, \\ 0 & \text{for } 3\delta/4 \leq r \leq \delta; \end{cases}$$

(2.2.2)
$$|\zeta'(r)| \leq C\delta^{-1},$$

where the constant C is independent of δ. Let $\alpha \in (-m, 2-m) \cup (2-m, 0]$ and $u_s(r), s = 0, 1, \ldots$, be smooth. Consider the following expressions

$$v_0(r) = -\int_r^\delta u_0'(\rho)\rho^\alpha d\rho, \tag{2.2.3}$$

$$v_s(r) = r^\alpha u_s(r)\zeta(r), s = 1, 2, \ldots \tag{2.2.4}$$

Consider also the integrals

$$I_s(u, z) = \int_0^\delta \left[u'z' + \frac{s(s+m-2)}{r^2} uz \right] r^{m-1} dr \tag{2.2.5}$$

and

$$J_{\alpha,s}(z) = \int_0^\delta \left[|z'|^2 + \frac{s(s+m-2)}{r^2} |z|^2 \right] r^{-\alpha+m-1} dr. \tag{2.2.6}$$

First we define integrals (2.2.5) and (2.2.6) on the regular functions $u(r)$ and $z(r)$, which satisfy the following conditions $z(\delta) = u(0) = 0$, for $\alpha \in (-m, 2-m)$, and $z(\delta) = u(\delta) = 0$, for $\alpha \in (2-m, 0]$. It follows from equalities (2.2.5), (2.2.6) that the functionals I_s and $J_{\alpha,s}$ can be extended to all $z \in W^{(1)}_{2,-\alpha+m-1}([0,\delta], 0)$ and $u \in W^{(1)}_{2,\alpha+m-1}([0,\delta], 0)$ under conditions $u(0) = z(\delta) = 0$ * for $s \geq 1$. In fact it is absolutely clear for $J_{\alpha,s}$. For such z and any $s \geq 0$ the right-hand side expression of (2.2.6) can be estimated with the help of inequality (2.1.8) by the squared norm of z in $W^{(1)}_{2,-\alpha+m-1}$. For I_s, this statement is also easy to prove. Multiply and divide the expression under the sign of the integral in (2.2.5) by $r^{\alpha/2}$ and apply the Hölder inequality. Then we come to the following relation

$$|I_s(u,z)|^2 \leq \int_0^\delta \left[|u'|^2 + \frac{s(s+m-2)}{r^2}|u|^2 \right] r^{\alpha+m-1} dr \times$$
$$\times \int_0^\delta \left[|z'|^2 + \frac{s(s+m-2)}{r^2}|z|^2 \right] r^{-\alpha+m-1} dr.$$

Estimate each factor on the right-hand side with the help of (2.1.9) and we get the required statement.

Lemma 2.2.1. *Suppose $\alpha \in (-m, 2-m) \cup (2-m, 0]$, the integral*

$$\int_0^\delta \left(|u_s'|^2 + |u_s|^2 \right) r^{\alpha+m-1} dr \tag{2.2.7}$$

is finite, $u_s(0) = 0$ for $\alpha \in (-m, 2-m)$, $u_s(\delta) = 0$ for $\alpha \in (2-m, 0]$ with $s \geq 1$. Then the following relations are true

*for $\alpha \in (-m, 2-m)$; if $\alpha \in (2-m, 0]$ then $z(\delta) = u(\delta) = 0$.

(2.2.8)
$$I_0(u_0, v_0) = J_{\alpha,0}(v_0);$$

(2.2.9)
$$\left(\min\left\{1, 1 - \frac{\alpha(\alpha+m-2)}{2(m-1)}\right\} - \eta\right) \int_0^\delta \left[|u_s'|^2 + \right.$$
$$\left. + \frac{s(s+m-2)}{r^2}|u_s|^2\right] r^{\alpha+m-1} dr -$$
$$-C \int_0^\delta \left[|u_s'|^2 + \frac{s(s+m-2)}{r^2}|u_s|^2\right] r^{m-1} dr \le I(u_s, v_s),$$

(2.2.10)
$$J_{\alpha,s}(v_s) \le$$
$$\left[1 - \frac{(m-2)\alpha}{m-1} + \eta\right] \int_0^\delta \left[|u_s'|^2 + \frac{s(s+m-2)}{r^2}|u_s|^2\right] r^{\alpha+m-1} dr +$$
$$+C \int_0^\delta \left[|u_s'|^2 + \frac{s(s+m-2)}{r^2}|u_s|^2\right] r^{m-1} dr,$$

where $s \ge 1$ and C is independent of s.

Proof. It is clear that we can only consider smooth functions with $u_s(0) = 0$ or $u_s(\delta) = 0$ $(s \ge 1)$. Then we can use the closure procedure in $W_{2,\alpha}^{(1)}$. For $s = 0$, we have
$$\int_0^\delta u_0' v_0' r^{m-1} dr = \int_0^\delta |u_0'|^2 r^{\alpha+m-1} dr,$$
$$\int_0^\delta |v_0'|^2 r^{-\alpha+m-1} dr = \int_0^\delta |u_0'|^2 r^{\alpha+m-1} dr$$
and (2.2.8) is proved. Consider now integral (2.2.5) for $z = v_s$ and $u = u_s$. Omit the index s and denote this integral by $I(u)$. Substitute v_s by its expression (2.2.4). After differentiation, we get

$$I(u) = \int_0^\delta \left[|u'|^2 + \frac{s(s+m-2)}{r^2}|u|^2\right] r^{\alpha+m-1} \zeta dr +$$
$$+\frac{\alpha}{2} \int_0^\delta \left(|u|^2\right)' r^{\alpha+m-2} \zeta dr + \int_0^\delta u' u r^{\alpha+m-1} \zeta' dr.$$

Integrating by parts in the middle term on the right-hand side and using the equalities $\zeta(\delta) = 0$ and $u(0) = 0$ (for $\alpha \in (-m, 2-m)$), we obtain

(2.2.11) $$I(u)=\int_0^\delta\left\{|u'|^2+\left[s(s+m-2)-\frac{\alpha(\alpha+m-2)}{2}\right]\frac{|u|^2}{r^2}\right\}r^{\alpha+m-1}\zeta dr+$$
$$+\int_0^\delta u'ur^{\alpha+m-1}\zeta' dr-\frac{\alpha}{2}\int_0^\delta |u|^2r^{\alpha+m-2}\zeta' dr.$$

Note that when integrating by parts, the term outside the integral vanishes. Now we estimate the second and the third integrals on the right-hand side of (2.2.11). Let

(2.2.12) $$S_\alpha(u)=\int_0^\delta |u|^2r^{\alpha+m-2}\zeta' dr.$$

From the definition of the cut-off function ζ it follows that $\zeta'(r)=0$, for $0\le r\le\delta/2$ and $3\delta/4\le r\le\delta$. Taking into account (2.2.2), we get the inequality

(2.2.13) $$|S_\alpha(u)|\le C\delta^{-2}\int_0^\delta |u|^2r^{\alpha+m-1}dr$$

where C is independent of δ, and the following relation

$$|S_\alpha(u)|\le C\int_0^\delta |u|^2r^{m-1}dr$$

is true.

Let us now estimate the integral

(2.2.14) $$T_\alpha(u)=\int_0^\delta u'ur^{\alpha+m-1}\zeta' dr.$$

Applying the Young's inequality

(2.2.15) $$|ab|\le\eta|a|^2+\frac{1}{4\eta}|b|^2$$

and taking into account that $\zeta'=0$, for $r<\delta/2$, we come to

$$|T_\alpha(u)|\le\eta\int_{\delta/2}^\delta |u'|^2r^{\alpha+m-1}|\zeta'|dr+C\int_{\delta/2}^\delta |u|^2r^{\alpha+m-1}|\zeta'|dr.$$

From the fact that $\eta>0$ is arbitrarily small and from (2.2.2), it follows that

(2.2.16) $$|T_\alpha(u)|\le\eta\int_0^\delta |u'|^2r^{\alpha+m-1}dr+C\int_{\delta/2}^\delta |u|^2r^{\alpha+m-1}dr.$$

Note that in (2.2.16) under the sign of the second integral on the right-hand side $r\ge\delta/2$ holds. Hence,

$$(2.2.17) \qquad |T_\alpha(u)| \le \eta \int_0^\delta |u'|^2 r^{\alpha+m-1} dr + C \int_0^\delta |u|^2 r^{m-1} dr.$$

Therefore it follows from inequalities (2.2.13) and (2.2.17) that

$$I(u) \ \ge \ \int_0^\delta \left\{ |u'|^2 + \left[s(s+m-2) - \frac{\alpha(\alpha+m-2)}{2} \right] \frac{|u|^2}{r^2} \right\} r^{\alpha+m-1} \zeta dr - $$
$$-\eta \int_0^\delta |u'|^2 r^{\alpha+m-1} dr - C \int_0^\delta |u|^2 r^{m-1} dr.$$

As $s \ge 1$, $\alpha(\alpha+m-2) > 0$ for $\alpha \in (2-m-2\gamma, 2-m)$, and $\alpha(\alpha+m-2) \le 0$, for $\alpha \in (2-m, 0]$, we have

$$I(u) \ge \min\left\{1, \min_{s\ge1}\left(1 - \frac{\alpha(\alpha+m-2)}{2s(s+m-2)}\right)\right\} \int_0^\delta \left[|u'|^2 + \frac{s(s+m-2)}{r^2}|u|^2 \right] r^{\alpha+m-1}\zeta dr -$$
$$-\eta \int_0^\delta |u'|^2 r^{\alpha+m-1} dr - C \int_0^\delta |u|^2 r^{m-1} dr.$$

The minimum on s is achieved at $s = 1$. Then

$$I(u) \ \ge \ \min\left\{1, 1 - \frac{\alpha(\alpha+m-2)}{2(m-1)}\right\} \int_0^\delta \left[|u'|^2 + \frac{s(s+m-2)}{r^2}|u|^2 \right] r^{\alpha+m-1}\zeta dr -$$
$$-\eta \int_0^\delta |u'|^2 r^{\alpha+m-1} dr - C \int_0^\delta |u|^2 r^{m-1} dr.$$

On the other hand, it is obvious that the following sequence of relations is valid

$$\int_0^\delta \left[|u'|^2 + \frac{s(s+m-2)}{r^2}|u|^2 \right] r^{\alpha+m-1}\zeta dr = \int_0^\delta \left[|u'|^2 + \frac{s(s+m-2)}{r^2}|u|^2 \right] r^{\alpha+m-1} dr +$$
$$+ \int_{\delta/2}^\delta \left[|u'|^2 + \frac{s(s+m-2)}{r^2}|u|^2 \right] (\zeta - 1) r^{\alpha+m-1} dr \ge$$
$$\ge \int_0^\delta \left[|u'|^2 + \frac{s(s+m-2)}{r^2}|u|^2 \right] r^{\alpha+m-1} dr - C \int_0^\delta \left[|u'|^2 + \frac{s(s+m-2)}{r^2}|u|^2 \right] r^{m-1} dr,$$

where C does not depend on s.
Comparing these relations with the last estimate for $I(u)$, we come to (2.2.9).
Inequality (2.2.10) can be proved in the similar way. In fact, omitting the index s for $s \ge 1$, after differentiation and squaring under the integral sign, we get

$$J_\alpha(v) = \int_0^\delta \left[|u'|^2 + \frac{s(s+m-2)}{r^2}|u|^2 \right] r^{\alpha+m-1}\zeta^2 dr +$$

$$+2\alpha \int_0^\delta u'ur^{\alpha+m-2}\zeta^2 dr + 2\int_0^\delta u'ur^{\alpha+m-1}\zeta\zeta' dr + \alpha^2 \int_0^\delta |u|^2 r^{\alpha+m-3}\zeta^2 dr +$$

$$+2\alpha \int_0^\delta |u|^2 r^{\alpha+m-2}\zeta\zeta' dr + \int_0^\delta |u|^2 r^{\alpha+m-1}(\zeta')^2 dr.$$

We can introduce the second term on the right-hand side in the following form

$$\alpha \int_0^\delta (u^2)' r^{\alpha+m-2}\zeta^2 dr.$$

Integrating by parts, we get

$$2\alpha \int_0^\delta u'ur^{\alpha+m-2}\zeta^2 dr = -\alpha(\alpha+m-2)\int_0^\delta |u|^2 r^{\alpha+m-3}\zeta^2 dr -$$

$$-\alpha \int_0^\delta |u|^2 r^{\alpha+m-2}(\zeta^2)' dr.$$

Using this expression on the right-hand side for $J_\alpha(v)$, we come to the equality

(2.2.18)
$$J_\alpha(v) = \int_0^\delta \left[|u'|^2 + \frac{s(s+m-2)-\alpha(m-2)}{r^2}|u|^2 \right] r^{\alpha+m-1}\zeta^2 dr +$$

$$+\int_0^\delta \left[u'u(\zeta^2)' + u^2(\zeta')^2 \right] r^{\alpha+m-1} dr.$$

It follows from the inequality $|\zeta| \le C$ that the modulus of the second integral can be estimated from above by $S_\alpha(u), T_\alpha(u)$ and the following integral

(2.2.19)
$$R_\alpha(u) = \int_0^\delta |u|^2 r^{\alpha+m-1}(\zeta')^2 dr.$$

From inequality (2.2.2) and the fact that $\zeta' = 0$, for $r < \delta/2$, follows the estimate for integral (2.2.19)

$$|R_\alpha(u)| \le C \int_0^\delta |u|^2 r^{m-1} dr.$$

Inequalities (2.2.13),(2.2.17) and the one introduced above give way to the estimate

$$J_\alpha(u) \leq \int_0^\delta \left[|u'|^2 + \frac{s(s+m-2)-\alpha(m-2)}{r^2}|u|^2\right] r^{\alpha+m-1}dr + $$
$$+\eta \int_0^\delta |u'|^2 r^{\alpha+m-1}dr + C\int_0^\delta |u|^2 r^{m-1}dr.$$

Therefore, we have

$$J_\alpha(u) \leq \frac{s(s+m-2)-\alpha(m-2)}{s(s+m-2)} \int_0^\delta \left[|u'|^2 + \frac{s(s+m-2)}{r^2}|u|^2\right] r^{\alpha+m-1}dr + $$
$$+\eta \int_0^\delta |u'|^2 r^{\alpha+m-1}dr + C\int_0^\delta |u|^2 r^{m-1}dr.$$

From the last inequality,(2.2.10) immediately follows. □

Take the function $\zeta_0(r)$ on $[0,\delta]$, with $\zeta_0(r)$ satisfying (2.2.1),(2.2.2) and for $\delta/2 \leq r \leq 3\delta/4$ taking the following form

$$\zeta_0(r) = \frac{16}{\delta^2}\left(r - \frac{3}{4}\delta\right)^2. \tag{2.2.20}$$

It is evident that the derivative of ζ_0 is bounded and the inequality (2.2.2) is valid. Let

$$\tilde{v}_0(r) = -\int_\delta^r u_0'(\rho)\rho^\alpha \zeta_0(\rho)d\rho$$

and $v_s(r)$ for $s \geq 1$ be defined by (2.2.4), where ζ is changed for ζ_0.

Lemma 2.2.2. *Let integral (2.2.7) be finite and $\alpha = 2-m-2\gamma (0<\gamma<1)$. Then the following relations*

$$\int_0^{3\delta/4} u_0'\tilde{v}_0' r^{m-1}dr = \int_0^{3\delta/4} \zeta_0^{-1}|\tilde{v}_0'|^2 r^{-\alpha+m-1}dr, \tag{2.2.21}$$

$$\left[1 - \frac{\alpha(\alpha+m-2)}{2(m-1)} - \eta\right]\int_0^\delta \left[|u_s'|^2 + \frac{s(s+m-2)}{r^2}|u_s|^2\right] r^{\alpha+m-1}\zeta_0 dr - \tag{2.2.22}$$
$$-C\delta^{-2}\int_0^\delta |u_s|^2 r^{\alpha+m-1}dr \leq I(u_s, v_s)$$

and

$$J^\circ_{\alpha,0}(v_s) \leq \left[1 - \frac{(m-2)\alpha}{m-1} + \eta\right] \times \tag{2.2.23}$$
$$\times \int_0^\delta \left[|u_s'|^2 + \frac{s(s+m-2)}{r^2}|u_s|^2\right] r^{\alpha+m-1}\zeta_0(r)dr + C\delta^{-2}\int_0^\delta |u_s|^2 r^{\alpha+m-1}dr$$

where

$$(2.2.24)\qquad J^{o}_{\alpha,0}(v_s)=\int_0^{3\delta/4}\zeta_0^{-1}\left[|v_s'|^2+\frac{s(s+m-2)}{r^2}|v_s|^2\right]r^{-\alpha+m-1}dr$$

hold true for $s\geq 1$, $u_s(0)=0$ *and* C *independent of either* δ *or* s.

Proof. We can assume at first that $u(s)(s\geq 0)$ is smooth and finite in the neighborhood of zero. The equality (2.2.21) is obvious because we have

$$u_0'\tilde{v}_0'r^{m-1}=\zeta_0^{-1}|\tilde{v}_0'|^2r^{-\alpha+m-1}.$$

To prove (2.2.22) we shall first present the same argument as in the previous lemma and obtain the equality (2.2.11) for $\zeta=\zeta_0$. The difference is only in the estimates for integrals (2.2.12), (2.2.14) and (2.2.19), where ζ is changed for ζ_0. For integral (2.2.12) use the estimate (2.2.13). We shall estimate the integral (2.2.14) in another way. Multiply and divide by $\sqrt{\zeta_0}$ under the sign of integral (2.2.14) and apply (2.2.15) we geting

$$|T_\alpha(u)|\leq\eta\int_0^{\delta}|u'|^2r^{\alpha+m-1}\zeta_0dr+\frac{1}{4\eta}\int_{\delta/2}^{3\delta/4}|u|^2r^{\alpha+m-1}|\zeta_0'|^2\zeta_0^{-1}dr.$$

From (2.2.20) we obtain $|\zeta_0'|^2\zeta_0^{-1}=64\delta^{-2}$ for $\delta/2<r<3\delta/4$ and come to the inequality

$$(2.2.25)\qquad |T_\alpha(u)|\leq\eta\int_0^{\delta}|u'|^2r^{\alpha+m-1}\zeta_0dr+C\delta^{-2}\int_0^{\delta}|u|^2r^{\alpha+m-1}dr,$$

where C is independent of δ. Now (2.2.22) directly follows from (2.2.11).
To prove (2.2.23) we substitute the expressions $v_s=u_sr^{\alpha}\zeta_0$ in (2.2.24). After simple calculations we obtain

$$\begin{aligned}J^{o}_{\alpha,0}(v_s)=&\int_0^{3\delta/4}\left[|u_s'|^2+\frac{s(s+m-2)}{r^2}|u_s|^2\right]r^{\alpha+m-1}\zeta_0dr+\\&+\alpha\int_0^{3\delta/4}\left(|u_s|^2\right)'r^{\alpha+m-2}\zeta_0dr+2\int_0^{3\delta/4}u_s'u_sr^{\alpha+m-1}\zeta_0'dr+\\&+\alpha^2\int_0^{3\delta/4}|u_s|^2r^{\alpha+m-3}\zeta_0dr+2\alpha\int_0^{3\delta/4}|u_s|^2r^{\alpha+m-2}\zeta_0'dr+\int_0^{3\delta/4}|u_s|^2r^{\alpha+m-1}\zeta_0^{-1}\zeta_0'^2dr.\end{aligned}$$

Integrating by parts under the sign of the second integral on the right-hand side. Taking into account that $u(0)=0$ and $\zeta_0=0$ for $r\geq 3\delta/4$ after elementary calculations we obtain

$$\begin{aligned}J^{o}_{\alpha,0}(v_s)=&\int_0^{3\delta/4}\left[|u_s'|^2+\frac{s(s+m-2)-\alpha(m-2)}{r^2}|u_s|^2\right]r^{\alpha+m-1}\zeta_0dr+\\&+\int_0^{3\delta/4}|u_s|^2\left[\zeta_0^{-1}(\zeta_0')^2+2\alpha r^{-1}\zeta_0'\right]r^{\alpha+m-1}dr+2T_\alpha(u_s).\end{aligned}$$

It is clear from (2.2.20) estimates

$$|\zeta_0'|^2\zeta_0^{-1} \le 64\delta^{-2}, |\zeta_0'| \le 8\delta^{-1} \tag{2.2.26}$$

hold true. Since $\zeta_0' = 0$ for $r < \delta/2$ we shall get

$$\left| \int_0^{3\delta/4} |u_s|^2(\zeta_0^{-1}\zeta_0'^2 + \alpha r^{-1}\zeta_0')r^{\alpha+m-1}dr \right| \le C\delta^{-2}\int_0^\delta |u_s|^2 r^{\alpha+m-1}dr,$$

which finally leads to (2.2.23). □

The last two lemmas were proved under the condition that α is negative. But later we shall need analogous estimates with a positive α. For $\alpha < 0$ the integral $\int_0^\delta |u|^2 r^{m-1}dr$ is comparatively weaker than the integral $\int_0^\delta |u|^2 r^{\alpha+m-1}dr$. But this is not true for $\alpha \ge 0$. Therefore in the inequalities of the type (2.2.9) and (2.2.10) instead of the above mentioned integrals the following weak integrals $\int_0^\delta |u'|^2 r^{\alpha'+m-1}dr$ and $\int_0^\delta |u|^2 r^{\alpha'+m-1}dr$, where $\alpha' > \alpha$ will be considered.

Lemma 2.2.3. *Let $\beta \in [0, m-2+2\gamma](0 < \gamma < 1)$, ζ satisfy (2.2.1) and (2.2.2) and the integral*

$$\int_0^\delta \left(|u'|^2 + \frac{|u|^2}{r^2}\right) r^{\beta+m-1}dr$$

be finite. Then the following statements are true:

$$1)\ I(u_0, v_0) = J_\beta(v_0) \tag{2.2.27}$$

2) *if $s \ge 1$, then*

$$\begin{aligned} I(u_s, v_s) &\ge \\ &\ge \int_0^\delta \left\{|u_s'|^2 + \left[s(s+m-2) - \frac{\beta(\beta+m-2)}{2}\right]\frac{|u_s|^2}{r^2}\right\} r^{\beta+m-1}dr - \\ &-C\left\{\int_0^\delta \left[|u_s'|^2 + \frac{s(s+m-2)}{r^2}|u_s|^2\right] r^{\beta'+m-1}dr\right\}, \end{aligned} \tag{2.2.28}$$

$$\begin{aligned} J_\beta(v_s) \le &\int_0^\delta \left[|u_s'|^2 + \frac{s(s+m-2) - \beta(m-2)}{r^2}|u_s|^2\right] r^{\beta+m-1}dr + \\ &+C\int_0^\delta \left[|u_s'|^2 + \frac{s(s+m-2)}{r^2}|u_s|^2\right] r^{\beta'+m-1}dr, \end{aligned} \tag{2.2.29}$$

where C is independent of either s or $\beta' > \beta$.

Proof. As in the previous considerations, we must estimate the last terms in equalities (2.2.11) and (2.2.18), where $\alpha < 0$ is changed to $\beta > 0$. Essentially in these equalities the integration runs only over the interval $\left[\frac{\delta}{2}, 3\delta/4\right]$. Hence, different powers of r can be estimated from above and below by different powers of δ to give

$$|S_\beta(u)| \leq C \int_0^\delta |u|^2 r^{\beta'+m-3} dr$$

and

$$|T_\beta(u)| \leq \eta \int_0^\delta |u'|^2 r^{\beta+m-1} dr + C \int_0^\delta |u|^2 r^{\beta'+m-1} dr.$$

The last terms in (2.2.18) is estimated in the same way to give the inequalities

(2.2.30)
$$I(u_s, v_s) \geq$$
$$\geq \int_0^\delta \left\{ |u_s'|^2 + \left[s(s+m-2) - \frac{\beta(\beta+m-2)}{2} \right] \frac{|u_s|^2}{r^2} \right\} r^{\beta+m-1} \zeta dr -$$
$$-\eta \int_0^\delta |u_s'|^2 r^{\beta'+m-1} dr - C \int_0^\delta |u_s|^2 r^{\beta'+m-1} dr,$$

(2.2.31)
$$J_\beta(v_s) \leq \int_0^\delta \left[|u_s'|^2 + \frac{s(s+m-2) - \beta(m-2)}{r^2} |u_s|^2 \right] \times$$
$$\times r^{\beta+m-1} \zeta dr + \eta \int_0^\delta |u_s'|^2 r^{\beta'+m-1} dr + C \int_0^\delta |u_s|^2 r^{\beta'+m-1} dr.$$

It is evident that

$$\int_0^\delta \left\{ |u_s'|^2 + \left[s(s+m-2) - \frac{\beta(\beta+m-2)}{2} \right] \frac{|u_s|^2}{r^2} \right\} r^{\beta+m-1} \zeta dr =$$
$$= \int_0^\delta \left\{ |u_s'|^2 + \left[s(s+m-2) - \frac{\beta(\beta+m-2)}{2} \right] \frac{|u_s|^2}{r^2} \right\} r^{\beta+m-1} dr +$$
$$+ \int_0^\delta \left\{ |u_s'|^2 + \left[s(s+m-2) - \frac{\beta(\beta+m-2)}{2} \right] \frac{|u_s|^2}{r^2} \right\} r^{\beta+m-1} (\zeta - 1) dr.$$

The second term on the right-hand side can be estimated in the following way

$$\left| \int_0^\delta \left\{ |u_s'|^2 + \left[s(s+m-2) - \frac{\beta(\beta+m-2)}{2} \right] \frac{|u_s|^2}{r^2} \right\} \times \right.$$
$$\left. \times r^{\beta+m-1} (\zeta - 1) dr \right| = \left| \int_{\delta/2}^\delta \left\{ |u_s'|^2 + \left[s(s+m-2) - \frac{\beta(\beta+m-2)}{2} \right] \times \right. \right.$$

$$\times \frac{|u_s|^2}{r^2}\Big\} r^{\beta+m-1}(\zeta-1)dr| \le C \int_{\delta/2}^{\delta} \left[|u_s'|^2 + \frac{s(s+m-2)}{r^2}|u_s|^2 \right] r^{\beta'+m-1} dr,$$

where C is independent of s. From this with the help of (2.2.30) we get (2.2.28). In the same manner (2.2.29) follows from (2.2.31). □

2.3 Hölder continuity for weak solutions of the nondegenerated elliptic second order systems with bounded nonlinearities in interior domain.

Let $u(x) = \{u^{(1)}(x), ..., u^{(N)}(x)\}$ be a vector-valued function from $C^\infty(\Omega)$ and let x_0 be an arbitrary point, the distance from $\partial\Omega$ more than $\delta > 0$. Take a ball $B_\delta(x_0)$ with the centre x_0 and radius δ and consider the spherical system of coordinates (r, θ),where $\theta = (\theta_1, \ldots, \theta_{m-1})$ belongs to unit sphere S, and r denotes the distance between $\forall x \in \Omega$ and x_0. Let $\{Y_{s,i}(\theta)\}$, $s = 0, 1, \ldots; i = 1, \ldots, k_s$, be a complete system of orthonormal spherical functions which are eigenfunctions of the Laplace -Beltrami operator on the unit sphere S. The eigenvalues of this operator are as follows $s(s+m-2)$. Expanding $u(x)$into a series in terms of $Y_{s,i}(\theta)$, we get

$$u(x) = \sum_{s=0}^{+\infty} \sum_{i=1}^{k_s} u_{s,i}(r) Y_{s,i}(\theta). \tag{2.3.1}$$

Define the function $v(x)$ by

$$v(x) = \sum_{s=0}^{+\infty} \sum_{i=1}^{k_s} v_{s,i}(r) Y_{s,i}(\theta), \tag{2.3.2}$$

where $u_{s,i}^{(k)}(r)$ and $v_{s,i}^{(k)}(r)$ are bounded together with (2.2.3) and (2.2.4). The function $v(x)$ will further play the role of test function in different integral identities. Notice, that $u_{s,i}(0) = 0$, for $s \ge 1$, which follows from the fact that $u_{s,i}(r), s = 0, 1 \ldots$, are the Fourier coefficients of $u(x)$ expanded in the terms of a system of spherical functions. In fact, for $s \ge 1$, we have

$$u_{s,i}(r) = \int_S [u(r,\theta) - u(x_0)]\, Y_{s,i}(\theta) dS,$$

which is true, because $Y_{s,i}(\theta)$, for $s \ge 1$, is orthogonal to $Y_{0,1}(\theta)$ =const. Let r go to zero to get the required relation.
Consider the integrals

$$X(u,v) = \int_{B_\delta(x_0)} DuDv dx, \tag{2.3.3}$$

$$Y_\alpha(v) = \int_{B_\delta(x_0)} |Dv|^2 r^{-\alpha} dx. \tag{2.3.4}$$

Lemma 2.3.1. *For integrals* (2.3.3),(2.3.4), $\alpha \in (2-m-2\gamma, 0], \alpha \neq 2-m$, *and sufficiently small* δ, *the inequalities*

$$X(u,v) \geq \left[\min\left\{1, 1-\frac{\alpha(\alpha+m-2)}{2(m-1)}\right\} - \eta\right] \int\limits_{B_\delta(x_0)} |Du|^2 r^\alpha dx - \tag{2.3.5}$$

$$-C \int\limits_{B_\delta(x_0)} |Du|^2 dx,$$

$$Y_\alpha(v) \leq \left[1 - \frac{(m-2)\alpha}{m-1} + \eta\right] \int\limits_{B_\delta(x_0)} |Du|^2 r^\alpha dx + C \int\limits_{B_\delta(x_0)} |Du|^2 dx \tag{2.3.6}$$

hold true.In addition we suppose that $u|_{\partial B_\delta} = 0$ *for* $\alpha \in (2-m, 0]$.

Proof. First, we shall establish (2.3.5). The function v under formulas (2.2.3) and (2.2.4) vanishes for $r = \delta$. Integrating by parts in (2.3.3), we get

$$X(u,v) = -\int\limits_{B_\delta(x_0)} \Delta uv dx + \int\limits_{B_\delta(x_0)} uv dx.$$

It is well known that

$$\Delta = r^{1-m}\frac{\partial}{\partial r}\left(r^{m-1}\frac{\partial}{\partial r}\right) - \frac{1}{r^2}\Delta', \tag{2.3.7}$$

where Δ' is the Beltrami operator on the unit sphere. For this operator we have

$$\Delta' Y_{s,i} = s(s+m-2)Y_{s,i}. \tag{2.3.8}$$

Substituting (2.3.1) and (2.3.2) in the integral (2.3.3) and taking into account the orthonormality of the system $\{Y_{s,i}(\theta)\}$ and (2.3.8), we come to

$$X(u,v) = \sum_{s,i}\int_0^\delta \left[-(u'_{s,i}r^{m-1})' + s(s+m-2)u_{s,i}r^{m-3}\right] v_{s,i} dr +$$

$$+\sum_{s,i}\int_0^\delta u_{s,i}v_{s,i}r^{m-1}dr.$$

The symbol $\sum\limits_{s,i}$ indicates that the summation is extended to all spherical functions, i.e. $\sum\limits_{s=0}^{+\infty}\sum\limits_{i=1}^{k_s}$. Integrating by parts once more, we get

$$X(u,v) = \sum_{s,i}\int_0^\delta \left[u'_{s,i}v'_{s,i} + \frac{s(s+m-2)}{r^2}u_{s,i}v_{s,i}\right] r^{m-1} dr +$$

$$+\sum_{s,i}\int_0^\delta u_{s,i}v_{s,i}r^{m-1}dr.$$

As we can see, $X(u,v)$ is expressed by integrals (2.2.5) and

$$X(u,v) = \sum_{s,i} I_s(u_{s,i}, v_{s,i}) + \sum_{s,i} \int_0^\delta u_{s,i} v_{s,i} r^{m-1} dr. \tag{2.3.9}$$

Consider now $Y_\alpha(v)$. Integrating by parts, we get

$$\begin{aligned} Y_\alpha(v) &= -\sum_{i=1}^m \int_{B_\delta(x_0)} D_i(r^{-\alpha} D_i v) v dx + \int_{B_\delta(x_0)} |v|^2 r^{-\alpha} dx = \\ &= -\int_{B_\delta(x_0)} \Delta v v r^{-\alpha} dx + \alpha \int_{B_\delta(x_0)} \frac{\partial v}{\partial r} v r^{-\alpha-1} dx + \int_{B_\delta(x_0)} |v|^2 r^{-\alpha} dx. \end{aligned}$$

Substitute expansion (2.3.2) on the right-hand side of this equality. Using the expression of the Laplace operator and formula (2.3.8), we get

$$\begin{aligned} Y_\alpha(v) &= \sum_{s,i} \int_0^\delta \Big[-\left(r^{m-1} v'_{s,i}\right)' v_{s,i} r^{-\alpha} + \alpha v'_{s,i} v_{s,i} r^{-\alpha+m-2} + \\ &+ s(s+m-2)|v_{s,i}|^2 r^{-\alpha+m-3} \Big] dr + \int_{B_\delta(x_0)} |v|^2 r^{-\alpha} dx. \end{aligned}$$

Integrating by parts in the first term of the right-hand side and cancelling similar terms, we come to

$$\begin{aligned} Y_\alpha(v) &= \sum_{s,i} \int_0^\delta \left[|v'_{s,i}|^2 + \frac{s(s+m-2)}{r^2} |v_{s,i}|^2 \right] r^{-\alpha+m-1} dr + \\ &+ \int_{B_\delta(x_0)} |v|^2 r^{-\alpha} dx. \end{aligned} \tag{2.3.10}$$

Similar considerations lead to the formula

$$\int_{B_\delta(x_0)} |D'u|^2 r^\alpha dx = \sum_{s,i} \int_0^\delta \left[|u'_{s,i}|^2 + \frac{s(s+m-2)}{r^2} |u_{s,i}|^2 \right] r^{\alpha+m-1} dr. \tag{2.3.11}$$

Apply to equalities (2.3.9) and (2.3.10) the estimates (2.2.9) amd (2.2.10). Then we have

$$\begin{aligned} X(u,v) &\geq \left(min \left\{ 1, 1 - \frac{\alpha(\alpha+m-2)}{2(m-1)} \right\} - \eta \right) \times \\ &\times \sum_{s,i} \int_0^\delta \left[|u'_{s,i}|^2 + \frac{s(s+m-2)}{r^2} |u_{s,i}|^2 \right] r^{\alpha+m-1} dr - \\ &- C \sum_{s,i} \int_0^\delta \left[|u'_{s,i}|^2 + \frac{s(s+m-2)}{r^2} |u_{s,i}|^2 \right] r^{m-1} dr + \int_{B_\delta(x_0)} uv dx, \end{aligned}$$

$$Y_\alpha(v) \le \left[1 - \frac{(m-2)\alpha}{m-1} + \eta\right] \sum_{s,i} \int_0^\delta \left[|u'_{s,i}|^2 + \frac{s(s+m-2)}{r^2}|u_{s,i}|^2\right] r^{\alpha+m-1} dr +$$

$$+C \sum_{s,i} \int_0^\delta \left[|u'_{s,i}|^2 + \frac{s(s+m-2)}{r^2}|u_{s,i}|^2\right] r^{m-1} dr + \int_{B_\delta(x_0)} |v|^2 r^{-\alpha} dx.$$

It follows from (2.3.11) that

(2.3.12)
$$X(u,v) \ge \left(min\left\{1, 1 - \frac{\alpha(\alpha+m-2)}{2(m-1)}\right\} - \eta\right) \times$$
$$\times \int_{B_\delta(x_0)} |D'u|^2 r^\alpha dx - C \int_{B_\delta(x_0)} |D'u|^2 dx + \int_{B_\delta(x_0)} uv dx,$$

(2.3.13)
$$Y_\alpha(v) \le \left[1 - \frac{(m-2)\alpha}{m-1} + \eta\right] \int_{B_\delta(x_0)} |D'u|^2 r^\alpha dx +$$
$$+C \int_{B_\delta(x_0)} |D'u|^2 dx + \int_{B_\delta(x_0)} |v|^2 r^{-\alpha} dx.$$

Now, we only have to estimate the following terms

$$\int_{B_\delta(x_0)} |v|^2 r^{-\alpha} dx \quad \text{and} \quad \int_{B_\delta(x_0)} uv dx.$$

Using the orthonormality of the system $\{Y_{s,i}(\theta)\}$, we find

$$\int_{B_\delta(x_0)} uv dx = \sum_{s,i} \int_0^\delta u_{s,i}(r) v_{s,i}(r) r^{m-1} dr.$$

Substitute the expressions $v_{s,i}(r)$ by $u_{s,i}(r)$ under formulas (2.2.3) and (2.2.4) in the last inequality. Then we get

$$\int_{B_\delta(x_0)} uv dx = -\int_0^\delta u_0(r) \int_r^\delta u'_0(\rho) \rho^\alpha d\rho r^{m-1} dr +$$
$$+\sum_{s=1}^{+\infty} \sum_{i=1}^{k_s} \int_0^\delta |u_{s,i}|^2 r^{\alpha+m-1} \zeta(r) dr$$

(by u_0 we shall denote $u_{0,1}$). First suppose that $\alpha = 2 - m - 2\gamma, 0 < \gamma < 1$. Adding and substracting the value $u_0(0)$ on the right-hand side and substituting the difference $u_0(r) - u_0(0)$ by the integral from the derivative $u'_0(r)$, we get

$$\int_{B_\delta(x_0)} uv dx = -\int_0^\delta \int_0^r u_0'(\rho)d\rho \int_r^\delta u_0'(\rho)\rho^\alpha d\rho r^{m-1} dr -$$

$$-u_0(0)\int_0^\delta \int_r^\delta u_0'(\rho)\rho^\alpha d\rho r^{m-1} dr + \sum_{s=1}^{+\infty}\sum_{i=1}^{k_s} \int_0^\delta |u_{s,i}|^2 r^{\alpha+m-1}\zeta(r)dr.$$

Integrating by parts on the right-hand side, we come to the following equality

$$\int_{B_\delta(x_0)} uv dx = -\int_0^\delta \int_0^r u_0'(\rho)d\rho \int_r^\delta u_0'(\rho)\rho^\alpha d\rho r^{m-1} dr - \tag{2.3.14}$$

$$-\frac{u_0(0)}{m}\int_0^\delta u_0'(r)r^{\alpha+m} dr + \sum_{s=1}^{+\infty}\sum_{i=1}^{k_s} \int_0^\delta |u_{s,i}|^2 r^{\alpha+m-1}\zeta(r)dr.$$

Now consider the first term on the right-hand side of (2.3.14) and express it in the form

$$I_1 = -\int_0^\delta \int_0^r u_0'(\rho)d\rho r^{\frac{\alpha+m-3}{2}} \int_r^\delta u_0'(\rho)\rho^\alpha d\rho r^{(-\alpha+m+1)/2} dr.$$

Note that $-(\alpha + m - 3) = 1 + 2\gamma > 1$ and $-(-\alpha + m + 1) = 1 - 2m - 2\gamma < 1$. Therefore, after using the Hölder inequality we can apply (2.1.9) to both factors to get

$$|I_1|^2 \le C\int_0^\delta |u_0'|^2 r^{\alpha+m-1} dr \int_0^\delta |u_0'|^2 r^{\alpha+m+3} dr,$$

where C is independent of δ. Splitting out the factor r^4 under the sign of the integral of the second factor on the right-hand side and estimating this factor from above by δ^4, after extracting the square root, we come to

$$|I_1| \le C\delta^2 \int_0^\delta |u_0'|^2 r^{\alpha+m-1} dr,$$

where C is independent of δ.
Let us now look at the second term of the right-hand side in (2.3.14)

$$I_2 = -\frac{u_0(0)}{m}\int_0^\delta u_0'(r)r^{\alpha+m} dr.$$

Applying (2.2.15) with $\eta = 1$, we get

$$|I_2| \leq |u_0(0)|^2 + \frac{1}{4m^2}|\int_0^\delta u_0'(r) r^{\alpha+m} dr|^2.$$

Estimate the second term on the right-hand side by introducing the integral of the subject expression in the following form

$$\int_0^\delta u_0'(r) r^{\alpha+m} dr = \int_0^\delta r^{(-\alpha+m+1)/2} \left[u_0'(r) r^{(-\alpha+m+1)/2}\right] dr$$

and apply the Hölder inequality. We get then

$$|\int_0^\delta u_0'(r) r^{\alpha+m} dr|^2 \leq \int_0^\delta r^{\alpha+m+1} dr \int_0^\delta |u_0'|^2 r^{\alpha+m-1} dr.$$

After a few simple calculations, taking into account that $\gamma < 1$, we come to the inequality

$$|\int_0^\delta u_0'(r) r^{\alpha+m} dr|^2 \leq C\delta^2 \int_0^\delta |u_0'|^2 r^{\alpha+m-1} dr,$$

with C independent of pendent of δ. Hence, we have

$$|I_2| \leq |u_0(0)|^2 + C\delta^2 \int_0^\delta |u_0|^2 r^{\alpha+m-1} dr.$$

Applying inequality (2.1.10) to the first term of the right-hand side, gives the following estimate

$$|I_2| \leq \left(\eta + C_2\delta^2\right) \int_0^\delta |u_0'|^2 r^{\alpha+m-1} dr + C \int_0^\delta |u_0|^2 r^{m-1} dr,$$

with C independent of δ.
Using the estimate for $|I_1|$, we get by means of (2.3.14) the following inequality

$$|\int_{B_\delta(x_0)} uv dx| \leq \left(\eta + C\delta^2\right) \int_0^\delta |u_0'|^2 r^{\alpha+m-1} dr + C_2 \int_{B_\delta(x_0)} |u|^2 r^\alpha dx +$$
$$+ C_3 \int_0^\delta |u_0|^2 r^{m-1} dr,$$

whith C_1 and C_2 independed of δ. Taking $\delta > 0$ sufficiently small we get

$$(2.3.15)\quad |\int_{B_\delta(x_0)} uvdx| \leq \eta \int_{B_\delta(x_0)} |D'u|^2 r^\alpha dx + C_1 \int_{B_\delta(x_0)} |u|^2 r^\alpha dx + C_2 \int_{B_\delta(x_0)} |u|^2 dx,$$

for $\alpha \in (-m, 2-m)$. If $\alpha \in (2-m, 0)$ then the relations(2.3.15) follow from the representation

$$u_0(r) = -\int_r^\delta u_0'(\rho)d\rho.$$

Now we need only estimate the integral $\int_{B_\delta(x_0)} |v|^2 r^{-\alpha} dx$ from above which can easily be done from expansion (2.3.1), and formulas (2.2.3) and (2.2.4) giving

$$\int_{B_\delta(x_0)} |v|^2 r^{-\alpha} dx = \sum_{s=1}^{+\infty} \sum_{i=1}^{k_s} \int_0^\delta |u_{s,i}|^2 r^{\alpha+m-1} \zeta^2 dr + \int_0^\delta |\int_r^\delta u_0'(\rho)\rho^\alpha d\rho|^2 r^{-\alpha+m-1} dr.$$

With $-(-\alpha+m-1) < 1$ the second term, as we have just seen, can be estimated as follows

$$\int_0^\delta |\int_r^\delta u_0'(\rho)\rho^\alpha d\rho|^2 r^{-\alpha+m-1} dr \leq C\delta^2 \int_0^\delta |u_0'|^2 r^{\alpha+m-1} dr,$$

with C independent of δ. Hence,

$$\int_{B_\delta(x_0)} |v|^2 r^{-\alpha} dx \leq C_1 \sum_{s=1}^{+\infty} \sum_{i=1}^{k_s} \int_0^\delta |u_{s,i}|^2 r^{\alpha+m-1} dr + C_2\delta^2 \int_0^\delta |u_0'|^2 r^{\alpha+m-1} dr,$$

whith C_2 independent of δ. Taking δ sufficiently small, we have

$$(2.3.16)\quad \int_{B_\delta(x_0)} |v|^2 r^{-\alpha} dx \leq \eta \int_{B_\delta(x_0)} |D'u|^2 r^\alpha dx + C \int_{B_\delta(x_0)} |u|^2 r^\alpha dx.$$

Inequalities (2.3.15),(2.3.16) along with relations (2.3.12), (2.3.13) and (2.1.17′) complete the proof of the lemmma for $\alpha \in (-m, 2-m)$. If $\alpha \in (2-m, 0)$, then instead of (2.1.17′) the conclusion of lemma 2.1.2 should be used. □

Corollary 2.3.1 For the integrals

$$(2.3.3')\quad X'(u,v) = \int_{B_\delta(x_0)} D'uD'vdx$$

and

(2.3.4')
$$Y'_\alpha(v) = \int\limits_{B_\delta(x_0)} |D'v|^2 r^{-\alpha} dx$$

the same inequalities (2.3.5) and (2.3.6) hold true. □

Remark 2.3.1. If $\alpha' \in [0, m), \alpha \in (2-m-2\gamma, 0], 2\alpha+\alpha' > -m$ and $u \in W^{(1)}_{2,2\alpha+\alpha'}(B_\delta(x_0), x_0)$ then inequality

(2.3.17)
$$||v||_{W^{(1)}_{2,\alpha'}} \leq C||u||_{W^{(1)}_{2,2\alpha+\alpha'}}$$

holds. This inequality follows immediately from (2.2.3), (2.2.4). □

Consider second order nondegenerated system (1.1.1) with bounded nonlineriaties and present this system in the form

(2.3.18)
$$L(u) \equiv -D_i a_i(x; D_j u) + a_0(x; D_j u) = 0$$

$$(i = 1, ..., m; j = 0, 1, ..., m).$$

Suppose that conditions 1) - 3) of 1.1 are satisfied. * As system (2.3.18) is a non-degenerated one with bounded nonlinearities, inequalities (1.1.2) , (1.1.3) should be introduced in the form

(2.3.19)
$$\mu_0|\xi|^2 \leq \frac{\partial a_i^{(k)}}{\partial p_i^{(k)}} \xi_i^{(k)} \xi_j^{(l)} \leq \nu_0 |\xi|^2,$$

$$|\frac{\partial a_i^{(k)}}{\partial p_i^{(k)}}| \leq \nu_0,$$

where the summation over the repeated indicies $i, j = 0, 1, ..., m; k, l = 1, ..., N$ is implied, μ_0 and ν_0 are positive constants.
The boundary conditions (1.1.13) can now be expressed in the form

(2.3.20)
$$u|_{\partial\Omega} = g,$$

where g is a trace of some $u_0 \in W^{(1)}_2(\Omega)$. We shall also consider the homogeneous condition

(2.3.21)
$$u|_{\partial\Omega} = 0.$$

Consider the iterative process for problem (2.3.18), (2.3.20)

---

*In this case $s = 2$

$$-\Delta u_{n+1} + u_{n+1} = -\Delta u_n + u_n - \varepsilon L(u_n), \quad u_{n+1}|_{\partial\Omega} = g. \tag{2.3.22}$$

The iterative equalities can be introduced as a system of integral identities

$$\int_\Omega Du_{n+1}Dvdx = \int_\Omega Du_nDvdx - \varepsilon\int_\Omega a_i(x;D_ju_n)D_ivdx \tag{2.3.23}$$

with boundary conditions

$$u_{n+1}|_{\partial\Omega} = g. \tag{2.3.24}$$

Here, as in the previous cases, the product of vector functions denotes the scalar product (in R^N).The integral identities should be satisfied by $\forall v \in \overset{\circ}{W}{}_2^{(1)}(\Omega)$.
When $a_i(x;D_ju) = 0$ and $a_i(x;D_ju), i = 1,...,m$, is independent of u, it is reasonable to consider the following process

$$-\Delta u_{n+1} = -\Delta u_n - \varepsilon L(u_n), \quad u_{n+1}|_{\partial\Omega} = g. \tag{2.3.25}$$

This process can also be introduced in the form of certain systems of integral identities

$$\int_\Omega D'u_{n+1}D'vdx = \int_\Omega D'u_nD'vdx - \varepsilon\int_\Omega a_i(x;D_ju_n)D_ivdx \tag{2.3.26}$$

with the boundary conditions

$$u_{n+1}|_{\partial\Omega} = g.$$

Let K be defined by (1.1.8) and (1.1.9).The corresponding values of ε are given in lemma 1.1.2. We now start with

Lemma 2.3.2. *Let the sequence of positive numbers σ_n, $n = 0,1,....$, satisfy inequality*

$$\sigma_{n+1}^2 \le A\sigma_n\sigma_{n+1} + Bq^n\sigma_n + Cq^{2n}, \tag{2.3.27}$$

where A, B, C and q are positive constants, $0 \le A < 1$ and $0 \le q < 1$.
Then there exists such a $Q(0 \le Q < 1)$, for which estimate

$$\sigma_n \le DQ^n \tag{2.3.28}$$

holds, where $D = const > 0$ and independent of n. If in (2.3.27) $q = 1$ then all σ_n are bounded.

Proof. It follows from (2.2.15) and from (2.3.27), that

$$\sigma_{n+1}^2 \leq \frac{A}{2}(\sigma_n^2 + \sigma_{n+1}^2) + \eta\sigma_n^2 + Cq^{2n}.$$

Therefore,

$$\sigma_{n+1}^2 \leq \frac{A-2\eta}{2-A}\sigma_n^2 + Cq^{2n}.$$

As $\sqrt{|a+b|} \leq \sqrt{|a|} + \sqrt{|b|}$, after extracting the square root, we get that

$$\sigma_{n+1} \leq \left(\frac{A-2\eta}{2-A}\right)^{\frac{1}{2}} \sigma_n + Cq^n.$$

Take $\eta > 0$ so small, that the inequality

$$\frac{A+2\eta}{2-A} < 1$$

holds.

Then $\sigma_{n+1} \leq t\sigma_n + Cq^n$, where $0 \leq t < 1$. As it is possible to increase q on the right-hand side of (2.3.27), we can suppose that $t < q < 1$. Let us denote this q by Q. By the method of mathematical induction we come to the following estimate

$$\sigma_{n+1} \leq t^{n+1}\sigma_0 + C\sum_{k=0}^{n} Q^k t^{n-k}.$$

Taking factor Q^n outside the sum, we get the following inequality

$$\sigma_{n+1} \leq t^{n+1}\sigma_0 + CQ^n \sum_{k=0}^{n} \left(\frac{t}{Q}\right)^{n-k}$$

As $tQ^{-1} < 1$, we have

$$\sigma_{n+1} \leq DQ^{n+1}.$$

If $q = 1$, then by the mathematical induction method we get

$$\sigma_{n+1} \leq t^{n+1}\sigma_0 + C\sum_{k=0}^{n} t^{n-k}.$$

From $0 \leq t < 1$ follows boundedness of the sequence $\{\sigma_n\}$. □

Theorem 2.3.1. *Let conditions 1)-3) of 1.1 for $s = 2$ be satisfied and* $\alpha \in (-m, 2-m)$. *Let g be a trace of some function belonging to $W_2^{(1)}(\Omega)$. If the inequalities*

$$1 - \frac{\alpha(\alpha+m-2)}{2(m-1)} > 0 \tag{2.3.29}$$

and

$$A_1^2 = K^2\left[1 - \frac{\alpha(m-2)}{m-1}\right]\left[1 - \frac{\alpha(\alpha+m-2)}{2(m-1)}\right]^{-1} < 1 \tag{2.3.30}$$

are satisfied, then the weak solution of system (2.3.18) *with boundary condition* (2.3.20) *is Hölder continious with an exponent* $\gamma = \frac{2-m-\alpha}{2}$ *in an arbitrary strictly interior subdomain* ω. *This solution can be found by means of process* (2.3.22) *with convergence in* $H_\alpha(\omega)$, *starting from a sufficiently smooth arbitrary initial iteration* $u_0(x)$, *at the rate of a geometric progression with the common ratio* A_1.

Proof. Let $\tilde{u}$ be harmonic in Ω, and satisfies boundary condition (2.3.20). If $u = u' + \tilde{u}$, the condition for u' on the boundary is reduced to a homogenous one. The system (2.3.18) becomes the form

$$D_i\tilde{a}_i(x; Du') - \tilde{a}_0(x; Du') = 0,$$

where $\tilde{a}_i(x; Du') = a_i(x; Du' + D\tilde{u})$. It is clear that all the coefficients $\tilde{a}_i$ satisfy conditions 1)-3). Now, we can suppose the boundary condition is homogeneous. Take any point $x_0 \in \omega$ and let $\delta > 0$ be less than the distance between x_0 and the boundary $\partial\Omega$. Take the test function v (2.3.2), where $u = u_{n+1} - u_n$. Subtract the two successsive integral identities (2.3.23) and introduce formula (1.2.14), for our case ($l = 1$), in the form

$$\int_\Omega D(u_{n+1} - u_n)Dv dx = \sum_{i=0}^{m}\int_\Omega \left[D_i(u_n - u_{n-1}) - \varepsilon\frac{\overline{\partial a_i}}{\partial p_k}D_k(u_n - u_{n-1})\right] D_i v dx.$$

Here i and k vary from zero to m , and the bar over the derivative indicates that its arguments are $\bar{p} = p_{n-1} + \theta(p_n - p_{n-1})$, where $0 < \theta < 1$. The vectors U_n and V, as in the proof of theorem 1.2.2 the last equality can be introduced in a form similar to (1.2.15) and we arrive at

$$\int_\Omega D(u_{n+1} - u_n)Dv dx = \int_\Omega (I - \varepsilon\overline{A})(U_n - U_{n-1})V dx. \tag{2.3.31}$$

As a result of the structure of v the integration on both sides of (2.3.31) is expanded only over $B_\delta(x_0)$. Let r be the distance between $x \in B_\delta(x_0)$ and x_0. Applying the Hölder inequality to the right-hand side of (2.3.31) and carrying out $\sup ||I - \varepsilon A||$ outside the sign of the integral, we come to

$$\int_\Omega D(u_{n+1} - u_n)Dv dx \leq \sup_{x,p} ||I - \varepsilon\overline{A}|| \int_\Omega |U_n - U_{n-1}||V| dx,$$

where $|.|$ denotes the length of the vector in $R^{(m+1)N}$. Multiplying and dividing by $r^{\frac{\alpha}{2}}$ under the sign of the integral on the right-hand side, we get the estimate

$$X(u_{n+1}-u_n,v) \le D(A)\left[\int\limits_{B_\delta(x_0)} |D(u_n-u_{n-1})|^2 r^\alpha dx\right]^{\frac{1}{2}} Y_\alpha^{\frac{1}{2}}(v).$$

Here with X and Y_α defined by formulas (2.3.3), (2.3.4), the function v given by (2.3.2) with $u = u_{n+1} - u_n$ and finally

$$D(A) = \inf_\varepsilon \sup_{x,\rho} ||I - \varepsilon \overline{A}||.$$

Let $0 < \delta < \delta_0$ and δ_0 be sufficiently small, by applying inequalities (2.3.5) and (2.3.6), we get the following estimate

$$(2.3.32) \quad \left[1 - \frac{\alpha(\alpha+m-2)}{2(m-1)} - \eta\right] \int\limits_{B_\delta(x_0)} |D(u_{n+1}-u_n)|^2 r^\alpha dx \le$$

$$\le D(A)\left[\int\limits_{B_\delta(x_0)} |D(u_n-u_{n-1})|^2 r^\alpha dx\right]^{\frac{1}{2}} \left\{\left[1 - \frac{(m-2)\alpha}{m-1} + \eta\right]^{\frac{1}{2}} \times\right.$$

$$\left.\times \left[\int\limits_{B_\delta(x_0)} |D(u_{n+1}-u_n)|^2 r^\alpha dx\right]^{\frac{1}{2}} + C||u_{n+1}-u_n||\right\} + C||u_{n+1}-u_n||^2.$$

Note that all constants are independent of $\tilde{u}$, i.e. on the boundary function g. In fact, $D(A)$ is calculated as an $\inf_\varepsilon \sup_{x,\rho}$ over expressions, which depend only on x and the eigenvalues of A^+ and C. The remaining constants depend on the boundary surface $\partial\Omega, \delta_0, \eta$ and the upper bounds of modulus of the elements A^+ and C. According to inequalities (2.3.19) these bounds can be estimated from above by constants which are independent of g. According to theorem 1.2.2 the inequality $||u_n - u_{n-1}|| < CK^n$ holds. Let

$$\sigma_n^2 = \int\limits_{B_\delta(x_0)} |D(u_n-u_{n-1})|^2 r^\alpha dx.$$

Then using inequalities (2.3.32) and $D(A) \le K$, we come to

$$\sigma_{n+1}^2 \le AK\sigma_n\sigma_{n+1} + C_1K^n\sigma_n + c_2K^{2n}.$$

From lemma 2.3.2 it follows that

$$\sigma_n \le CA_1^n \quad (A_1 = AK),$$

where C is independent of x_0. Then for any $x_0 \in \omega$, we have

$$\int_{B_\delta(x_0)} |D(u_n - u_{n-1})|^2 r^\alpha dx \le C A_1^{2n}.$$

Taking the supremum over all $x_0 \in \omega$ on the left-hand side, we get

$$\sup_{\delta<\delta_0, x_0\in\omega} \int_{B_\delta(x_0)} |D(u_n - u_{n-1})|^2 r^\alpha dx \le C A_1^{2n}.$$

Next using formula (2.1.5) we obtain

$$||u_n - u_{n-1}||'_\alpha \le C A_1^n.$$

Using the equivalence of the norm in $H_\alpha(\omega)$, we come to

$$||u_n - u_{n-1}||_\alpha \le C A_1^n.$$

where the norm for w is defined by formula(2.1.4). From the embedding H_α in $C^{0,\gamma}$ (theorem 2.1.2) it follows that the solution satisfies the Hölder condition. □

Remark 2.3.2. Note the important case, where system (2.3.18) is symmetric ($A^- = 0$) and γ is small. In this case by means of formula (1.1.9), we have $K = (\Lambda - \lambda)(\Lambda + \lambda)^{-1}$.Thus, for symmetric second order systems, the weak solution will satisfy the Hölder condition when

$$\frac{\Lambda - \lambda}{\Lambda + \lambda}\sqrt{1 + \frac{(m-2)^2}{m-1}} < 1. \tag{2.3.33}$$

It is trivial that for $m = 2$, this condition is satisfied for any nondegenerating second order elliptic system with bounded nonlinearities. Further (section 2.5) we shall show that condition (2.3.33) is precise, which means that if the strong inequality in (2.3.33) is replaced by an equality, the solution may become discontinuous.

It is not at all difficult to prove a theorem similar to 2.3.1 for systems with coefficients independent of u, and for the iterative process (2.3.26).

Theorem 2.3.2. *Let g be a trace of some function, belonging to $W_2^{(1)}(\Omega)$ and the coefficients $a_i(x; D_j u)(i,j = 1, ..., m)$,be independent of u, with $a_0 = 0$. If inequalities* (2.3.19) *hold for $\xi_0^k = 0, k = 1, ..., N,$ and relation*

$$A_1 < 1, \tag{2.3.34}$$

with A_1 denoted by (2.3.30) *is valid, then the weak solution of system* (2.3.18) *with boundary conditions* (2.3.21) *satisfies the Hölder condition in any subdomain ω, which lies strictly inside Ω. Iterative process* (2.3.26) *converges to this solution in the norm H_α with $\alpha = 2 - m - 2\gamma$. The process converges beginning with an arbitrary $u_0 \in W_q^{(1)}(\omega) \cap W_2^{(1)}(\Omega)$ and $q > 2m(m+\alpha)^{-1}$, satisfying* (2.3.21) *for ε, which is defined in 1.2.2, by* (1.2.12) *and* (1.2.13) *at the rate of a geometric progression, with the common ratio $A_1 K'$.*

The proof is similar to that of the previous theorem. The only difference occurs in replacing the matrix A by the matrix A', which is defined by (1.2.18) for $l = 1$. In the symmetric case (and small $\gamma > 0$) the condition (2.3.33) should be introduced formally in the form

$$\text{(2.3.35)} \qquad \frac{\Lambda' - \lambda'}{\Lambda' + \lambda'}\sqrt{1 + \frac{(m-2)^2}{m-1}} < 1.$$

□

We now shall obtain a result which is analogyous to lemma 2.3.1 but with positive exponent $\beta = -\alpha$.

Lemma 2.3.3. *Let $\beta \in [0, m-2+2\gamma)$ $(0 < \gamma < 1)$. Then for integrals* (2.3.3′) *and* (2.3.4′) *(corollary* 2.3.1) *the estimates*

$$\text{(2.3.5′)} \qquad X'(u,v) \geq \frac{m^2 - \beta^2}{4(m-1) + (\beta + m - 2)^2} \int\limits_{B_\delta(x_0)} |D'u|^2 |x - x_0|^\beta dx - \\ -C \int\limits_{B_\delta(x_0)} |Du|^2 r^{\beta'} dx$$

and

$$\text{(2.3.6′)} \qquad Y'_\beta(v) \leq (1+\eta) \int\limits_{B_\delta(x_0)} |D'u|^2 |x - x_0|^\beta dx + C \int\limits_{B_\delta(x_0)} |Du|^2 r^{\beta'} dx,$$

where u and v are bounded by relatitions (2.2.3),(2.2.4),(2.3.1) *and* (2.3.2) *(where β is substituted for α).*

Proof. It is clear in this case that the relations analogous to (2.3.9) and (2.3.10) hold true. Now apply inequalities (2.2.30) and (2.2.31) for these relations. Taking into account $\beta \geq 0$ we get (2.3.6′) and

$$X'(u,v) \geq \int_0^\delta |u_0'|^2 r^{\beta+m-1} dr + \sum_{s=1}^{+\infty}\sum_{i=1}^{k_s} \int_0^\delta \{|u'_{s,i}|^2 + \\ + \left[s(s+m-2) - \frac{\beta(\beta+m-2)}{2}\right] \frac{|u_{s,i}|^2}{r^2}\Big\} r^{\beta+m-1} dr - \\ -C \int\limits_{B_\delta(x_0)} |Du|^2 r^{\beta'} dx.$$

Considering separately each term

$$I_1 = \int_0^\delta \left\{ |u'|^2 + \left[s(s+m-2) - \frac{\beta(\beta+m-2)}{2}\right] \frac{|u|^2}{r^2} \right\} r^{\beta+m-1} dr.$$

(We omit temporarily the index s). Let $0 < t_s < 1$. Then

$$I_1 = \int_0^\delta \left\{(1-t_s)|u'|^2 + t_s|u'|^2 + \left[s(s+m-2) - \frac{\beta(\beta+m-2)}{2}\right]\frac{|u|^2}{r^2}\right\} r^{\beta+m-1}dr.$$

Estimate $\int_0^\delta |u'|^2 r^{\beta+m-1}dr$ from below. Applying Hardy inequality(2.1.8) and taking into account the relation

$$u(r)\zeta(r) = -\int_r^\delta [u(\rho)\zeta(\rho)]' d\rho,$$

we get

$$\int_0^\delta |u\zeta|^2 r^{\beta+m-3}dr \le \frac{4}{(m-2+\beta)^2}\int_0^\delta |(u\zeta)'|^2 r^{\beta+m-1}dr$$

and immediately

$$\int_0^\delta |u|^2 r^{\beta+m-3}dr \le \frac{4}{(m-2+\beta)^2}\int_0^\delta |u'|^2 r^{\beta+m-1}dr + C\int_0^\delta (|u'|^2+|u|^2) r^{\beta'+m-1}dr.$$

In fact

$$\int_0^\delta |u|^2\zeta^2 r^{\beta+m-3}dr = \int_0^\delta |u|^2 r^{\beta+m-3}dr + \int_0^\delta |u|^2(\zeta^2-1) r^{\beta+m-3}dr =$$

$$= \int_0^\delta |u|^2 r^{\beta+m-3}dr + \int_{\frac{\delta}{2}}^\delta |u|^2(\zeta^2-1) r^{\beta+m-3}dr.$$

Since $\frac{\delta}{2} \le r \le \delta$ we obtain

$$\int_0^\delta |u|^2\zeta^2 r^{\beta+m-3}dr \ge \int_0^\delta |u|^2 r^{\beta+m-3}dr - C\int_{\frac{\delta}{2}}^\delta |u|^2 r^{\beta'+m-3}dr.$$

On the other side we have

$$\int_0^\delta |(u\zeta)'|^2 r^{\beta+m-1}dr = \int_0^\delta |u'|^2 r^{\beta+m-1}dr + \int_0^\delta uu'(\zeta^2)' r^{\beta+m-1}dr +$$

$$+\int_0^\delta |u|^2|\zeta'|^2 r^{\beta+m-1}dr + \int_0^\delta |u'|^2 r^{\beta+m-1}(\zeta^2-1) r^{\beta+m-1}dr.$$

All integrals excepting the first are integrated over the interval $[\frac{\delta}{2},\delta]$. Therefore

$$\int_0^\delta |(u\zeta)'|^2 r^{\beta+m-1}dr \le \int_0^\delta |u'|^2 r^{\beta+m-1}dr + C\int_{\frac{\delta}{2}}^\delta (|u'|^2+|u|^2)r^{\beta'+m-1}dr$$

and we obtain the desired estimate, giving

$$I_1 \ge \int_0^\delta \left[(1-t_s)|u'|^2 + \frac{4^{-1}(\beta+m-2)^2 t_s + s(s+m-2) - 2^{-1}\beta(\beta+m-2)}{r^2}|u|^2\right] \times$$

$$\times r^{\beta+m-1}dr - C\int_{\frac{\delta}{2}}^\delta (|u'|^2+|u|^2)r^{\beta'+m-1}dr.$$

We obtain t_s from the equality

$$s(s+m-2)(1-t_s) = \frac{(\beta+m-2)^2}{4}t_s + s(s+m-2) - \frac{\beta(\beta+m-2)}{2}.$$

Then

$$t_s = \frac{2^{-1}\beta(\beta+m-2)}{s(s+m-2)+4^{-1}(\beta+m-2)^2}$$

and

$$1-t_s \ge 1-t_1 = \frac{m^2-\beta^2}{4(m-1)+(\beta+m-2)^2}.$$

While $1-t_1$ is a decreasing function on β we get

$$\int_0^\delta \left\{|u'|^2 + \left[s(s+m-2) - \frac{\beta(\beta+m-2)}{2}\right]\frac{|u|^2}{r^2}\right\} r^{\beta+m-1}dr \ge$$

$$\ge (1-t_1)\int_0^\delta \left[|u'|^2 + \frac{s(s+m-2)}{r^2}|u|^2\right] r^{\beta+m-1}dr - C\int_{\frac{\delta}{2}}^\delta (|u'|^2+|u|^2)r^{\beta'+m-1}dr.$$

This leads us to (2.3.5′).

□

Remark 2.3.3. If we add in (2.3.4′) under the sign of the integral the factor $\zeta(r)$ (2.3.6′) also holds. In this case the integrals on the right-hand side of (2.3.6′) will have this same factor $\zeta(r)$. This follows from the properties formulated by (2.2.1),(2.2.2) and the inequalities (2.2.26). With the help of these relations it is possible to prove the necessary estimates. For example

$$\int_0^\delta |u|^2 r^{\beta+m-1}|\zeta'|^2 dr = \int_{\delta/2}^\delta |u|^2 r^\beta |\zeta'|^2 dr \le C\int_{\delta/2}^\delta |u|^2 r^{\beta'+m-1}\zeta dr \le C\int_0^\delta |u|^2 r^{\beta'+m-1}\zeta dr.$$

Consider now inside $B_\delta(x_0)$ the problem

$$\Delta u = \operatorname{div} f, \quad u|_{\partial B_\delta} = 0, \tag{2.3.36}$$

where f is at first a sufficiently smooth function.

Lemma 2.3.4. *Let $\alpha \in (-m, 2-m) \cup (2-m, 0]$ with δ sufficiently small and α satisfying inequality (2.3.29). If $f \in \mathcal{L}_{2,\alpha}(B_\delta(x_0), x_0)$, then the weak solution of (2.3.35) satisfies the inequality*

$$\int_{B_\delta(x_0)} |D'u|^2 |x - x_0|^\alpha dx \leq \left[1 - \frac{(m-2)\alpha}{m-1} + \eta\right] \times \\ \times \left[min\left\{1 - \frac{\alpha(\alpha + m - 2)}{2(m-1)}; 1\right\}\right]^{-1} \times \\ \times \int_{B_\delta(x_0)} |f|^2 |x - x_0|^\alpha dx + C \int_{B_\delta(x_0)} |Du|^2 dx. \tag{2.3.37}$$

If $f \in \mathcal{L}_{2,-\alpha}(B_\delta(x_0), x_0)$ then the inequality

$$\int_{B_\delta(x_0)} |D'u|^2 |x - x_0|^{-\alpha} dx \leq \left[\frac{4(m-1) + (-\alpha + m - 2)^2}{m^2 - \alpha^2} + \eta\right] \times \\ \times \int_{B_\delta(x_0)} |f|^2 |x - x_0|^\alpha dx + C \int_{B_\delta(x_0)} |Du|^2 |x - x_0|^{\alpha'} dx, \tag{2.3.38}$$

where $\alpha < \alpha'$ holds.

Proof. Without loss of generality we can assume, that f is a smooth function. Problem (2.3.36) can be written in the weak form

$$\int_{B_\delta(x_0)} D'uD'v dx = \int_{B_\delta(x_0)} f \nabla v dx.$$

After multiplying and dividing by $|x - x_0|^\alpha$ under the sign of the right-hand side integral, we apply the Hölder inequality and come to the following estimate

$$|\int_{B_\delta(x_0)} D'uD'v dx|^2 \leq \int_{B_\delta(x_0)} |f|^2 |x - x_0|^{\pm\alpha} dx \int_{B_\delta(x_0)} |D'v|^2 |x - x_0|^{\mp\alpha} dx.$$

In the case for α take for v function (2.3.2) and apply inequalities (2.3.5′) and (2.3.6′). In the case $-\alpha$ the inequalities analogous to (2.3.5) and (2.3.6) should be used. □

Later in section 2.5 we shall prove (2.3.37) is sharp.
The results of this section were obtained in a series of papers by the author (Koshelev [3]-[11]) and in his monography [15]).

2.4 Hölder continuity in the entire domain

We shall first consider a singular integral operator

$$(2.4.1) \qquad J(f) = \begin{cases} \frac{1}{(m-2)|S|} D_k \int\limits_\Omega f^{(k)}(y)|x-y|^{2-m}dy, & m \geq 3, \\ \frac{1}{2\pi} D_k \int\limits_\Omega f^{(k)}(y) ln|x-y|dy, & m = 2, \end{cases}$$

where $f \in L_{2,\alpha}(\Omega; x_0)$ for $\forall x_0 \in \Omega$ ($\Omega \subset R^m$) and $\alpha \in (-m, 2-m) \cup (2-m, 0]$ and S is the area of the unit sphere.

Lemma 2.4.1. *Let Ω and ω be two bounded domains in R^m and $\omega \subset \Omega$. Suppose that $x_0 \in \omega$ and $dist\{\omega, \partial\Omega\} > 0$. Then for $\alpha \in (-m, 2-m) \cup (2-m, 0)$, satisfying inequality* (2.3.29) *the estimate*

$$(2.4.2) \int\limits_\omega |\nabla J|^2 |x-x_0|^\alpha dx \leq \left[1 - \frac{\alpha(m-2)}{m-1} + \eta\right]\left[min\{1 - \frac{\alpha(\alpha+m-2)}{2(m-1)}, 1\}\right]^{-1} \times$$
$$\times \int\limits_\Omega |f|^2 |x-x_0|^\alpha dx + C \int\limits_\Omega |f|^2 dx$$

holds true.

Proof. We shall consider only the case $m \geq 3$ For $m = 2$ the proof is identical. Let us first prove the lemma for the case when $\omega = B_{\frac{\delta}{2}}$ and $\Omega = B_\delta$. Consider problem (2.3.36) inside B_δ. This problem can be written in the following way

$$\int\limits_{B_\delta} \nabla u \nabla v dx = \int\limits_{B_\delta} f \nabla v dx,$$

where v is an arbitrary smooth function finite in B_δ. The solution of problem (2.3.36) can be written in the form

$$u(x) = \frac{1}{(m-2)|S|} \int\limits_{B_\delta} \operatorname{div} f \left[\frac{1}{|x-y|^{m-2}} - \left(\frac{\delta}{|x|}\right)^{m-2} \frac{1}{|x'-y|^{m-2}}\right] dy,$$

where x' is the conjugate to point x. Integrating by parts and taking into account that the expression in the square brackets is equal to zero on ∂B_δ we come to

$$u(x) = -\frac{1}{(m-2)|S|} \int\limits_{B_\delta} f^{(k)}(y) \frac{\partial}{\partial y_k} |x-y|^{2-m} dy + \int\limits_{B_\delta} f(y) g(x,y) dy,$$

where $g(x, y)$ is smooth inside $B_{\delta/2}$. Stating that

$$\frac{\partial}{\partial y_k}(|x-y|^{2-m}) = -\frac{\partial}{\partial x_k}(|x-y|^{2-m}),$$

we get

$$u(x) = \frac{1}{(m-2)|S|} D_k \int_{B_\delta} f^{(k)}(y) \frac{dy}{|x-y|^{m-2}} + \int_{B_\delta} f(y) g(x,y) dy.$$

After differentiating once more we obtain

$$\frac{1}{(m-2)|S|} D_k D_i \int_{B_\delta} f^{(k)}(y) \frac{dy}{|x-y|^{m-2}} = D_i u - \int_{B_\delta} f(y) D_i g(x,y) dy.$$

Then

$$D_i J = D_i u - \int_{B_\delta} f(y) D_i g(x,y) dy.$$

Taking into account that $D_i g$ are bounded for $|x - x_0| < \frac{\delta}{2}$ after squaring, multiplying by $|x - x_0|^\alpha$ and integrating over $B_{\frac{\delta}{2}}$ we get

$$\sum_{i=1}^{m} \int_{B_{\frac{\delta}{2}}} |D_i J|^2 |x - x_0|^\alpha dx \leq (1+\eta) \sum_{i=1}^{m} \int_{B_{\frac{\delta}{2}}} |D_i u|^2 |x - x_0|^\alpha dx +$$

$$+C \int_{B_{\frac{\delta}{2}}} |f(y)|^2 dy.$$

Applying (2.3.37) to the right-hand side we come to the inequality

(2.4.3)
$$\int_{B_{\frac{\delta}{2}}} |D' J|^2 |x - x_0|^\alpha dx \leq \left[1 - \frac{\alpha(m-2)}{m-1} + \eta\right]$$

$$\times \left[\min\left\{1 - \frac{\alpha(\alpha+m-2)}{2(m-1)}, 1\right\}\right]^{-2} \times$$

$$\times \int_{B_\delta} |f|^2 |x - x_0|^\alpha dx + \left(C \int_{B_\delta} |Du|^2 dx + \int_{B_\delta} |f|^2 dx\right).$$

Since u is a weak solution of (2.3.36) we get (2.4.2) for $\omega = B_{\frac{\delta}{2}}$ and $\Omega = B_\delta$.
Let us now consider the common case and first extend f by zero for entire space R^m. Take the balls $B_{\frac{\delta}{2}} \supset \omega$ and $B_\delta \subset \Omega$. Apply inequality (2.4.3) for this case and reduce the left-hand side integral to the domain ω. This provides the proof of the whole lemma. □

Consider in the half space $R^m_- = \{x : x_m < 0\}$ the problem

(2.4.4)
$$\Delta u = \operatorname{div} f, \quad u|_{x_m=0},$$

where the finite $f \in \mathcal{L}_{2,\alpha}(R^m_-; x_0)$. As usual we can start with a sufficiently smooth f. Let $D_d = \{x : 0 > x_m > -d\}$, with $d > 0$ sufficiently small. Take the point $x_0 \in D_d$ and the ball $B_\delta(x_0)$. The Green function $G(x,y)$ of (2.4.4) has the form

$$G(x,y) = \frac{1}{(m-2)|S|}\left(\frac{1}{|x-y|^{m-2}} - \frac{1}{|x^*-y|^{m-2}}\right), \tag{2.4.5}$$

where x^* is symmetric to x with respect to the plane $x_m = 0$ (we consider for the sake of brevity only $m > 2$; for $m = 2$ the expressions for $G(x,y)$ contain the logarithm but the result will be identical).
First let $B_{\frac{\delta}{2}} \cap R^m_+ = \emptyset$, with $R^m_+ = \{x : x_m > 0\}$. Multiply the differential equation of (2.4.4) by ζG, where ζ is the cut-off function (2.2.1) and integrate over the domain $B_\delta^- = B_\delta \cap (x_m < 0)$ If we integrate once by parts and use the boundary condition (2.4.4) for $G(x,y)$ we come to the equality

$$\int\limits_{B_\delta^-} \nabla u \nabla(\zeta G) dy = \int\limits_{B_\delta^-} f \nabla(\zeta G) dy.$$

Let $B^-_{\frac{\delta}{2},\delta} = B_\delta^- \backslash B^-_{\frac{\delta}{2}}$. Integrating by parts on the left-hand side once more and taking in account that $\zeta' = 0$ inside $B_{\frac{\delta}{2}}$ we get

$$u(x) = -\int\limits_{B_\delta^-} \zeta f \nabla_y G dy - \int\limits_{B^-_{\frac{\delta}{2},\delta}} G f \nabla_y \zeta dy - \int\limits_{B^-_{\frac{\delta}{2},\delta}} u(\Delta_y \zeta G + 2\nabla_y \zeta \nabla_y G) dy, \tag{2.4.6}$$

where ∇_y denotes the differentiation with respect to y.

The right-hand side can be divided in two units: one, containing the regular part of Green function in (2.4.6) and the other unit, containing the singular part.
Then (2.4.6) can be written in the form

$$\begin{aligned} u(x) = -\frac{1}{(m-2)|S|}\Bigg[& \int\limits_{B_\delta^-} \zeta f \nabla_y (|x-y|^{2-m}) dy + \\ & + \int\limits_{B^-_{\frac{\delta}{2},\delta}} |x-y|^{2-m} f \nabla_y \zeta dy + \int\limits_{B^-_{\frac{\delta}{2},\delta}} u(y)|x-y|^{2-m} \, \Delta_y \, \zeta dy + \\ & + 2 \int\limits_{B^-_{\frac{\delta}{2},\delta}} u \nabla_y \zeta \nabla_y |x-y|^{2-m} dy \Bigg] + \int\limits_{B_\delta^-} f(y) g(x,y) dy + \\ & + \int\limits_{B_\delta^-} u(y) h(x,y) dy, \end{aligned} \tag{2.4.7}$$

where g and h are smooth functions for $x \in B_{\frac{\delta}{2}}$. Notice the second, third and fourth integrals on the right-hand side are taken over $B^-_{\frac{\delta}{2},\delta}$. So if $x \in B^-_{q\frac{\delta}{2}}$ with $0 < q < 1$ then the derivatives under the signs of these integrals will be bounded. Notice also the first integral on the right-hand side of (2.4.7) coincides with integral (2.4.1), where f is changed to $f\zeta$. This is clear, because we can take the differentiation over x_k outside the sign of the integral. The integrals which contain the derivatives of ζ are regular for $|x| < q\frac{\delta}{2}$. Therefore all the integrals after differentiating with respect to x can be estimated in $L_{2,\alpha}(B^-_{q\frac{\delta}{2}})$ through the norms in $L_2(B_\delta)$ for u and f. To estimate the derivatives of the first integral we apply (2.4.2), and we have

$$(2.4.8)\quad \int_{B^-_{q\frac{\delta}{2}}} |\nabla u|^2 r^\alpha dx \le \left[1 - \frac{\alpha(m-2)}{m-1} + \eta\right]\left[min\{1 - \frac{\alpha(\alpha+m-2)}{2(m-1)}, 1\}\right]^{-2} \times$$

$$\times \int_{B^-_\delta} |f|^2 r^\alpha dx + C\left(\int_{B^-_\delta} |f|^2 dx + \int_{B^-_\delta} |Du|^2 dx\right).$$

Suppose now that $B_{\frac{\delta}{2}} \cap R^m_+ \neq \emptyset$. Multiplying the differential equation of (2.4.4) by $\zeta G(x,y)$ and integrating by parts we come to (2.4.6), which can be written in the form

$$(2.4.9)\quad u(x) = -\frac{1}{(m-2)|S|}\left\{\int_{B^-_\delta} \zeta f(y)[\nabla_y|x-y|^{2-m} - \right.$$

$$\left. -\nabla_y|x^* - y|^{2-m}]dy\right\} +$$

$$+ \int_{B^-_{\frac{\delta}{2},\delta}} H(x,y)f(y)dy + \int_{B^-_{\frac{\delta}{2},\delta}} K(x,y)u(y)dy$$

where $H(x,y)$ and $K(x,y)$ for $|x| < q\frac{\delta}{2}(0 < q < 1)$ are regular kernels. So, it is necessary to estimate the first term

$$(2.4.10)\quad R(x) = -\frac{1}{(m-2)|S|}\int_{B^-_\delta} \zeta f(y)\left[\nabla_y|x-y|^{2-m} - \nabla_y|x^*-y|^{2-m}\right]dy.$$

After taking the differentiation outside the integral, we can write $R(x)$ in the form

$$(2.4.11)\quad R(x) = \frac{1}{(m-2)|S|}\left[\sum_{i=1}^{m} \frac{\partial}{\partial x_i}\int_{B^-_\delta} \zeta f^{(i)}(y)|x-y|^{2-m}dy - \right.$$

$$\left. -\sum_{i=1}^{m-1} \frac{\partial}{\partial x_i}\int_{B^-_\delta} \zeta f^{(i)}(y)|x^*-y|^{2-m}dy + \frac{\partial}{\partial x_m}\int_{B^-_\delta} \zeta f^{(m)}(y)|x^*-y|^{2-m}dy\right].$$

Here we have used the identities

$$\frac{\partial}{\partial y_i}|x^* - y|^{2-m} = -\frac{\partial}{\partial x_i}|x^* - y|^{2-m} \quad (i = 1, ..., m-1)$$

and

$$\frac{\partial}{\partial y_m}|x^* - y|^{2-m} = \frac{\partial}{\partial x_m}|x^* - y|^{2-m}.$$

Let us extend the functions $f^{(i)}(x)(i = 1, ..., m-1)$ antysimmetrically and the function $f^{(m)}(x)$ symmetrically on $x_m > 0$. Denote by B_δ^+ the symmetric to B_δ^- domain with respect to $x_m = 0$. Then (2.4.11) can be written in the following way

$$R(x) = \frac{1}{(m-2)|S|} \sum_{i=1}^{m} \frac{\partial}{\partial x_i} \int_{B_\delta^- \cup B_\delta^+} \zeta f^{(i)}(y)|x - y|^{2-m} dy$$

and $R(x)$ is an odd function on x_m. Applying inequality (2.4.2) we come to the estimate

(2.4.12)
$$\int_{B_{q\frac{\delta}{2}}^- \cup B_{q\frac{\delta}{2}}^+} |\nabla R|^2 r^\alpha dx \leq \left[1 - \frac{(m-2)\alpha}{m-1} + \eta\right] \times$$
$$\times \left[min\{1 - \frac{\alpha(\alpha + m - 2)}{2(m-1)}, 1\}\right]^{-2} \int_{B_\delta^- \cup B_\delta^+} |f|^2 r^\alpha dx + C \int_{B_\delta^-} |f|^2 dx.$$

Return now to (2.4.9) and extend $u(x)$ to B_δ^+ in an asymmetric way. As we have mentioned, the last two terms of (2.4.9) can be estimated through $\int_{B_\delta^-} |u|^2 dx$ and $\int_{B_\delta^-} |f|^2 dx$. So, for the weak solution of (2.4.4) we have the estimate

(2.4.13)
$$\int_{B_\delta^- \cup B_\delta^+} |\nabla u|^2 r^\alpha dx \leq \left[1 - \frac{(m-2)\alpha}{m-1} + \eta\right] \times$$
$$\times \left[min\{1 + \frac{\alpha(\alpha + m - 2)}{2(m-1)}, 1\}\right]^{-1} \int_{B_\delta^- \cup B_\delta^+} |f|^2 r^\alpha dx + C \left(\int_{B_\delta^-} |Du|^2 dx + \int_{B_\delta^-} |f|^2 dx \right)$$

where α satisfies (2.3.29).

Consider a slightly more general problem

(2.4.14)
$$\Delta u - u = \text{div} f + \varphi, \ u|_{x_m} = 0.$$

Under the same conditions with respect to f we come to

Lemma 2.4.2. *If $\varphi \in \mathcal{L}_{2,\alpha}(B_g^-(x_0);x_0)(x_0 \in \overline{\Omega})$, the functions u, f and φ are extended to $x_m > 0$ respectively in symmetric and asymmetric ways, then the weak solution of* (2.4.14) *satisfies inequality*

$$\int_{B_\delta^- \cup B_\delta^+} |D'u|^2 r^\alpha dx \le (A_1^2 + \eta) \int_{B_\delta^- \cup B_\delta^+} |f|^2 r^\alpha dx + \eta \int_{B_\delta^- \cup B_\delta^+} |\varphi|^2 r^\alpha dx \tag{2.4.15}$$

$$+C\left[\int_{B_\delta^-} |Du|^2 dx + \int_{B_\delta^-} |f|^2 dx + \int_{B_\delta^-} |\varphi|^2 dx\right],$$

where A_1 is defined by (2.3.30).

Proof. Multiply differential equation (2.4.14) by ζG, where G is defined by (2.4.5). Integrating over $B_\delta^-(x_0)(x_0 \in R_-^m)$ we get

$$u(x) = -\int_{B_\delta^-} \zeta f \nabla_y G dy - \int_{B_{\frac{\delta}{2},\delta}^-} G f \nabla_y \zeta dy - \tag{2.4.16}$$

$$-\int_{B_{\frac{\delta}{2},\delta}^-} u\left[\Delta_y \zeta G + 2\nabla_y \zeta \nabla_y G\right] dy + \int_{B_\delta^-} u\zeta G dy + \int_{B_\delta^-} \varphi G \zeta dy$$

which is analogous to (2.4.6).

After differentiating with respect to x the first three terms can be estimated by the right-hand side of (2.4.13). Therefore we consider only the last two terms. It is enough to show how to estimate one of these terms, for example, the last one. Extend this term on $x_m > 0$ in an even way. Since the first derivatives are operators with weak singlularity, we get

$$\int_{B_{\frac{\delta}{2}}^-} |\nabla_x \int_{B_\delta^-} \varphi\ \zeta G dy|^2 dx \le C \int_{B_\delta^-} |\varphi|^2 dx. \tag{2.4.17}$$

The second derivatives of the terms under consideration can be represented through singular integrals. Therefore, applying Stein [1] estimates and (2.1.17′) we get

$$\int_{B_{\frac{\delta}{2}}^-} |\nabla \int_{B_\delta^-} \varphi \zeta G dy|^2 r^\alpha dx \le \eta \int_{B_{\frac{\delta}{2}}^-} |D'^2 \int_{B_\delta^-} \varphi G \zeta dy|^2 r^\alpha dx +$$

$$+C \int_{B_{\frac{\delta}{2}}^-} |D' \int_{B_\delta^-} \varphi G \zeta dy|^2 dx \le \eta \int_{B_\delta^-} \varphi^2 r^\alpha dx + C \int_{B_\delta^-} \varphi^2 dx.$$

Here we used the inequality $\zeta \le 1$. So, the two last terms on the right-hand side of (2.4.16) are estimated and we come to (2.4.15). □

Now consider problem (2.3.18),(2.3.20).

Theorem 2.4.1. *Let $\partial\Omega \in C^{1,æ}$, $\alpha = 2 - m - 2\gamma (0 < \gamma < 1)$, K satisfy* (2.3.29) *and* (2.3.30), *g be a trace of the function from $H_\alpha(\bar\Omega)$ and conditions 1) - 3) of section 1.1 are satisfied . Then the weak solution of problem* (2.3.18), (2.3.20) *belongs to $H_\alpha(\bar\Omega)$ and therefore satisfies the Hölder condition with the exponent $(2 - m - \alpha)/2$. Concerning the convergency of methods* (2.3.23) and (2.3.24) *the same statements as in theorem 2.3.1 are true but for the whole of the domain $\bar\Omega$.*

Proof. Due to the same reasons as in theorem 2.3.1 we can suppose that the boundary condition is homogenious. Let $w_{n+1} = u_{n+1} - u_n$, where u_n is the iteration of the above mentioned process. We can write

(2.4.18)
$$\int_\Omega Dw_{n+1}Dv dx = \sum_{i=0}^m \int_\Omega \left[D_i w_n - \varepsilon \overline{\frac{\partial a_i}{\partial p_k}} D_k w_n\right] D_i v dx.$$

Denote

(2.4.19)
$$D_i w_n - \varepsilon \overline{\frac{\partial a_i}{\partial p_k}} D_k w_n = f_i$$

(2.4.20)
$$w_n - \varepsilon \overline{\frac{\partial a_0}{\partial p_k}} D_k w_n = \varphi.$$

Then integral identity (2.4.18) can be written in the form

(2.4.21)
$$\int [D' w_{n+1} D' v + w_{n+1} v]\, dx = \sum_{i=1}^m \int (f_i D_i v + \varphi v) dx.$$

Straighten the piece of $\partial\Omega$, lying inside $B_\delta(x_0)$ with the help of the transformation

$$x_i' = x_i \quad (i = 1, ..., m-1),$$

$$x_m' = x_m - \psi(x_1, ..., x_{m-1}),$$

where $x_m = \psi(x_1, ..., x_{m-1})$ is the equation of boundary surface, lying inside $B_\delta(x_0)$ with a sufficiently small $\delta > 0$. Since $\partial\Omega \in C^{1,æ}$ we can suppose that all $|D\psi|$ are also sufficiently small. After rewriting the integral identity (2.4.21) into new coordinates we shall have

(2.4.22)
$$\sum_{i=1}^m \int_{\Omega'} \left[\frac{\partial w_{n+1}}{\partial x_i'} + (1 - \delta_{im}) \frac{\partial \psi}{\partial x_i} \frac{\partial w_{n+1}}{\partial x_m'}\right] \times$$
$$\times \left[\frac{\partial v}{\partial x_i'} + (1 - \delta_{im}) \frac{\partial \psi}{\partial x_i} \frac{\partial v}{\partial x_m'}\right] |J| dx +$$
$$+ \int_{\Omega'} w_{n+1} v |J| dx' = \sum_{i=1}^m \int_{\Omega'} f_i \left[\frac{\partial v}{\partial x_i'} + (1 - \delta_{im}) \frac{\partial \psi}{\partial x_i} \frac{\partial v}{\partial x_m'}\right] |J| dx' + \int_{\Omega'} \varphi v |J| dx'.$$

Take x_0' such, that after passing to the new coordinates the piece of the plane $x_m' = 0$ will be inside $B_\delta'(x_0')$ where $B_\delta'(x_0')$ is a ball in new coordinate space.
It is obvious (2.4.22) can be represented in the form

$$\int_{\Omega'} (D'w_{n+1}D'v + w_{n+1}v)dx' = \int_{\Omega'} (f'D'v + \varepsilon'v)dx',$$

where $f' = f + F, \varphi' = \varphi + \Phi$. Here F and Φ are linear homogeneous forms on w_{n+1} and its first derivatives with small coefficients (depending on $D\phi$).
Then

$$\int_{B_\delta'^{-}(x_0')} |F|^2|x' - x_0'|^\alpha dx' < \eta \int_{B_\delta'^{-}(x_0')} \left(|D'w_{n+1}|^2 + |w_{n+1}|^2\right)|x' - x_0'|^\alpha dx'$$

and

$$\int_{B_\delta'^{-}(x_0')} |\Phi|^2|x' - x_0'|^\alpha dx' < \eta \int_{B_\delta'^{-}(x_0')} \left(|D'w_{n+1}|^2 + |w_{n+1}|^2\right)|x' - x_0'|^\alpha dx'.$$

After extending w' in an odd way for $x_m' > 0$ we get from the previous lemma that

$$\text{(2.4.23)} \quad \int_{B_\delta^{-'}(x_0')\cup B_\delta'^{+}(x_0')} (|D'w_{n+1}|^2 + |w_{n+1}|^2)|x' - x_0'|^\alpha dx' \le (A_1^2 + \eta) \times$$
$$\times \int_{B_\delta^{-'}(x_0')\cup B_\delta'^{+}(x_0')} |f'|^2|x' - x_0'|^\alpha dx' + \eta \int_{B_\delta^{-'}(x_0')\cup B_\delta'^{+}(x_0')} |\varphi'|^2|x' - x_0'|^\alpha dx' +$$
$$+C\int_{\Omega'} (|Dw_n|^2 + |w_{n+1}|^2 + |\varphi_{n+1}'|^2)dx'.$$

Now we can return to the old coordinates and consider the problem identical to theorem 2.3.1.

□

2.5 The sharpness of the regularity conditions for solutions of second order systems

As was mentioned in the introduction, there exists a non-smooth (for example, non-Hölder) solution of the boundary value problem (2.3.18),(2.3.20) for cases where the dimension is sufficiently large, even through the given natural data are smooth (Giusti, Miranda[1], Mazja [1]). This fact also holds for some linear elliptic problem of divergent form (De Giorgi [2]). In the previous chapter we have obtained some conditions sufficient for the smoothness of weak solutions. Therefore, there arises a question, of how sharp these conditions are. It turns out that the conditions on dispersion of A^+ eigenvalues for the smoothness of the weak solution in some cases are sharp. For precise classes of elliptic systems, singled out by inequalities (2.3.29),(2.3.32) and (2.3.33)

, these conditions are necessary in order that the desired smoothness of the solutions hold for each system of the class. These results were published in the author's paper (Koshelev [12]).

Consider system (2.3.18) with boundary conditions (2.3.20), where g is a trace in $W_2^{(1)}$. Let Ω be a strictly interior subdomain of Ω. Notice the theorem 2.3.1 establishes that when (2.3.33) for a symmetric A' is satisfied, the weak solution of the problem will satisfy the Hölder condition.

Let us show that this condition is sharp. For this purpose, consider the following system

$$D_j\left[A_{hk}^{ij}(x)D_iu^{(h)}\right] = 0 \quad (i,j,h,k = 1,...,m), \tag{2.5.1}$$

where

$$A_{hk}^{ij}(x) = \delta_{hk}\delta_{ij} + \left(c\delta_{ih} + d\frac{x_ix_h}{|x|^2}\right)\left(c\delta_{jk} + d\frac{x_jx_k}{|x|^2}\right). \tag{2.5.2}$$

For $c = m-2$ and $d = m$, system (2.5.1) is that of De Giorgi [2]. Equations (2.5.1) will be considered in a m-dimensional unit ball $|x| < 1$, on the boundary of which the condition

$$u|_{|x|=1} = x \tag{2.5.3}$$

is given. After simple calculations, one can see that problem (2.5.1), (2.5.3) has the solution $u = x|x|^{-b}$, where

$$b = m/2 - \sqrt{\frac{m^2}{4} - \frac{m(m-1)cd + (m-1)d^2}{1+(c+d)^2}}. \tag{2.5.4}$$

System (2.5.1) is the Euler equation for the variational problem with the functional

$$\int_{|x|<1}\left\{|D'u|^2 + \left[\left(c\delta_{ij} + d\frac{x_ix_j}{|x|^2}\right)D_iu^{(j)}\right]^2\right\}dx.$$

So, the matrix A', defined by (1.2.18)., is symmetric. Then

$$A'\xi\xi = A'^{+}\xi\xi = \sum_{i,k=1}^{m}\left\{|\xi_i^{(k)}|^2 + \left[\left(c\delta_{ik} + d\frac{x_ix_k}{|x|^2}\right)\xi_i^{(k)}\right]^2\right\} \tag{2.5.5}$$

and it is evident that $\lambda' = 1$. In this case $K' = (\Lambda' - \lambda')(\Lambda' + \lambda')^{-1}$. It is easy to find Λ'. From (2.2.5) it follows that

$$\Lambda' = 1 + \sup_{|x|<1,|\xi|=1}\left[\left(c\delta_{ik} + d\frac{x_ix_k}{|x|^2}\right)\xi_i^{(k)}\right]^2. \tag{2.5.6}$$

Thus we have to find the largest eigenvalues of the following matrix

$$(2.5.7) \qquad B = \left\{ \left(c\delta_{ih} + d\frac{x_i x_h}{|x|^2} \right) \left(c\delta_{jk} + d\frac{x_j x_k}{|x|^2} \right) \right\}.$$

From the Hölder inequality it follows that the greatest eigenvalue of the matrix $\{a_j a_l\}$, $j,l = 1,...,M$, where a_j are arbitrary numbers, is $\sum_{j=1}^{M} a_j^2$. Hence the greatest eigenvalue of the matrix B is $\sum_{i,k=1}^{m} \left(c\delta_{ik} + d\frac{x_i x_k}{|x|^2} \right)^2 = c^2(m-1) + (c+d)^2$. According to (2.5.6) we have $\Lambda' = 1 + c^2(m-1) + (c+d)^2$. Take

$$(2.5.8) \qquad c = (m-1)^{-\frac{1}{2}} \left[1 + \frac{(m-2)^2}{m-1} \right]^{-\frac{1}{4}}, \quad d = \frac{c^2+1}{(m-2)c}.$$

For such c and d, from (2.5.4), we get $b = 1$. So, problem (2.5.1),(2.5.3) has an unique weak solution $u = x|x|^{-1}$ belonging to $W_2^{(1)}$ and discontinuous at $x = 0$. Meanwhile, after some calculations, we come to the equality

$$\frac{\Lambda' - \lambda'}{\Lambda' + \lambda'} \sqrt{1 + \frac{(m-2)^2}{m-1}} = 1.$$

In fact,

$$\frac{\Lambda' - \lambda'}{\Lambda' + \lambda'} = \frac{(m-1)c^2 + (c+d)^2}{2 + (m-1)c^2 + (c+d)^2}.$$

It is not difficult to see that

$$(m-1)c^2 + (c+d)^2 = \frac{(m-1)^2 \left[1 + \frac{(m-2)^2}{m-1} \right] c^4 + 2(m-1)c^2 + 1}{(m-2)^2 c^2}.$$

Substititing here the value of c, we have

$$(m-1)c^2 + (c+d)^2 = 2\frac{1 + (m-1)c^2}{(m-2)^2 c^2}.$$

Then

$$\frac{\Lambda' - \lambda'}{\Lambda' + \lambda'} = \frac{1 + (m-1)c^2}{1 + (m-1)\left[1 + \frac{(m-2)^2}{m-1} \right] c^2} =$$

$$= \left[1 + \frac{(m-2)^2}{m-1} \right]^{-\frac{1}{2}}.$$

Let us now return to the problem (2.3.36) and show that for $\alpha = 2 - m - \varepsilon$ and small $\varepsilon > 0$ the (2.3.37) is sharp. We shall consider only the differential equation

$$(2.5.9) \qquad \Delta u = div f$$

in a bounded domain Ω and assume there exists a weak solution $u(x)$ of (2.5.9), which belongs to $W_2^{(1)}(\Omega)$. Let $x_0 \in \Omega$ and $\delta < dist(x_0, \delta\Omega)$.

Theorem 2.5.1. *If $f \in \mathcal{L}_{2,\alpha}(\Omega, x_0)$ with $\alpha = 2 - m - 2\gamma$ and small $\gamma = \varepsilon/2 > 0$ then the weak solution $u \in W_2^{(1)}(\Omega)$ of (2.5.9) satisfies the inequality*

(2.5.10)
$$\int_\Omega |D'u|^2 |x - x_0|^\alpha dx \le \left[1 + \frac{(m-2)^2}{m-1} + O(\varepsilon)\right] \times$$
$$\times \int_\Omega |f|^2 |x - x_0|^\alpha dx + C(\varepsilon) \int_\Omega \left(|f|^2 + |Du|^2\right) dx.$$

The constant before the first integral of the right-hand side integral is sharp.

Proof. As in the proof of Lemma 2.3.4 we come immediately to 2.5.10. To prove the sharpness consider $\Omega = B_1(o)$ and take

(2.5.11)
$$u_\varepsilon(x) = a x_m r^{\varepsilon - 1}$$

with

(2.5.12)
$$a = \frac{m^2 - 3m + 3 + m\varepsilon - 2\varepsilon}{(1-\varepsilon)(m - 1 + \varepsilon)}.$$

Consider also

(2.5.13)
$$f_\varepsilon = \nabla\left(x_m r^{1-m}\right) r^{m-2+\varepsilon}.$$

Simple calculations lead to equalities $div\, f_\varepsilon = -(m^2 - 3m + 3 + m\varepsilon - 2\varepsilon)x_m$ and

$$\Delta\left(x_m r^{-1+\varepsilon}\right) = (\varepsilon - 1)(m - 1 + \varepsilon) x_m r^{-3+\varepsilon}.$$

From (2.5.11) - (2.5.13) we have $\Delta u_\varepsilon = div\, f_\varepsilon$. The relation (2.5.13) gives $f_\varepsilon^{(i)}(x) = (1-m)x_i x_m r^{-3+\varepsilon} + \delta_{im} r^{-1+\varepsilon} (i = 1, ..., m)$. So

$$\sum_{i=1}^m \int_B |f_\varepsilon^{(i)}|^2 r^{2-m-\varepsilon} dx = (1-m)(3-m)\int_B x_m^2 r^{-2-m+\varepsilon} dx + \int_B r^{-m+\varepsilon} dx.$$

Taking into account the equality

(2.5.14)
$$m \int_B x_m^2 r^{-2-m+\varepsilon} dx = \int_B r^{-m+\varepsilon} dx = \frac{1}{\varepsilon}|S|$$

where $|S|$ is the area of a unit sphere in R_m we get

(2.5.15)
$$m \int_B |f_\varepsilon|^2 r^{2-m-\varepsilon} dx = \frac{m^2 - 3m + 3}{m} \frac{|S|}{\varepsilon}.$$

From (2.5.11) after differentiation we obtain

(2.5.16)
$$(\nabla u_\varepsilon)^{(i)} = a\left[(\varepsilon - 1)x_i x_m + \delta_{im} r^2\right] r^{\varepsilon - 3}$$

and

$$\int_B |\nabla u_\varepsilon|^2 r^{2-m-\varepsilon} dx = a^2 \left[(\varepsilon^2 - 1)\int_B r^{-m-2-\varepsilon} x_m^2 dx + \int_B r^{-m+\varepsilon} dx\right].$$

From (2.5.14) follows

$$\int_B |\nabla u_\varepsilon|^2 r^{2-m-\varepsilon} dx = a^2 \frac{m-1+\varepsilon^2}{m} \frac{|S|}{\varepsilon}. \tag{2.5.17}$$

Comparing (2.5.15) and (2.5.17) we obtain

$$\int_B |\nabla u_\varepsilon|^2 r^{2-m-\varepsilon} dx = \left[1 + \frac{(m-2)^2}{m-1} + O(\varepsilon)\right] \int_B |f_\varepsilon|^2 r^{2-m-\varepsilon} dx. \tag{2.5.18}$$

Remark 2.5.1. *The optimal function can be written in the form*

$$u_\varepsilon = \frac{m^2 - 3m + 3 + m\varepsilon - 2\varepsilon}{(1-\varepsilon)(m-1)} r^\varepsilon \cos\theta \tag{2.5.19}$$

where θ is the angle between axes x_m and direction r. Obviously, when ε passes through $\varepsilon = 0$ from positive to negative values the function v_ε looses the continuity at $\varepsilon = 0$ but still preserves the boundedness. For $\varepsilon < 0$ the function (2.5.11) became unbounded.

□

Chapter 3

Some properties and applications of regular solutions for quasilinear ellliptic systems

3.1 The Liouville theorem

The Liouville theorem, universally known from the theory of analytic functions, was further generalized by many authors. The investigation was started by S. Bernstein. Later much research was aimed at expanding the theorem to the general elliptic equations.
Important results were obtained by Landis[1] for elliptic equations, satisfying the so-called Cordes condition. Many significant results belong to Serrin [1], Peletier and Serrin[1], Ivanov[1]. Some valuable results concerning the Liouville theorem for elliptic systems with linear principal part were achieved by Hildebrandt and Widman[1]. For a broad class of linear and quasilinear elliptic systems the Liouville theorem was proved by Nechas and Oleinik [1]. In Nechas's [1] lectures the validity of the Liouville theorem was tied to the regularity of weak solution for elliptic systems.
In this section we will prove the Liouville theorem for a broad class of quasilinear elliptic systems. This result was first claimed in the author's article [12] and later proved in our monograph [15].

Let u and v be represented by series (2.3.1), (2.3.2), where $v_{0,1}(r) = \int_{\delta}^{r} u'_{0,1}(\rho)\rho^{\alpha}\zeta_0(s)d\rho$, $v_{s,1}(r) = u_{s,i}(r)r^{\alpha}\zeta_0(r)(s \geq 1)$ and $\zeta_0(r)$ is a cut-off function satisfying (2.2.1), (2.2.2) and (2.2.20), $\alpha = 2 - m - 2\gamma, (0 < \gamma < 1), r = |x - x_0|$. Similar to (2.2.4) and (2.3.3′) denote the integrals

$$X'(u,v) = \int_{B_\delta(x_0)} D'uD'vdx, \tag{3.1.1}$$

$$Y^{o'}_{\alpha,0}(v) = \int_{B_{3\delta/4}(x_0)} \zeta_0^{-1}|D'v|^2 r^{-\alpha}dx. \tag{3.1.2}$$

Lemma 3.1.1. *If $u \in W^{(1)}_{2,\alpha}(B_\delta)$, then for integrals* (3.1.1), (3.1.2) *hold the following inequalities*

$$X'(u,v) \geq \left[1 - \frac{\alpha(\alpha+m-2)}{2(m-1)} - \eta\right] \times \tag{3.1.3}$$

$$\times \int\limits_{B_\delta(x_0)} |D'u|^2 r^\alpha \zeta_0 dx - C\delta^{-2} \int\limits_{B_\delta(x_0)} |u|^2 r^\alpha dx,$$

$$Y^{o'}_{\alpha,0}(v) \leq \left[1 - \frac{\alpha(m-2)}{(m-1)} + \eta\right] \int\limits_{B_\delta(x_0)} |D'u|^2 r^\alpha \zeta_0 dx + C\delta^{-2} \int\limits_{B_\delta(x_0)} |u|^2 r^\alpha dx, \tag{3.1.4}$$

where C is independent of δ.

Proof. Using expansions (2.3.1),(2.3.2), analagously to lemma 2.3.1 we get

$$X'(u,v) = \sum_{s,i} I_s(u_{s,i}, v_{s,i}) = \sum_{s=0}^{+\infty}\sum_{i=1}^{k_s} \int\limits_0^\delta \left[u'_{s,i}v'_{s,i} + \frac{s(s+m-2)}{r^2} u_{s,i}v_{s,i}\right] r^{m-1} dx$$

and

$$Y^{o'}_{\alpha,0}(v) = \sum_{s,i} J^o_{\alpha,0}(v_{s,i}) = \sum_{s=0}^{+\infty}\sum_{i=1}^{k_s} \int\limits_0^{3\delta/4} \left[|v'_{s,i}|^2 + \frac{s(s+m-2)}{r^2}|v_{s,i}|^2\right] r^{-\alpha+m-1}\zeta_0^{-1} dr,$$

where $I_s(u_{s,i}, v_{s,i})$ and $J^o_{\alpha,0}(v_{s,i})$ are determined by (2.2.5) with $\zeta = \zeta_0$ and (2.2.24). Then with the help of (2.2.21),(2.2.22), (2.2.23) (lemma 2.2.2) we come to (3.1.3) and (3.1.4).

□

Consider in the entire space R^m a system with coefficients independent of u

$$L'(u) \equiv -\sum_{i=1}^m D_i a_i(x; D'u) = \varphi(x). \tag{3.1.5}$$

Suppose $\varphi(x)$ is smooth and the coefficients $a_i(x;p)$ satisfy conditions 1) -3) from section 1.1.1 for $s = 2$ and $\xi_0^{(k)} = 0 (k, l = 1, .., N)$.Consider the matrix

$$A' = \left\{\frac{\partial a_i^{(k)}}{\partial p_j^{(l)}}\right\} \quad (i,j = 1,...,m; k,l = 1,...,N), \tag{3.1.6}$$

which differs from (1.1.4) in the absence of the rows and columns containing derivatives with respect to u and with $a_0 = 0$. (compare (1.2.18)).
Let $\lambda'_1, ..., \lambda'_{mN}$ be the eigenvalues of A'^+ (symmetric part of A'), $\lambda' = \inf_{i,x,p} \lambda'_i(x,p)$, $\Lambda' = \sup_{i,x,p} \lambda'_i(x,p)$ and σ' be the sup of eigenvalues for the matrix $C' = A'^+A'^- - A'^-A'^+ - (A'^+)^2$ with A'^- being the skew symmetric part of A'. Changing λ_i for skew-symmetric λ'_i and σ to σ', we can denote K'_ε by (1.1.7). Using the finite difference theorem, we come to

$$(3.1.7)\qquad \int_\Omega \left\{D_i(u^{(k)}-z^{(k)})-\varepsilon\left[a_i^{(k)}(x,D'u)-a_i^{(k)}(x,D'z)\right]\right\}D_iv^{(k)}dx=$$

$$=\int_\Omega\left[D_i(u^{(k)}-z^{(k)})-\varepsilon\frac{\overline{\partial a_i^{(k)}}}{\partial p_j^{(l)}}D_j(u^{(l)}-z^{(l)})\right]D_iv^{(k)}dx,$$

where the bar over derivatives denotes that they are calculated for some arguments, lying between u and z.

Since $v=0$ outside $B_\delta=B_\delta(0)$, we have

$$(3.1.8)\qquad \inf_{\varepsilon>0}\left|\int_\Omega\left\{D_i(u^{(k)}-z^{(k)})-\varepsilon\left[a_i^{(k)}(x,D'u)-a_i^{(k)}(x,D'z)\right]\right\}D_iv^{(k)}dx\right|^2\leq$$
$$\leq K'^2Y'_{-\alpha}(u-z)Y'_\alpha(v),$$

where Y'_α is defined by (2.3.4') and

$$K'=\inf_{\varepsilon>0}K'_\varepsilon<1.$$

Consider the iterative process (2.3.26) for the system (3.1.5) with the condition $u|_{\partial B_\delta}=0$. Let $w_n=u_n-u_{n-1}$. Then from the last inequality we get

$$|X'(w_{n+1},v)|<K'\left[Y'_{-\alpha}(w_n)\right]^{1/2}\left[Y'_\alpha(v)\right]^{1/2}.$$

If we take $\alpha=0$ and $v=w_{n+1}$ then we obtain the following estimate

$$||w_{n+1}||_{W_2^{(1)}(B_\delta)}\leq K'||w_n||_{W_2^{(1)}(B_\delta)}.$$

This gives the convergence of process (2.3.26) in $W_2^{(1)}(B_\delta)$ and proves the existence of weak solution in the same space.
Let $\alpha=2-m-2\gamma$ and $\gamma>0$ be sufficiently small. With the help of inequalities (2.3.5) and (2.3.6) for integrals (3.1.1) and (3.1.2) as in theorem 2.3.1, we get

$$K'^2\left[1+\frac{(m-2)^2}{(m-1)}\right]<1$$

which follows the Hölder continuity for the weak solution in an arbitrary interior domain.
Assume that the coefficients of system (3.1.5) satisfy the conditions

$$(3.1.9)\qquad a_i(x,0)\equiv 0\quad (i=1,..,m).$$

Take $z\equiv 0$. Then analogously to (3.1.8) we get

$$(3.1.10)\qquad \left|\int_{B_\delta}[D'u-\varepsilon a(x;D'u)]\,D'vdx\right|^2\leq K'^2\int_{B_\delta}|D'u|^2r^\alpha\zeta_0dxY^{o'}_{\alpha,0}(v),$$

where $Y^{o'}_{\alpha,0}(v)$ is determined by (3.1.2).

Theorem 3.1.1. *If the coefficients of* (3.1.5) *satisfy conditions 1)-3) of section 1.1 for* $\xi_0^{(k)} = 0 (k = 1, ..., N), s = 2, \varphi(x) \equiv 0$, *relations* (3.1.9) *and inequality*

$$K'\sqrt{1 + \frac{(m-2)^2}{m-1}} < 1 \tag{3.1.11}$$

holds true, then any solution bounded in the entire R^m *and belonging to* $W^{(1)}_{2,loc}$ *is a constant.*

Proof. According to theorem 2.3.2 our solution will belong to $H_\alpha(B_g)$ with $\alpha = 2 - m - 2\gamma$ and $\gamma > 0$, but small. Let us write system (3.1.5) in the form

$$\int_{B_\delta} D'uD'vdx = \int_{B_\delta} [D'u - \varepsilon a(x; D'u)] D'vdx.$$

Applying (3.1.10) we come to

$$|\int_{B_\delta} D'uD'vdx|^2 \leq K'^2 \int_{B_\delta} |D'u|^2 r^\alpha \zeta_0 dx Y'^{\circ}_{\alpha,0}(v).$$

Apply inequalities (3.1.3) and (3.1.4). We come then to

$$\left[\int_{B_\delta} |D'u|^2 r^\alpha \zeta_0 dx - C\delta^{-2} \int_{B_\delta} |u|^2 r^\alpha dx\right]^2 \leq$$
$$\leq K'^2 \left[1 + \frac{(m-2)^2}{m-1} + O(\gamma) + \eta\right] \int_{B_\delta} |D'u|^2 r^\alpha \zeta_0 dx \times$$
$$\times \left[\int_{B_\delta} |D'u|^2 r^\alpha \zeta_0 dx + C_2\delta^{-2} \int_{B_\delta(x_0)} |u|^2 r^\alpha dx\right],$$

where C is independent of δ. If u is bounded then

$$\int_{B_\delta} |u|^2 r^\alpha dx \leq C \int_{B_\delta} r^\alpha dx \leq C\delta^{2-2\gamma}. \tag{3.1.12}$$

Applying (2.2.15) after elementary calculations we get

$$\int_{B_\delta} |D'u|^2 r^\alpha \zeta_0 dx \leq \left[K'\sqrt{1 + \tfrac{(m-2)^2}{m-1}} + \eta + O(\gamma)\right] \int_{B_\delta} |D'u|^2 r^\alpha \zeta_0 dx + C\delta^{-2\gamma}.$$

At the same time condition (3.1.11) gives us the following inequality

$$\int_{B_\delta} |D'u|^2 r^\alpha \zeta_0 dx \leq C\delta^{-2\gamma}.$$

Taking into account that $\zeta_0 \geq 0$ and $\zeta_0 = 1$ for $r \leq \delta/2$, we get

$$\int_{B_{\delta/2}} |D'u|^2 r^\alpha dx \leq C\delta^{-2\gamma}$$

which leads to

$$\int_{R^m} |D'u|^2 r^\alpha dx = 0$$

and to u =const. □

System (2.5.1) shows condition (3.1.11) is exact.

3.2 The Korn inequality in weighted spaces

There exists a number of important problems for which the inequalities on the left-hand side of (2.3.19) are not valid.
For example, the equilibrium system for the linear elasticity is not strongly elliptic. In fact, let $S = \{\sigma_i^{(j)}\}$ and $E = \{\varepsilon_{jl}\}$ be respectively the symmetric tensors of stresses and strains. In the orthotropic case the elements of these tensors are bounded by equations

(3.2.1) $$\sigma_{ii} = \lambda divu\delta_{ik} + \mu\varepsilon_{ik}$$

and the system of the equilibrium equations

$$D_i\sigma_{ik} = f^{(k)}(x) \quad (i, k = 1, ..., m)$$

takes the form

(3.2.2) $$L(u) \equiv \mu \bigtriangleup u + (\lambda + \mu)\nabla(divu) = f(x).$$

Let u denotes the displacement vector, $f(x)$ the external mass forces, μ and ν so-called Lame constants $(\mu, \nu > 0)$. Then

(3.2.3) $$\varepsilon_{ij} = D_i u^{(j)} + D_j u^{(i)}.$$

System (3.2.2) can be written in the form

(3.2.4) $$\frac{\partial}{\partial x_i}\left\{\left[[\mu\delta_{ij}\delta_{kl} + (\lambda + \mu)\delta_{il}\delta_{jk}]\, D_j u^{(l)}\right]\right\} = f(x).$$

The matrix of ellipticity (1.2.18) for this system looks as follows

(3.2.5) $$A' = \{\mu\delta_{ij}\delta_{kl} + (\lambda + \mu)\delta_{il}\delta_{jk}\} \quad (i, j, k, l = 1, ..., m).$$

The quadratic form in (2.3.19) can be written in the following way

$$\sum \frac{\partial a_i^{(k)}}{\partial p_j^{(l)}} \xi_i^{(k)} \xi_j^{(l)} = \sum \left[\mu |\xi_i^{(k)}|^2 + (\lambda + \mu) \xi_i^{(k)} \xi_k^{(i)} \right].$$

It is clear that this form is not positively defined. In fact, take all $\xi_i^{(k)}$ except $\xi_1^{(2)}$ and $\xi_2^{(1)}$ equal to zero. Then

$$\sum \frac{\partial a_i^{(k)}}{\partial p_j^{(l)}} \xi_i^{(k)} \xi_j^{(l)} = \mu \left(|\xi_1^{(2)}|^2 + |\xi_2^{(1)}|^2 \right) + 2(\lambda + \mu) \xi_1^{(2)} \xi_2^{(1)} =$$

$$= \mu \left[\xi_1^{(2)} + \frac{\lambda + \mu - \sqrt{\lambda(\lambda + 2\mu)}}{\mu} \xi_2^{(1)} \right] \left[\xi_1^{(2)} + \frac{\lambda + \mu + \sqrt{\lambda(\lambda + 2\mu)}}{\mu} \xi_2^{(1)} \right].$$

It is now obvious that the right-hand side can be equal to zero, in spite of the fact that $\xi_1^{(2)}$ and $\xi_2^{(1)}$ are not equal to zero. Besides, it is known the system is elliptic in the general sense.

Suppose the solution of (3.2.2) satisfies boundary condition

$$u|_{\partial\Omega} = 0. \tag{3.2.6}$$

The positiveness of the operator $L(u)$ in (3.2.2) with condition (3.2.6) follows from the Korn inequality

$$\sum_{j,k=1}^{m} \int_{\Omega} |D_j u^{(k)} + D_k u^{(j)}|^2 dx \geq 2 \int_{\Omega} |D' u|^2 dx. \tag{3.2.7}$$

The existence of the weak solution for the problem (3.2.2), (3.2.4) is proved with the help of (3.2.7). The energetic space coincides here with $\overset{\circ}{W}{}_2^{(1)}(\Omega)$. But this inequality is not sufficient if you want to prove the regularity of the weak solution; a stronger inequality is needed. For example inequality

$$\int_{\Omega} |D_j u^{(k)} + D_k u^{(j)}|^p dx \geq C_p \int_{\Omega} |D' u|^p dx \tag{3.2.8}$$

with $p > m$ will lead to the fact that the weak solution for $f \in \mathcal{L}_p (p > m)$ belongs to $C^{0,\gamma}$.

Inequality (3.2.8) was obtained by Mosolov and Myasnikov [1] by way of using the boundedness of a singular integral in $\mathcal{L}_p$. It is worthwile to remark that the constant C_p is bounded with the norm of the above mentioned integral. The exact expression of this constant is unknown.

In this section we shall represent the Korn inequality in a weighted space. Originally it was proved in our papers [13],[14] and [15]. The result for $m = 2$ was published by Kavrajskaja [1]. These statements can also be obtained in the same way as (3.2.8), using the boundedness of a singular integral in $\mathcal{L}_{2,\alpha}$ as proved by Stein[1]. However, an explicit sharp constant can not be obtained in this way.

Let $r = |x - x_0|$ and $0 \geq \alpha > -m$.

Lemma 3.2.1. *If $u \in W^{(1)}_{2,\alpha}(B_\delta(x_0), x_0)$, vanishes for $r = \delta$, satisfies (3.2.6) and for $-m < \alpha < 2 - m$ the condition*

$$u(x_0) = 0, \tag{3.2.9}$$

then the inequality

$$\begin{aligned} W_\alpha(u) \equiv \sum_{i,k=1}^{m} \int_{B_\delta(x_0)} |D_k u^{(i)} + D_i u^{(k)}|^2 r^\alpha dx = \\ = 2 \int_{B_\delta(x_0)} |D'u|^2 r^\alpha dx + 2 \int_{B_\delta(x_0)} \left(divu + \alpha u_r r^{-1}\right)^2 r^\alpha dx - \\ -4\alpha \int_{B_\delta(x_0)} u_r^2 r^{\alpha-2} dx + 2\alpha \int_{B_\delta(x_0)} |u|^2 r^{\alpha-2} dx \end{aligned} \tag{3.2.10}$$

is true. Here u_r is the projection of u in the direction from x_0 to x .
If $u \in W^{(1)}_{2,-\alpha}(B_\delta(x_0), x_0)$ and u satisfies only (3.2.6) then (3.2.10) is valid after replacing α by $-\alpha$.

Proof. It is sufficient to prove the lemma for smooth u and $\delta = 1$. Square under the sign of the left-hand side integral. We get

$$\begin{aligned} \sum_{i,k=1}^{m} \int_B \left[D_k u^{(i)} + D_i u^{(k)}\right]^2 r^{\pm\alpha} dx = 2 \sum_{i,k=1}^{m} \int_B |D_k u^{(i)}|^2 r^{\pm\alpha} dx + \\ +2 \int_B D_k u^{(i)} D_i u^{(k)} r^{\pm\alpha} dx, \end{aligned}$$

where $B = B_1(x_0)$.
Integrate twice by parts under the sign of of the second integral on the left-hand side term and use the condition (3.2.9) for $\alpha < 2 - m$. We have then

$$\begin{aligned} \int_B D_k u^{(i)} D_i u^{(k)} r^{\pm\alpha} dx = -\left(\int_B u^{(i)} D_i D_k u^{(k)} r^{\pm\alpha} dx + \right. \\ \left. + \int_B u^{(i)} D_i u^{(k)} D_k r^{\pm\alpha} dx\right) = \int_B D_i u^{(i)} D_k u^{(k)} r^{\pm\alpha} dx + \\ + \int_B u^{(i)} D_k u^{(k)} D_i r^{\pm\alpha} dx + \int_B D_i u^{(i)} u^{(k)} D_k r^{\pm\alpha} dx + \int_B u^{(i)} u^{(k)} D_i D_k r^{\pm\alpha} dx. \end{aligned}$$

Taking into account

$$\begin{aligned} D_i r^{\pm\alpha} &= \pm\alpha r^{\pm\alpha-1} cos(x_i, r), \\ D_i D_k r^{\pm\alpha} &= \pm\alpha r^{\pm\alpha-2} \delta_{ik} + (\pm\alpha)(\pm\alpha - 2) r^{\pm\alpha-2} cos(x_i, r) cos(x_k, r) \end{aligned}$$

and

$$u_r = u^{(i)} cos(r, x_i),$$

we come to (3.2.10). □

As in (2.3.1) consider the expansions

$$u^{(l)}(x) = \sum_{s,k} u^{(l)}_{s,k}(r) Y_{s,k}(\theta). \tag{3.2.11}$$

Let

$$v(x) = u(x) - u_0(r) Y_{0,s}(\theta) \tag{3.2.12}$$

(we define $u_{0,1}(r) = u_0(r)$). As we noticed at the very beginning of section 2.3 the functions $u^{(l)}_{s,k_s}(r)$ are equal to zero for $r = 0$ and $s \geq 1$. Hence $v(x_0) = 0$.

Lemma 3.2.2. *If $u \in W^{(1)}_{2,\pm\alpha}(B_\delta(x_0), x_0)$ u satisfies (3.2.9) for the case where α vanishes for $r = \delta$, then*

$$W_{\pm\alpha}(u) = W_{\pm\alpha}(v) + 4 \int_{B_\delta} divv div u_0 r^{\pm\alpha} dx + W_{\pm\alpha}(u_0) \tag{3.2.13}$$

holds.

Proof. We start as in the previous lemma with $\delta = 1$ and smooth $u(x)$. Then from the orthogonality of $Y_{s,k}(\theta)$ follows the relation $\int_S v'_r dS = 0$. Taking into account that u_0 depends only on r, we get

$$\int_B D_i v^{(k)} D_i u_0^{(k)} r^{\pm\alpha} dx = \int_B \left(v^{(k)}\right)'_r \left(u_0^{(k)}\right)'_r r^{\pm\alpha} dx =$$

$$= \int_0^1 \left(u_0^{(k)}\right)'_r r^{\pm\alpha+m-1} dr \int_S \left(v^{(k)}\right)'_r dS = 0.$$

Therefore

$$W_{\pm\alpha}(u) = W_{\pm\alpha}(v) + W_{\pm\alpha}(u_0) + 4 \int_B D_i v^{(k)} D_k u_0^{(i)} r^{\pm\alpha} dx.$$

Consider the last term on the right-hand side. Integrating twice by parts under the sign of the integral we arrive at

$$\int_B D_i v^{(k)} D_k u_0^{(i)} r^{\pm\alpha} dx = -\int_B v^{(k)} D_i D_k u_0^{(i)} r^{\pm\alpha} dx -$$

$$-\int_B v^{(k)} D_k u_0^{(i)} D_i r^{\pm\alpha} dx = \int_B D_k v^{(k)} D_i u_0^{(i)} r^{\pm\alpha} dx +$$

$$+\int_B v^{(k)} D_i u_0^{(i)} D_k r^{\pm\alpha} dx - \int_B v^{(k)} D_k u_0^{(i)} D_i r^{\pm\alpha} dx.$$

Taking into account that u_0 depends only on r we come to

$$\int_B v^{(k)} D_i u_0^{(i)} D_k r^{\pm\alpha} dx = \pm\alpha \int_B v^{(k)} \left(u_0^{(i)}\right)'_r cos(x_i, r) cos(x_k, r) r^{\pm\alpha-1} dx$$

and

$$\int_B v^{(k)} D_k u_0^{(i)} D_i r^{\pm\alpha} dx = \pm\alpha \int_B v^{(k)} \left(u_0^{(i)}\right)'_r cos(x_k, r) cos(x_i, r) r^{\pm\alpha-1} dx.$$

So, the last two right-hand side expressions can differ from zero only for $i = k$. Hence

$$\int_B D_i v^{(k)} D_k u_0^{(i)} r^{\pm\alpha} dx = \int_B divv div u_0 r^{\pm\alpha} dx.$$

□

Lemma 3.2.3. *If the conditions of lemma 3.2.2 are satisfied then the inequality*

$$W_{\pm\alpha}(u) \geq C_{\pm\alpha} \int_{B_\delta} |D'u|^2 r^{\pm\alpha} dx \tag{3.2.14}$$

holds, where

$$C_\alpha = min\left\{\frac{2m+\alpha}{m}, \frac{2(\alpha+m)^2}{(\alpha+m)^2-4\alpha}\right\} \quad (-m < \alpha \leq 0), \tag{3.2.15}$$

$$C_{-\alpha} = \begin{cases} min\left\{2\frac{m+\alpha}{m}, 2\frac{(\alpha-m)^2+12\alpha}{(\alpha-m)^2+4\alpha}\right\} (-m < \alpha \leq 0, m \geq 3), \\ min\left\{2\frac{m+\alpha}{m}, 2\frac{(\alpha-m)^2+12\alpha}{(\alpha-m)^2+4\alpha}\right\} (\sqrt{12} - 4 < \alpha \leq 0, m = 2). \end{cases} \tag{3.2.16}$$

Proof. It is sufficient to prove the lemma for $\delta = 1$. Consider at first the case with α. Take the right- hand side of (3.2.13) and apply lemma 3.2.1 to $W_\alpha(v)$. This is possible because $v(x_0) = v|_{\partial B} = 0$. We get

$$W_\alpha(u) = 2\int_B |D'v|^2 r^\alpha dx + 2\int_B \left(divv + \alpha r^{-1} v_r\right)^2 r^\alpha dx -$$
$$-4\alpha \int_B v_r^2 r^{\alpha-2} dx + 2\alpha \int_B |v|^2 r^{\alpha-2} dx + 4\int_B divv div u_0 r^\alpha dx + W_\alpha(u_0).$$

After elementary calculations we come to the following relation

$$\begin{aligned} W_\alpha(u) &= 2\int_B |D'v|^2 r^\alpha dx + 2\int_B \left(divv + \alpha r^{-1} v_r + div u_0\right)^2 r^\alpha dx - \\ &-4\alpha \int_B \left(r^{-1} v_r + \frac{1}{2} div u_0\right)^2 r^\alpha dx + 2\alpha \int_B |v|^2 r^{\alpha-2} dx + \\ &+(\alpha - 2)\int_B (div u_0)^2 r^\alpha dx + W_\alpha(u_0). \end{aligned} \tag{3.2.17}$$

Taking into account that $\alpha \leq 0$ and omitting some positive terms we arrive to

$$W_\alpha(u) \geq 2\left[\int |D'v|^2 r^\alpha dx + \alpha \int_B |v|^2 r^{\alpha-2} dx\right] +$$
$$+W_\alpha(u_0) + (\alpha - 2)\int_B (div u_0)^2 r^\alpha dx.$$

Evidently

$$W_\alpha(u_0)+(\alpha-2)\int_B (divu_0)^2 r^\alpha dx=\sum_{i,k=1}^m\int_B\left[D_i u_0^{(k)}+D_k u_0^{(i)}\right]^2 r^\alpha dx+$$

$$+(\alpha-2)\int_B\left(D_i u_0^{(i)}\right)^2 r^\alpha dx=\sum_{i,k=1}^m\int_B\left[(u_0^{(i)})'_r cos(x_k,r)+\right.$$

$$\left.+(u_0^{(k)})'_r cos(x_i,r)\right]^2 r^\alpha dx+(\alpha-2)\int_B\sum_{i=1}^m\left[(u_0^{(i)})'_r cos(x_i,r)\right]^2 r^\alpha dx.$$

Taking into account that $\int_S cos(x_i,r)cos(x_k,r)dS=0$ for $i\neq k$ we obtain the following relations

$$W_\alpha(u_0)+(\alpha-2)\int_B (divu_0)^2 r^\alpha dx=2\sum_{i=1}^m\int_B[(u_0^{(i)})'_r]^2 r^\alpha dx+$$

$$+\alpha\sum_{i=1}^m\int_B[(u_0^{(i)})'_r]^2 r^\alpha cos^2(x_i,r)dx=$$

$$=\sum_{i=1}^m\int_0^1[(u_0^{(i)})'_r]^2 r^{\alpha+m-1}dr\int_S\left[2+\alpha cos^2(x_i,r)\right]dS.$$

Since the integral $\int_S\left[2+\alpha cos^2(x_i,r)\right]dS$ is independent of i, we get then

$$\int_S\left[2+\alpha cos^2(x_i,r)\right]dS=2|S|+\alpha\int_S cos^2\theta_1 dS=$$

$$=\left[2+\alpha\int_0^\pi sin^{m-2}\theta_1 cos^2\theta_1 d\theta_1\left(\int_0^\pi sin^{m-2}\theta_1 d\theta_1\right)^{-1}\right]|S|=(2+\alpha/m)\,|S|,$$

where $|S|$ is the area of the unit sphere. We have then

$$W_\alpha(u_0)+(\alpha-2)\int_B (divu_0)^2 r^\alpha dx=\frac{2m+\alpha}{m}\sum_{k=1}^m\int_B|\nabla u_0^{(k)}|^2 r^\alpha dx.$$

and

$$W_\alpha(u)\geq 2\left(\int_B|D'v|^2 r^\alpha dx+\alpha\int_B|v|^2 r^{\alpha-2}dx\right)+\frac{2m+\alpha}{m}\sum_{k=1}^m\int_B|\nabla u_0^{(k)}|^2 r^\alpha dx.$$

From

$$\sum_{l=1}^m\int_B|\nabla v^{(l)}|^2 r^\alpha dx=\sum_{l=1}^m\sum_{s,k\geq 0}\int_0^1\left[|v_{s,k}^{(l)\prime}(r)|^2+s(s+m-2)|v_{s,k}^{(l)}(r)|^2 r^{-2}\right]r^{\alpha+m-1}dr,$$

we get

$$(3.2.18)\quad W_\alpha(u) \geq 2\sum_{l=1}^{m} \sum_{s,k;s\geq 1} \int_0^1 \left\{ |v_{s,k}^{(l)\prime}(r)|^2 + [s(s+m-2)+\alpha]\, |v_{s,k}^{(l)}(r)|^2 r^{-2} \right\} \times$$

$$\times r^{\alpha+m-1} dr + \frac{2m+\alpha}{m} \sum_{k=1}^{m} \int_B |\nabla u_0^{(k)}|^2 r^\alpha dx,$$

where the summation runs over $s \geq 1$ and the dash denotes the derivative with respect to r.

Consider all the terms separately. Taking into account the inequality (2.1.9) we obtain the following relation (we omit for the moment the indicies)

$$\int_0^1 \left\{ |v^{(l)\prime}|^2 + [s(s+m-2)+\alpha]\, |v^{(l)}|^2 r^{-2} \right\} r^{\alpha+m-1} dr \geq$$

$$\geq q \int_0^1 |v^{(l)\prime}|^2 r^{\alpha+m-1} dr +$$

$$+ \left[(1-q)\frac{(\alpha+m-2)^2}{4} + s(s+m-2) + \alpha \right] \int_0^1 |v^{(l)}|^2 r^{\alpha+m-3} dr,$$

where $0 < q < 1$. Take q such that

$$qs(s+m-2) = (1-q)\left[\frac{\alpha+m-2}{2}\right]^2 + s(s+m-2) + \alpha.$$

Hence for all $s \geq 1$ we have the following relation

$$q \geq 1 + \alpha \left[m - 1 + \left(\frac{\alpha+m-2}{2}\right)^2 \right]^{-1}.$$

This leads to

$$\int_0^1 \left\{ |v^{(l)\prime}|^2 + [s(s+m-2)+\alpha]\, |v^{(l)}|^2 r^{-2} \right\} r^{\alpha+m-1} dr \geq$$

$$\geq \left\{ 1 + \alpha \left[m - 1 + \left(\frac{\alpha+m-2}{2}\right)^2 \right]^{-1} \right\} \times$$

$$\times \int_0^1 \left[|v^{(l)\prime}|^2 + s(s+m-2)|v^{(l)}|^2 r^{-2} \right] r^{\alpha+m-1} dr =$$

$$= \left\{ 1 + \alpha \left[m - 1 + \left(\frac{\alpha+m-2}{2}\right)^2 \right] \right\} \int_B |D'v|^2 r^\alpha dx.$$

and we come to (3.2.14) with the help of (3.2.18).

If we take $m = 2, u^{(1)} = u(r) sin\varphi$ and $u^{(2)} = -u(r) cos\varphi$, then $u_r = v_r = divu =$

$divv = u_0 = 0$, $u_\varphi = -u(r)$. From (3.2.17) it follows

$$W_\alpha(u) = 2\left[\sum_{l=1}^{2}\int_B |\nabla v^{(l)}|^2 r^\alpha dx + \alpha\int_B |v|^2 r^{\alpha-2}dx\right].$$

The function v contains only one harmonic. Therefore

$$W_\alpha(u) = 2\pi\int_0^1 \left[u'^2 + (\alpha+1)u^2\right] r^{\alpha+1}dr.$$

So, the constant (3.2.15), which was received from the sharp inequality (2.1.9) is also sharp.

Consider now the case, where the weight is equal to $r^{-\alpha}$. Applying (3.2.10) and (3.2.13) for $-\alpha$ we get the following equality

$$W_{-\alpha}(u) = 2\int_B \Big[\sum_{l=1}^{m} |\nabla v^{(l)}|^2 + \left(divv - \alpha v_r r^{-1}\right)^2 + 2\alpha v_r^2 r^{-2} -$$
$$-\alpha|v|_r^2 r^{-2} + 2divvdivu_0\Big] r^{-\alpha}dx + W_{-\alpha}(u_0).$$

Considering $\alpha \le 0$ we find the following relation

$$W_{-\alpha}(u) \ge 2\int_B \Big[\sum_{l=1}^{m} |\nabla v^{(l)}|^2 + \left(divv - \alpha v_r r^{-1}\right)^2 + 2\alpha|v|^2 r^{-2} -$$
$$-\alpha|v|^2 r^{-2} + 2divvdivu_0\Big] r^{-\alpha}dx + W_{-\alpha}(u_0).$$

After elementary calculations we arrive at

$$W_{-\alpha}(u) \ge 2\int_B \Big[\sum_{l=1}^{m} |\nabla v^{(l)}|^2 + \left(divv - \alpha r^{-1}v_r + divu_0\right)^2 -$$
$$-\alpha(divu_0 - v_r r^{-1})^2 + 2\alpha|v|^2 r^{-2} + (\alpha-1)(divu_0)^2\Big] r^{-\alpha}dx + W_{-\alpha}(u_0).$$

Omitting the second and the third terms under the sign of the integral we obtain

$$W_{-\alpha}(u) \ge 2\int_B \left[\sum_{l=1}^{m} |\nabla v^{(l)}|^2 + 2\alpha|v|^2 r^{-2}\right] r^{-\alpha}dx + W_{-\alpha}(u_0) + 2(\alpha-1)\int_B (divu_0)^2 r^{-\alpha}dx.$$

Repeating the arguments which we applied for the case r^α we get

$$W_{-\alpha}(u_0) + 2(\alpha-1)\int_B (divu_0)^2 r^{-\alpha}dx = 2\frac{m+\alpha}{m}\int_B |\nabla u_0|^2 r^{-\alpha}dx.$$

After some calculations and after using (2.3.5′) we obtain

$$2\int_B \left[\sum_{l=1}^{m} |\nabla v^{(l)}|^2 + 2\alpha|v|^2 r^{-2}\right] r^{-\alpha}dx \ge 2\frac{(\alpha-m)^2 + 12\alpha}{(\alpha-m)^2 + 4\alpha}\sum_{l=1}^{m}\int_B |\nabla v^{(l)}|^2 r^{-\alpha}dx.$$

From this follows (3.2.14). The condition $\alpha > \sqrt{12} - 4$ provides for $m = 2$ positiveness of the constant. □

We can now prove the following theorem

Theorem 3.2.1. *If $u \in W^{(1)}_{2,\pm\alpha}(\Omega, x_0)$, where $-m < \alpha \leq 0$ satisfies (3.2.6) and x_0 is inside the bounded domain Ω, then the inequality*

$$(3.2.19)\quad \sum_{i,k=1}^{m} \int_{\Omega} |D_k u^{(i)} + D_i u^{(k)}|^2 r^{\pm\alpha} dx \geq C_{\pm\alpha} \int_{\Omega} |D'u|^2 r^{\pm\alpha} dx - C \int_{\Omega} |D'u|^2 r^{a_{\pm\alpha}} dx,$$

holds, where $C_{\pm\alpha}$ is defined by (3.2.15) *and* (3.2.16) *, $a_\alpha = 0$ and $a_{-\alpha} = \beta$, where β is arbitrary.*

Proof. Let a smooth u be defined in Ω and $\delta > 0$ is less than the distance between $x_0 \in \Omega$ and $\partial\Omega$. Take the cut-off function$\zeta(r)$ (2.2.1) , which is extended for $r > \delta$ by zero. Then

$$\sum_{i,k=1}^{m} \int_{\Omega} \left[D_i u^{(k)} + D_k u^{(i)}\right]^2 r^\alpha \zeta^2(r) dx =$$

$$= \sum_{i,k=1}^{m} \int_{\Omega} \left[D_i \left(\zeta u^{(k)}\right) + D_k \left(\zeta u^{(i)}\right) - D_i \zeta u^{(k)} - D_k \zeta u^{(i)}\right]^2 r^\alpha dx =$$

$$= \sum_{i,k=1}^{m} \left\{ \int_{\Omega} \left[D_i \left(\zeta u^{(k)}\right) + D_k \left(\zeta u^{(i)}\right)\right]^2 r^\alpha dx - 2 \int_{\Omega} \left[u^{(k)} D_i \zeta + u^{(i)} D_k \zeta\right] \left[D_i(\zeta u^{(k)}) + \right.\right.$$

$$\left.\left. + D_k(\zeta u^{(i)})\right] r^\alpha dx + \int_{\Omega} \left[u^{(k)} D_i \zeta + u^{(i)} D_k \zeta\right]^2 r^\alpha dx \right\}.$$

Omitting the last term on the right-hand side and taking into account that $D_k \zeta \equiv 0$ for $r \leq \delta/2$ and for $r \geq 3\delta/4$ we get

$$\sum_{i,k=1}^{m} \int_{\Omega} \left[D_i u^{(k)} + D_k u^{(i)}\right]^2 r^\alpha \zeta^2(r) dx \geq \sum_{i,k=1}^{m} \left\{ \int_{B_\delta} \left[D_i \left(\zeta u^{(k)}\right) + D_k \left(\zeta u^{(i)}\right)\right]^2 r^\alpha dx - \right.$$

$$\left. -2 \int_{r \geq \delta/2} \left[u^{(k)} D_i \zeta + u^{(i)} D_k \zeta\right] \left[D_i(\zeta u^{(k)}) + D_k(\zeta u^{(i)})\right] r^\alpha dx \right\}.$$

From the lemma 3.2.3 follows

$$(3.2.20)\quad \sum_{i,k=1}^{m} \int_{\Omega} \left[D_i u^{(k)} + D_k u^{(i)}\right]^2 r^\alpha \zeta^2(r) dx \geq C_\alpha \sum_{k=1}^{m} \int_{B_\delta} |\nabla(\zeta u^{(k)})|^2 r^\alpha dx -$$

$$-2 \sum_{i,k=1}^{m} \int_{r \geq \delta/2} \left[u^{(k)} D_i \zeta + u^{(i)} D_k \zeta\right] \left[D_i \left(\zeta u^{(k)}\right) + D_k \left(\zeta u^{(i)}\right)\right] r^\alpha dx.$$

Considering the smoothness of $\zeta(r)$ it is easy to see that the inequality

$$|\sum_{i,k=1}^{m}\int_{r\geq\delta/2}\left[u^{(k)}D_i\zeta+u^{(i)}D_k\zeta\right]\left[D_i\left(\zeta u^{(k)}\right)+D_k\left(\zeta u^{(i)}\right)\right]r^\alpha dx|\leq$$
$$\leq C\int_\Omega\left(\sum_{k=1}^{m}|\nabla u^{(k)}|^2+|u|^2\right)dx,$$

holds.
Therefore

$$\sum_{i,k=1}^{m}\int_\Omega\left[D_iu^{(k)}+D_ku^{(i)}\right]^2 r^\alpha\zeta^2(r)dx\geq$$
$$\geq C_\alpha\sum_{k=1}^{m}\int_{B_\delta}|\nabla(\zeta u_k)|^2r^\alpha dx-C\int_\Omega\left(\sum_{k=1}^{m}|\nabla u^{(k)}|^2+|u|^2\right)dx.$$

Applying the same arguments we come to inequality

$$\sum_{k=1}^{m}\int_{B_\delta(x_0)}|\nabla\left(\zeta u^{(k)}\right)|^2r^\alpha dx\geq\sum_{k=1}^{m}\int_{B_{\delta/2}}|\nabla u^{(k)}|^2r^\alpha dx-C\int_\Omega\left(\sum_{k=1}^{m}|\nabla u^{(k)}|^2+|u|^2\right)dx.$$

Estimate the first term on the right-hand side in the following manner

$$\sum_{k=1}^{m}\int_{B_{\delta/2}}|\nabla u^{(k)}|^2r^\alpha dx=\sum_{k=1}^{m}\left[\int_\Omega|\nabla u^{(k)}|^2r^\alpha dx-\int_{\Omega\setminus B_{\delta/2}}|\nabla u^{(k)}|^2r^\alpha dx\right]\geq$$
$$\geq\sum_{k=1}^{m}\left[\int_\Omega|\nabla u^{(k)}|^2r^\alpha dx-(\delta/2)^\alpha\int_{\Omega\setminus B_{\delta/2}}|\nabla u^{(k)}|^2dx\right]\geq$$
$$\geq\sum_{k=1}^{m}\left[\int_\Omega|\nabla u^{(k)}|^2r^\alpha dx-C\int_\Omega|\nabla u^{(k)}|^2dx\right].$$

So, we come to the inequality

$$\sum_{k=1}^{m}\int_{B_\delta}|\nabla\left(\zeta u^{(k)}\right)|^2r^\alpha dx\geq\sum_{k=1}^{m}\left[\int_\Omega|\nabla u^{(k)}|^2r^\alpha dx-C\int_\Omega|\nabla u^{(k)}|^2dx\right].$$

With the help of this estimate and relation (3.2.20) we'll arrive at

$$\sum_{i,k=1}^{m}\int_\Omega\left[D_iu^{(k)}+D_ku^{(i)}\right]^2r^\alpha\zeta^2(r)dx\geq$$
$$\geq C_\alpha\sum_{k=1}^{m}\int_\Omega|\nabla u^{(k)}|^2r^\alpha dx-C\int_\Omega\left(\sum_{k=1}^{m}|\nabla u^{(k)}|^2+|u|^2\right)dx.$$

Now it is sufficient to estimate from the above the left-hand side of the last inequality . It is clear that

$$\sum_{i,k=1}^{m}\int_{\Omega}\left[D_i u^{(k)} + D_k u^{(i)}\right]^2 r^{\alpha}\zeta^2(r)dx \le$$

$$\le \sum_{i,k=1}^{m}\int_{B_{\delta/2}}\left[D_i u^{(k)} + D_k u^{(i)}\right]^2 r^{\alpha}dx + C\sum_{i,k=1}^{m}\int_{\Omega\setminus B_{\delta/2}}\left[D_i u^{(k)} + D_k u^{(i)}\right]^2 dx \le$$

$$\le \sum_{i,k=1}^{m}\int_{\Omega}\left[D_i u^{(k)} + D_k u^{(i)}\right]^2 r^{\alpha}dx + C\sum_{k=1}^{m}\int_{\Omega}|\nabla u^{(k)}|^2 dx.$$

To show this we take into account $\zeta = 1$ for $r \le \delta/2$ and apply the elementary estimate

$$\sum_{i,k=1}^{m}\int_{\Omega}\left[D_i u^{(k)} + D_k u^{(i)}\right]^2 dx \le C\sum_{k=1}^{m}\int_{\Omega}|\nabla u^{(k)}|^2 dx.$$

So, the estimate is proved in the case where the weight is equal to r^{α}. In the case when the weight is equal to $r^{-\alpha}$ you must apply the same considerations . The only difference appears when under the sign of integrals the factors containing ζ' are multiplied and divided by r with an arbitrary exponent. □

3.3 Hölder continuity of displacements for elasto-plastic media with hardening

The equilibrium system for elasto-plastic media can be written in the form

$$L(u) \equiv \sum_{k=1}^{m}\frac{\partial a_k(x; D'u)}{\partial x_k} - a_0(x) = 0 \qquad (3.3.1)$$

where u is the unknown displacement vector. We consider this system with the boundary condition

$$u|_{\partial\Omega} = 0. \qquad (3.3.2)$$

Here $u = (u^{(1)}, ..., u^{(m)})$ and $a_k = (a_k^{(1)}, ..., a_k^{(m)}), k = 0, 1, ..m$, are vector valued functions in R^m. The most interesting cases are $m = 2$ and $m = 3$.

We suppose that the $a_k(x,p)$ are measurable, continuously differentiable with respect to p in any bounded domain, and for some $q > m$ and all $u \in W_q^{(1)}(\Omega)$ the coefficients a_k belong to $\mathcal{L}_q(\Omega)$ $(k = 0, 1, ..., m)$. We have seen that even for linear orthotropic elastisity (3.2.1) the equlibrium system (3.2.2) does not satisfy the conditions of strong ellipticity (1.1.2). Therefore we must consider the conditions which guarantee regular solvability of the problem under some other assumptions. Suppose that for any differentiable u and v the inequalities

$$\sum_{i,k=1}^{m}\left[a_k^{(i)}(x; D'u) - a_k^{(i)}(x; D'v)\right]\frac{\partial(u^{(i)} - v^{(i)})}{\partial x_k} \ge \qquad (3.3.3)$$

$$\ge a\sum_{i,k=1}^{m}\left[\frac{\partial(u^{(i)} - v^{(i)})}{\partial x_k} + \frac{\partial(u^{(k)} - v^{(k)})}{\partial x_i}\right]^2;$$

$$(3.3.4)\qquad \sum_{i,k=1}^{m}\left[a_k^{(i)}(x;D'u)-a_k^{(i)}(x;D'v)\right]^2 \le$$

$$\le b\sum_{i,k=1}^{m}\left[a_k^{(i)}(x;D'u)-a_k^{(i)}(x;D'v)\right]\frac{\partial(u^{(i)}-v^{(i)})}{\partial x_k}$$

where a and b are positive constants are satisfied. Let functions $a_k^{(i)}$, connecting the components $\sigma_k^{(i)}$ of the thensor of stresses and the components ε_{jl} of the thensor of strains be bounded with relations $\sigma_k^{(i)} = a_k^{(i)}(x,\varepsilon_{jl})$, which satisfy (3.3.3) and (3.3.4). Apply now the iterative process

$$(3.3.5)\qquad \Delta u_{n+1} = \Delta u_n - \varepsilon L(u_n), \quad u_{n+1}|_{\partial\Omega} = 0.$$

In section 1.2 under conditions 1)-3) the convergence of (3.3.5) was proved in the energetic norm. In section 2.3 under more restrictive assumptions of theorem 2.3.1 the convergence of (3.3.5) in $H_\alpha \subset C^{0,\gamma}$ was obtained. We shall now prove analogous results for the systems satisfying both (3.3.3) and (3.3.4)

Theorem 3.3.1. *If inequalities* (3.3.3) *and* (3.3.4) *are satisfied, then the iterative process* (3.3.5) ($u_0 = 0$) *converges in the norm of* $W_2^{(1)}(\Omega)$ *to a solution of the problem* (3.3.1),(3.3.2) *for* $0 < \varepsilon < 2b^{-1}$.

Proof. The weak equations of (3.3.5) can be written in the form

$$(3.3.6)\int_\Omega D'u_{n+1}D'v\,dx = \int_\Omega\left[D_k u_n - \varepsilon a_k(x;D'u_n)\right]D_k v\,dx - \varepsilon\int_\Omega a_0 v\,dx, \quad u_{n+1}|_{\partial\Omega} = 0.$$

Subtracting the two successive equations and applying Hölder inequality we get

$$(3.3.7)\qquad \left|\int_\Omega D'(u_{n+1}-u_n)D'v\,dx\right|^2 \le \left\{\int_\Omega |D'u_n - D'u_{n-1}|^2dx - \right.$$

$$-2\varepsilon\int_\Omega\left[a_k^{(i)}(x;D'u_n)-a_k^{(i)}(x;D'u_{n-1})\right]D_k(u_n^{(i)}-u_{n-1}^{(i)})dx+$$

$$\left.+\varepsilon^2\sum_{k=1}^{m}\int_\Omega |a_k(x;D'u_n)-a_k(x;D'u_{n-1})|^2dx\right\}\int_\Omega |D'v|^2dx.$$

Take $v = u_{n+1} - u_n$. Then from (3.3.4) follows

$$\int_\Omega |D'u_{n+1} - D'u_n|^2dx \le \int_\Omega |D'(u_n - u_{n-1})|^2dx -$$

$$-\varepsilon(2-\varepsilon b)\int_\Omega\left[a_k^{(i)}(x;D'u_n)-a_k^{(i)}(x;D'u_{n-1})\right]D_k(u_n^{(i)}-u_{n-1}^{(i)})dx.$$

Let $0 < \varepsilon < 2b^{-1}$. Then from (3.3.3) we obtain

$$\int_\Omega |D'u_{n+1} - D'u_n|^2dx \le \int_\Omega |D'(u_n - u_{n-1})|^2dx -$$

$$-\varepsilon(2-\varepsilon b)a\sum_{i,k=1}^{m}\int_\Omega |D_k(u_n^{(i)}-u_{n-1}^{(i)}) + D_i(u_n^{(k)}-u_{n-1}^{(k)})|^2dx.$$

After applying the Korn inequality (3.2.7), we arrive at the following estimate

$$\int_\Omega |D'(u_{n+1} - u_n)|^2 dx \le (1 - 4a\varepsilon + 2ab\varepsilon^2) \int_\Omega |D'(u_n - u_{n-1})|^2 dx.$$

For $\varepsilon = b^{-1}$ we have

$$\int_\Omega |D'(u_{n+1} - u_n)|^2 dx \le (1 - 2\frac{a}{b}) \int_\Omega |D'(u_n - u_{n-1})|^2 dx.$$

Therefore

$$2a \le b \tag{3.3.8}$$

and we reach the end of the proof. □

Theorem 3.3.2. *If the conditions of the previous theorem are satisfied and the inequality*

$$\left(1 - \frac{2}{m-1}\frac{a}{b}\right)\left[1 + \frac{(m-2)^2}{m-1}\right] < 1 \tag{3.3.9}$$

holds, then the weak solution of the problem (3.3.1), (3.3.2) *is Hölder continuous in any strongly interior to the domain* Ω.

Proof. Let $x_0 \in \Omega$ and $\delta > 0$ be so small enough that $B_\delta(x_0) \cap \Omega \neq 0$. Take $\alpha = 2 - m - 2\gamma$ where $\gamma > 0$ is sufficiently small. Assume that the test function $v(x)$ is taken with the help of (2.3.1),(2.3.2), (2.2.3) and (2.2.4), where $u(x) = u_{n+1}(x) - u_n(x)$. Then using the same arguments as in theorem 2.3.1 we get

$$\begin{aligned} \int_\Omega |D'(u_{n+1} - u_n)|^2 r^\alpha dx \le \left[1 + \frac{(m-2)^2}{m-1} + O(\gamma) + \eta\right] \times \\ \times \left\{ \int_{B_\delta(x_0)} |D'(u_n - u_{n-1})|^2 r^\alpha dx - 2\varepsilon \int_{B_\delta(x_0)} \left[a_k^{(i)}(x; D'u_n) - a_k^{(i)}(x; D'u_{n-1})\right] \times \right. \\ \left. \times D_k(u_n^{(i)} - u_{n-1}^{(i)}) r^\alpha dx + \varepsilon^2 \sum_{k=1}^m \int_{B_\delta(x_0)} |a_k(x; D'u_n) - a_k(x; D'u_{n-1})|^2 r^\alpha dx \right\} + \\ + C\left[\int_\Omega |D'(u_n - u_{n-1})|^2 dx + \int_\Omega |D'(u_{n+1} - u_n)|^2 dx\right]. \end{aligned} \tag{3.3.10}$$

Since $v = 0$ for $r \ge \delta$, we obtain from (3.3.6) the estimate

$$\left|\int_\Omega D'(u_{n+1} - u_n) D'v dx\right|^2 \le \sum_{k=1}^m \int_{B_\delta(x_0)} |D_k(u_n - u_{n+1}) - \varepsilon[a_k(x; D'u_n) - a_k(x; D'u_{n-1})]|^2$$

$$r^\alpha dx \times \int_{B_\delta(x_0)} |D'v|^2 r^{-\alpha} dx.$$

The relation (3.3.10) follows from the inequalities (2.3.5), (2.3.6) for the integrals (2.3.3') and (2.3.4') (corollary (2.3.1)). Applying (3.3.3), (3.3.4) and the Korn inequality (3.2.19) in weighted spaces for $\alpha = 2 - m - 2\gamma$ with a small positive γ, we arrive at

$$\int_\Omega |D'(u_{n+1} - u_n)|^2 r^\alpha dx \leq \left(1 - 4\varepsilon\frac{a}{m-1} + 2\varepsilon^2\frac{ab}{m-1}\right) \times$$
$$\times \left[1 + \frac{(m-2)^2}{m-1} + O(\gamma) + \eta\right] \int_{B_\delta} |D'(u_n - u_{n-1})|^2 r^\alpha dx +$$
$$+C\left[\int_\Omega |D'(u_n - u_{n-1})|^2 dx + \int_\Omega |D'(u_{n+1} - u_n)|^2 dx\right].$$

Taking $\varepsilon = b^{-1}$ we come to (3.3.9) which guaranties, as in theorem 2.3.1 the convergence of (3.3.6) (or (3.3.5)) in $H_\alpha(\Omega')$.

□

Remark 3.3.1. It can be seen from (3.3.9) , that for $m = 2$ under the assumptions of the theorems 3.3.1 and 3.3.2 the weak solution is always Hölder continuous inside Ω.

The inequality (3.3.9) can hold only for $m \leq 3$. In fact for $m = 3$ from (3.3.9) we obtain

$$b < 3a, \tag{3.3.11}$$

which does not contradict (3.3.8), but for $m \geq 4$ we get from (3.3.9)

$$b < \frac{7}{6}a, \tag{3.3.12}$$

which is uncompatible with (3.3.8).

Now we apply our results to the theory of small elasto-plastic deformations for hardening media (see, for example, Ily'ushin[1]). In this case the components of the stresses and strains tensors are bounded with the equations

$$a_k^{(i)} = \sigma_{ik} = \left(K - \frac{2}{m}G\right) divu\delta_{ik} + G\left(\frac{\partial u^{(i)}}{\partial x_k} + \frac{\partial u^{(k)}}{\partial x_i}\right) + \tag{3.3.13}$$
$$+G\omega(e)\left[\frac{2}{m}divu\delta_{ik} - \left(\frac{\partial u^{(i)}}{\partial x_k} + \frac{\partial u^{(k)}}{\partial x_i}\right)\right],$$

where K and G are elastic constants, ω is a function characterizing the hardening, and e is the intensity of the deviator for the deformation tensor.

To solve problem (3.3.1) and (3.3.2) where $a_k^{(i)}$ is defined by (3.3.13) the well-known method of elastic solutions has been applied. The solution for the problem with a clamped boundary consists in the following iterative process:

$$\sum_{k=1}^{3}\frac{\partial}{\partial x_k}\left[\left(K-\frac{2}{m}G\right)divu_{n+1}\delta_{ik}+G\left(\frac{\partial u_{n+1}^{(i)}}{\partial x_k}+\frac{\partial u_{n+1}^{(k)}}{\partial x_i}\right)\right]= \tag{3.3.14}$$

$$=-G\sum_{k=1}^{3}\frac{\partial}{\partial x_k}\left\{\omega(e_n)\left[\frac{2}{m}divu_n\delta_{ik}-\left(\frac{\partial u_n^{(i)}}{\partial x_k}+\frac{\partial u_n^{(k)}}{\partial x_i}\right)\right]\right\}+f^{(i)}(x),$$

$$u_{n+1}|_{\partial\Omega}=0,$$

where$f(x)$ is the vector of mass forces. Here u_n and e_n , respectively, denote the succesive approximation u_n of the displacement vector and the value of the intensity on the $n-$ approximation. Vorovich and Krasovskiĭ [1] proved convergence of this process in the energy norm for the basic problems. In contrast to this process, where at each step it is necessary to solve a linear problem of elasticity theory depending on the elastic constants, in the iterative process (3.3.5) at each step the Poisson equation was solved. In the next section and in 4.7 we shall also consider the process of elastic solutions.

Consider the intensity e_0 of the deviator for tensor of strains

$$e_0=\frac{\sqrt{2}}{m}\sqrt{\sum_{h<s}\left[(D_hu^{(h)}-D_su^{(s)})^2+\frac{m}{2}(D_hu^{(s)}+D_su^{(h)})^2\right]} \tag{3.3.15}$$

and the deviator for tensor of strains

$$\sigma_0=\frac{\sqrt{2}}{2}\sqrt{\sum_{j<k}(\sigma_{jj}-\sigma_{kk})^2+2m\sigma_{jk}^2}. \tag{3.3.16}$$

In the considered model

$$\sigma_0=\Phi(e_0), \tag{3.3.17}$$

where

$$\Phi(e_0)=mG\left[1-\omega(e_0)\right]e_0. \tag{3.3.18}$$

Then, according to Hencky theory, the differentiable function ω should satisfy the following inequalities

$$1>\omega+e_0\omega'\geq\omega\geq 0,\quad \omega'\geq 0. \tag{3.3.19}$$

If we substitute the expressions (3.3.13) into the equilibrium equations

$$D_k\sigma_{ik}=f^{(i)}(x) \tag{3.3.20}$$

we have then the system (3.3.1) with $a_0(x)=f(x)$. Considering this system inside a bounded domain Ω with condition (3.3.2) the existence proof of a Hölder continuous solution (displacement) of the problem (3.3.20),(3.3.2) will follow from theorem 3.3.2. Therefore we must verify that the inequalities (3.3.3) and (3.3.4) hold true under the conditions (3.3.19).

Lemma 3.3.1. *If ω has a bounded derivative and* (3.3.19) *holds, then the functions* (3.3.13) *satisfy the inequalities* (3.3.3), (3.3.4) *with*

$$a = \frac{G}{2}\left[1 - \sup_{e_0}(\omega + \omega' e_0)\right], \tag{3.3.21}$$

$$b = 2\max\left\{\frac{\sup\limits_{e_0}\left|m\left[K - \frac{2}{m}(1-\omega)G\right]^2 - \frac{4G^2(\omega' e_0)^2}{m}\right|}{\min\left\{\inf\limits_{e_0}\left[K - \frac{2}{m}(1-\omega)G\right], \inf\limits_{e_0}\frac{G}{2}\left[1 - (\omega + \omega' e_0)\right]\right\}}, \frac{\sup\limits_{e_0} G^2\left[(1-\omega)^2 + (\omega' e_0)^2\right]}{\min\left\{\inf\limits_{e_0}\left[K - \frac{2}{m}(1-\omega)G\right], \inf\limits_{e_0}\frac{G}{2}\left[1 - (\omega + \omega' e_0)\right]\right\}}\right\},$$

where the inf and sup are taken over $[0, +\infty)$.

Proof. Obviously

$$e_0 = \frac{\sqrt{2}}{m}\sqrt{\frac{m}{4}\sum_{j;k=1}^{m}[D_j u^{(k)} + D_k u^{(j)}]^2 - (divu)^2}. \tag{3.3.22}$$

In fact, the expression under the square root in (3.3.15) can also be written in the following form

$$\sum_{j,k=1}^{m}\frac{1}{2}\left[D_j u^{(j)} - D_k u^{(k)}\right]^2 + \frac{m}{4}\sum_{j\neq k}\left[D_j u^{(k)} + D_k u^{(j)}\right]^2. \tag{3.3.23}$$

Squaring in the first sum, we get

$$-\sum_{j=1}^{m} D_j u^{(j)}\sum_{k=1}^{m} D_k u^{(k)} + m\sum_{j=1}^{m}\left[D_j u^{(j)}\right]^2 + \frac{m}{4}\sum_{j\neq k}\left[D_j u^{(k)} + D_k u^{(j)}\right]^2,$$

which coincides with (3.3.22). Differentiating (3.3.22) with respect to $p_j^{(l)} = D_j u^{(l)}$ we get

$$\frac{\partial e_0}{\partial p_j^{(l)}} = \frac{2}{m^2 e_0}\left[\frac{m}{2}\sum_{i,k=1}^{m}\left(D_i u^{(k)} + D_k u^{(i)}\right)\delta_{ij}\delta_{kl} - divu\delta_{jl}\right]. \tag{3.3.24}$$

Considering the expression

$$\Omega(\alpha, u) = \omega(e_0)\left[\frac{2}{m}divu\delta_{ik} - \left(D_i u^{(k)} + D_k u^{(i)}\right)\right]\alpha_{ik},$$

where α_{ik} are functional parameters and the summation runs from 1 to m for i and k, then applying the finite difference theorem, we get

$$\Omega(\alpha,u)-\Omega(\alpha,v)=\frac{\overline{\partial\Omega}}{\partial p_j^{(l)}}D_j(u^{(l)}-v^{(l)}).$$

After differentiating we get

$$\text{(3.3.25)}\qquad \frac{\overline{\partial\Omega}}{\partial p_j^{(l)}}=\overline{\omega}\left(\frac{2}{m}\delta_{jl}\delta_{ik}-\delta_{ij}\delta_{kl}-\delta_{il}\delta_{jk}\right)\alpha_{ik}-\frac{2\overline{\omega'}}{m^2\overline{e_0}}\Big[-\frac{2}{m}\overline{divu}\delta_{ik}+$$
$$+\left(\overline{D_iu^{(k)}}+\overline{D_ku^{(i)}}\right)\Big]\left[\frac{m}{2}\left(\overline{D_ju^{(l)}}+\overline{D_lu^{(j)}}\right)-\overline{divu}\delta_{jl}\right]\alpha_{ik},$$

where the bar over the derivatives indicates that they are calculated for intermediate arguments. If $\alpha_{ik}=D_i\left(u^{(k)}-v^{(k)}\right)$, then

$$\text{(3.3.26)}\qquad \Omega\left(D'(u-v),u\right)-\Omega\left(D'(u-v),v\right)=$$
$$=\overline{\omega}\left[\frac{2}{m}|div(u-v)|^2-\frac{1}{2}\sum_{i,k=1}^{m}|D_i(u^{(k)}-v^{(k)})+D_k(u^{(i)}-v^{(i)})|^2\right]-$$
$$-\frac{4}{m^3}\frac{\overline{\omega'}}{\overline{e_0}}\left|\left[\frac{m}{2}\left(\overline{D_iu^{(k)}}+\overline{D_ku^{(i)}}\right)-\overline{divu}\delta_{ik}\right]D_i(u^{(k)}-v^{(k)})\right|^2.$$

From (3.3.13), (3.3.24) and (3.3.25) we have

$$\left[a_k^{(i)}(x;D'u)-a_k^{(i)}(x;D'v)\right]D_i\left(u^{(k)}-v^{(k)}\right)=$$
$$=\left(K-\frac{2}{m}G\right)|div(u-v)|^2+\frac{G}{2}\sum_{i,k=1}^{m}|D_i\left(u^{(k)}-v^{(k)}\right)+D_k\left(u^{(i)}-v^{(i)}\right)|^2+$$
$$+G\left[\Omega\left(D'(u-v),u\right)-\Omega\left(D'(u-v),v\right)\right].$$

Applying (3.3.26) we come to the equality

$$\text{(3.3.27)}\qquad \left[a_k^{(i)}(x;D'u)-a_k^{(i)}(x;D'v)\right]D_i\left(u^{(k)}-v^{(k)}\right)=\left(K-\frac{2}{m}G(1-\overline{\omega})\right)\times$$
$$\times|div(u-v)|^2+\frac{G(1-\overline{\omega})}{2}\sum_{i,k=1}^{m}|D_i\left(u^{(k)}-v^{(k)}\right)+D_k\left(u^{(i)}-v^{(i)}\right)|^2-$$
$$-\frac{4G\overline{\omega'}}{m^3\overline{e_0}}\left|\left[\frac{m}{2}\left(\overline{D_iu^{(k)}}+\overline{D_ku^{(i)}}\right)-\overline{divu}\delta_{ik}\right]D_i(u^{(k)}-v^{(k)})\right|^2.$$

Let us estimate the last term on the right-hand side. First we have

$$\left|\left[\frac{m}{2}\left(\overline{D_iu^{(k)}}+\overline{D_ku^{(i)}}\right)-\overline{divu}\delta_{ik}\right]D_i(u^{(k)}-v^{(k)})\right|=$$
$$=\frac{1}{2}\left|\left[\frac{m}{2}\left(\overline{D_iu^{(k)}}+\overline{D_ku^{(i)}}\right)-\overline{divu}\delta_{ik}\right]\left[D_i(u^{(k)}-v^{(k)})+D_k(u^{(i)}-v^{(i)})\right]\right|.$$

Applying the Hölder inequality we get

$$\left|\left[\frac{m}{2}\left(\overline{D_iu^{(k)}}+\overline{D_ku^{(i)}}\right)-\overline{divu}\delta_{ik}\right]D_i(u^{(k)}-v^{(k)})\right|^2\le$$
$$\le\frac{m^3}{8}\sum_{i,k=1}^{m}|D_i(u^{(k)}-v^{(k)})+D_k(u^{(i)}-v^{(i)})|^2e_0^2.$$

From (3.3.27) it now follows that

$$\left[a_k^{(i)}(x; D'u) - a_k^{(i)}(x; D'v)\right] D_i\left(u^{(k)} - v^{(k)}\right) \geq \tag{3.3.28}$$
$$\geq \left(K - \frac{2}{m}G(1-\overline{\omega})\right) |div(u-v)|^2 +$$
$$+\frac{G}{2}[1 - (\overline{\omega} + \overline{\omega' e_0})] \sum_{i,k=1}^{m} \left|D_i\left(u^{(k)} - v^{(k)}\right) + D_k\left(u^{(i)} - v^{(i)}\right)\right|^2.$$

Hence

$$\left[a_k^{(i)}(x; D'u) - a_k^{(i)}(x; D'v)\right] D_i\left(u^{(k)} - v^{(k)}\right) \geq$$
$$\geq \frac{G}{2}[1 - \sup_{e_0}(\omega + \omega' e_0)] \sum_{i,k=1}^{m} \left|D_i\left(u^{(k)} - v^{(k)}\right) + D_k\left(u^{(i)} - v^{(i)}\right)\right|^2,$$

thus a is determined by the first inequality (3.3.21).
Let us now prove the second relation (3.3.21). Consider the following expression

$$\left[a_k^{(i)}(x; D'u) - a_k^{(i)}(x; D'v)\right] \alpha_{ik} =$$
$$= \left\{\left(K - \frac{2}{m}G\right) div(u-v)\delta_{ik} + G\left[D_i(u^{(k)} - v^{(k)}) + D_k(u^{(i)} - v^{(i)})\right]\right\} \alpha_{ik} +$$
$$+G[\Omega(\alpha, u) - \Omega(\alpha, v)],$$

where α_{ik} are functional parameters. Taking into account (3.3.25) we get

$$\frac{\partial \Omega}{\partial p_j^{(l)}} = \omega\left(\frac{2}{m}\delta_{jl}\delta_{ik} - \delta_{ij}\delta_{kl} - \delta_{il}\delta_{jk}\right)\alpha_{ik} -$$
$$-\frac{4}{m^3}\frac{\omega'}{e_0}\left[\sum_{s<h}^{m}\left(D_s u^{(s)} - D_h u^{(h)}\right)(\delta_{js}\delta_{ls} - \delta_{jh}\delta_{lh}) +\right.$$
$$\left.+\frac{m}{2}\sum_{s<h}^{m}\left(D_h u^{(s)} + D_s u^{(h)}\right)(\delta_{jh}\delta_{sl} + \delta_{js}\delta_{hl})\right] \times$$
$$\times \left[\frac{m}{2}\left(D_i u^{(k)} + D_k u^{(i)}\right) - divu\delta_{ik}\right]\alpha_{ik}.$$

Applying the finite difference theorem we obtain

$$\Omega(\alpha, u) - \Omega(\alpha, v) = \omega\left\{\frac{2}{m}div(u-v)\delta_{ik} - \left[D_i(u^{(k)} - v^{(k)}) + D_k(u^{(i)} - v^{(i)})\right]\alpha_{ik}\right\} \cdot$$
$$-\frac{4}{m^3}\frac{\overline{\omega'}}{\overline{e_0}}\left\{\sum_{s<h}^{m}\left(\overline{D_s u^{(s)}} - \overline{D_h u^{(h)}}\right)\left[D_s(u^{(s)} - v^{(s)}) - D_h(u^{(h)} - v^{(h)})\right] +\right.$$
$$\left.+\frac{m}{2}\sum_{s<h}^{m}\left[\overline{D_h u^{(s)}} + \overline{D_s u^{(h)}}\right]\left[D_h(u^{(s)} - v^{(s)}) + D_s(u^{(h)} - v^{(h)})\right]\right\} \times$$
$$\times \left[\frac{m}{2}\left(\overline{D_i u^{(k)}} + \overline{D_k u^{(i)}}\right) - \overline{divu}\delta_{ik}\right]\alpha_{ik}.$$

From this follows

$$\left[a_k^{(i)}(x;D'u)-a_k^{(i)}(x;D'v)\right]\alpha_{ik}=$$
$$=\Big\{\left[K-\frac{2}{m}G(1-\overline{\omega})\right]div(u-v)\delta_{ik}+G(1-\overline{\omega})\left[D_i(u^{(k)}-v^{(k)})+D_k(u^{(i)}-v^{(i)})\right]-$$
$$-\frac{4G\overline{\omega'}}{m^3\overline{e_0}}\Big\{\sum_{s<h}^{m}\left[\overline{D_su^{(s)}}-\overline{D_hu^{(h)}}\right]\left[D_s(u^{(s)}-v^{(s)})-D_h(u^{(h)}-v^{(h)})\right]+$$
$$+\frac{m}{2}\sum_{s<h}^{m}\left[\overline{D_hu^{(s)}}+\overline{D_su^{(h)}}\right]\left[D_h(u^{(s)}-v^{(s)})+D_h(u^{(h)}-v^{(h)})\right]\Big\}\times$$
$$\times\left[\frac{m}{2}\left(\overline{D_iu^{(k)}}+\overline{D_ku^{(i)}}\right)-\overline{divu}\delta_{ik}\right]\Big\}\alpha_{ik}.$$

Substitute $\alpha_{ik}=a_k^{(i)}(x;D'u)-a_k^{(i)}(x;D'v)$ and apply Hölder inequality. After simple calculations we come to the following inequality

$$\sum_{i,k=1}^{m}\left|a_k^{(i)}(x;D'u)-a_k^{(i)}(x;D'v)\right|^2\le$$
$$\le\sum_{i,k=1}^{m}\Big|\left[K-\frac{2}{m}G(1-\overline{\omega})\right]div(u-v)\delta_{ik}+G(1-\overline{\omega})\left[D_i(u^{(k)}-v^{(k)})+\right.$$
$$\left.+D_k(u^{(i)}-v^{(i)})\right]-\frac{4G\overline{\omega'}}{m^3\overline{e_0}}\Big\{\sum_{s<h}\left[\overline{D_su^{(s)}}-\overline{D_hu^{(h)}}\right]\left[D_s(u^{(s)}-v^{(s)})-D_h(u^{(h)}-v^{(h)})\right]+$$
$$+\frac{m}{2}\sum_{s<h}^{m}\left[\overline{D_hu^{(s)}}+\overline{D_su^{(h)}}\right]\left[D_h(u^{(s)}-v^{(s)})+D_s(u^{(h)}-v^{(h)})\right]\Big\}\times$$
$$\times\left[\frac{m}{2}\left(\overline{D_iu^{(k)}}+\overline{D_ku^{(i)}}\right)-\overline{divu}\delta_{ik}\right]\Big|^2.$$

Applying the inequality $(a+b)^2\le 2(a^2+b^2)$ we get

$$\sum_{i,k=1}^{m}\left|a_k^{(i)}(x;D'u)-a_k^{(i)}(x;D'v)\right|^2\le$$
$$\le 2G^2(1-\overline{\omega})^2\sum_{i,k=1}^{m}\left[D_i(u^{(k)}-v^{(k)})+D_k(u^{(i)}-v^{(i)})\right]^2+$$
$$+2\sum_{i,k=1}^{m}\Big|\left[K-\frac{2}{m}G(1-\overline{\omega})\right]|div(u-v)|\delta_{ik}-\frac{4G\overline{\omega'}}{m^3\overline{e_0}}\Big\{\sum_{s<h}^{m}\left[\overline{D_su^{(s)}}-\overline{D_hu^{(h)}}\right]\times$$
$$\times\left[D_s(u^{(s)}-v^{(s)})-D_h(u^{(h)}-v^{(h)})\right]+$$
$$+\frac{m}{2}\sum_{s<h}^{m}\left[\overline{D_hu^{(s)}}+\overline{D_su^{(h)}}\right]\left[D_h(u^{(s)}-v^{(s)})+D_s(u^{(h)}-v^{(h)})\right]\Big\}\times$$
$$\times\left[\frac{m}{2}\left(\overline{D_iu^{(k)}}+\overline{D_ku^{(i)}}\right)-\overline{divu}\delta_{ik}\right]\Big|^2.$$

Taking into account the equality

$$\sum_{i,k=1}^{m}\delta_{ik}\left[\frac{m}{2}(D_iu^{(k)}+D_ku^{(i)})-divu\delta_{ik}\right]=0,$$

after squaring we obtain

$$\sum_{i,k=1}^{m}\left|a_k^{(i)}(x;D'u)-a_k^{(i)}(x;D'v)\right|^2\leq$$
$$\leq 2\sum_{i,k=1}^{m}\left\{\left[K-\frac{2}{m}G(1-\omega)\right]^2[div(u-v)]^2\delta_{ik}+\right.$$
$$\left.+G^2(1-\omega)^2\left[D_i(u^{(k)}-v^{(k)})+D_k(u^{(i)}-v^{(i)})\right]^2\right\}+$$
$$+\frac{32G^2}{m^6}\frac{\overline{\omega'}^2}{\overline{e_0}^2}\left\{\sum_{s<h}^{m}\left[\overline{D_su^{(s)}}-\overline{D_hu^{(h)}}\right]\left[D_s(u^{(s)}-v^{(s)})-D_h(u^{(h)}-v^{(h)})\right]+\right.$$
$$\left.+\frac{m}{2}\sum_{s<h}^{m}\left[\overline{D_hu^{(s)}}+\overline{D_su^{(h)}}\right]\left[D_h(u^{(s)}-v^{(s)})+D_s(u^{(h)}-v^{(h)})\right]\right\}^2\times$$
$$\times\sum_{i,k=1}^{m}\left[\frac{m}{2}\left(\overline{D_iu^{(k)}}+\overline{D_ku^{(i)}}\right)-\overline{divu}\delta_{ik}\right]^2.$$

Applying the Hölder inequality to the second braces and using (3.3.15) and (3.3.22) we calculate the following set of inequalities

$$\left\{\sum_{s<h}^{m}\left[\overline{D_su^{(s)}}-\overline{D_hu^{(h)}}\right]\left[D_s(u^{(s)}-v^{(s)})-D_h(u^{(h)}-v^{(h)})\right]+\right.$$
$$\left.+\frac{m}{2}\sum_{s<h}^{m}\left[\overline{D_hu^{(s)}}+\overline{D_su^{(h)}}\right]\left[D_h(u^{(s)}-v^{(s)})+D_s(u^{(h)}-v^{(h)})\right]\right\}^2\leq$$
$$\leq\left[\sum_{s<h}^{m}\left(|\overline{D_su^{(s)}}-\overline{D_hu^{(h)}}|^2+\frac{m}{2}|\overline{D_hu^{(s)}}+\overline{D_su^{(h)}}|^2\right)\right]\times$$
$$\times\left[\sum_{s<h}^{m}\left(|D_s(u^{(s)}-v^{(s)})-D_h(u^{(h)}-v^{(h)})\right)|^2+\right.$$
$$\left.+\frac{m}{2}\sum_{s<h}^{m}|D_h(u^{(s)}-v^{(s)})+D_s(u^{(h)}-v^{(h)})|^2\right]=$$
$$=\frac{m^2}{2}e_0^2\left[\frac{m}{4}\sum_{h,s=1}^{m}|D_h(u^{(s)}-v^{(s)})+D_s(u^{(h)}-v^{(h)})|^2-|div(u-v)|^2\right].$$

Hence

$$\sum_{i,k=1}^{m}\left|a_k^{(i)}(x;D'u)-a_k^{(i)}(x;D'v)\right|^2\leq$$
$$\leq 2\left\{m\left[K-\frac{2}{m}G(1-\overline{\omega})\right]^2[div(u-v)]^2+\right.$$
$$\left.+G^2(1-\overline{\omega})^2\sum_{i,k=1}^{m}\left[D_i(u^{(k)}-v^{(k)})+D_k(u^{(i)}-v^{(i)})\right]^2\right\}+$$
$$+\frac{8G^2(\overline{\omega'}\overline{e_0})^2}{m}\left\{\frac{m}{4}\sum_{i,k=1}^{m}\left[D_i(u^{(k)}-v^{(k)})+D_k(u^{(i)}-v^{(i)})\right]^2-|div(u-v)|^2\right\}.$$

From this follows

$$\sum_{i,k=1}^{m} \left|a_k^{(i)}(x;D'u) - a_k^{(i)}(x;D'v)\right|^2 \le$$
$$\le 2\left\{m\left[K - \frac{2}{m}G(1-\overline{\omega})\right]^2 - \frac{4G^2(\overline{\omega}'\overline{e_0})^2}{m}\right\}|div(u-v)|^2 +$$
$$+2G^2\left[(1-\overline{\omega})^2 + (\overline{\omega}'\overline{e_0})^2\right]\sum_{i,k=1}^{m}\left[D_i(u^{(k)}-v^{(k)}) + D_k(u^{(i)}-v^{(i)})\right]^2$$

and we obtain

$$\sum_{i,k=1}^{m} \left|a_k^{(i)}(x;D'u) - a_k^{(i)}(x;D'v)\right|^2 \le \tag{3.3.29}$$
$$\le 2max\left\{\sup_{e_0}\left[m|K - \frac{2}{m}G(1-\omega)|^2 - \frac{4G^2(\omega'e_0)^2}{m}\right], \sup_{e_0} G^2\left[(1-\omega)^2 + (\omega'e_0)^2\right]\right\}\times$$
$$\times\left\{|div(u-v)|^2 + \sum_{i,k=1}^{m}\left|D_i(u^{(k)}-v^{(k)}) + D_k(u^{(i)}-v^{(i)})\right|^2\right\}.$$

From (3.3.28) we find

$$\left(a_k^{(i)}(x;D'u) - a_k^{(i)}(x;D'v)\right)D_i(u^{(k)}-v^{(k)}) \ge$$
$$\ge min\left\{\inf_{e_0}\left[K - \frac{2}{m}G(1-\omega)\right], \inf_{e_0}\frac{G}{2}[1-(\omega+\omega'e_0)]\right\}\times$$
$$\times\left\{|div(u-v)|^2 + \sum_{i,k=1}^{m}\left|D_i(u^{(k)}-v^{(k)}) + D_k(u^{(i)}-v^{(i)})\right|^2\right\}.$$

With the help of (3.3.29) we derive the following estimate

$$\sum_{i,k=1}^{m} \left|a_k^{(i)}(x;D'u) - a_k^{(i)}(x;D'v)\right|^2 \le$$
$$\le 2max\left\{\frac{\sup_{e_0}\left[m|K - \frac{2}{m}G(1-\omega)|^2 - \frac{4G^2(\omega'e_0^2)^2}{m}\right]}{min\left\{\inf_{e_0}\left[K-\frac{2}{m}G(1-\omega)\right], \inf_{e_0}\frac{G}{2}[1-(\omega+\omega'e_0)]\right\}},\right.$$
$$\left.\frac{\sup_{e_0} G^2\left[(1-\omega)^2 + (\omega'e_0)^2\right]}{min\left\{\inf_{e_0}\left[K-\frac{2}{m}G(1-\omega)\right], \inf_{e_0}\frac{G}{2}[1-(\omega+\omega'e_0)]\right\}}\right\}\times$$
$$\times\left[a_k^{(i)}(x;D'u) - a_k^{(i)}(x;D'v)\right]D_i(u^{(k)}-v^{(k)})$$

establishing the second inequality of (3.3.21). □

Theorem 3.3.2 with the help of lemma 3.3.1 leads us to the following result.

Theorem 3.3.3. *If the conditions of lemma* 3.3.1 *are satisfied,* $f \in \mathcal{L}_q(\Omega)(q > m)$ *and inequality* (3.3.9) *holds, then the elasto-plastic problem* (3.3.20), (3.3.2) *has inside*

Ω a Hölder continuous solution (displacement), which can be obtained with the help of iterative process (3.3.14). *This process converges in $H_\alpha(\Omega')(\Omega' \ll \Omega)$ as a geometric progression with the rate of convergence*

$$q = \left(1 - \frac{2}{m-1}\frac{a}{b}\right)^{1/2}\left[1 + \frac{(m-2)^2}{m-1}\right]^{1/2}.$$

For the system (3.2.2) we get

$$a = \mu/2$$

and

$$b = \lambda m + 4\mu.$$

This shows, that the last theorem can not be applied for $m \geq 3$. We shall investigate in the next section a different method from (3.3.5), the method of elastic solutions.

3.4 Method of elastic solutions; the appearance of cracks

In the previous section we considered boundary value problems satisfying the conditions (3.3.13). These equations are true for the orthotropic case. We shall now suppose that the media is characterized by more general relations

(3.4.1) $$\sigma_{ik} = a_k^{(i)}(x; \varepsilon_{jl}), \quad (i, j, k, l = 1, ..., m).$$

This situation can appear, for example in the following cases:
1). Anisotropic nonhomogenious elastic media with

(3.4.2) $$\sigma_{i,k} = \sum_{pq=1}^{m} c_{ikpq}\varepsilon_{p,q};$$

2). An isotropic media with hardening, satisfying the conditions (3.3.13), (3.3.18) and (3.3.19).
It appears that occationally the functions $a_k^{(i)}$ are not smooth. For example, if Φ in relations (3.3.18) is determined by a polygonal line then $a_k^{(i)}$ are nondifferentiable with respect to ε_{pq}.
We shall start our consideration under following assumptions about the coefficients $a_k^{(i)}$:
1) $a_k^{(i)}$ satisfy Caratheodory conditions
2) the symmetric matrix

(3.4.3) $$A = \left\{\frac{\partial a_k^{(i)}}{\partial \varepsilon_{jl}}\right\}$$

is positive determined and all its elements are bounded, or, equally, the inequalities

(3.4.4)
$$\frac{\partial a_k^{(i)}}{\partial \varepsilon_{jl}} \xi_i^{(k)} \xi_j^{(l)} \geq \mu_0 \sum_{i=1}^{m} |\xi_i|^2$$

(3.4.5)
$$\left| \frac{\partial a_k^{(i)}}{\partial \varepsilon_{jl}} \right| \leq \nu_0 \quad .$$

hold, where $\mu_0, \nu_0 = \text{const} > 0$ and $\xi_i = (\xi_i^{(1)}, ..., \xi_i^{(m)},)$ $(i = 1, .., m)$ is an arbitrary set of vectors from R^m.
3) For any $q > 1$ and any displacement $u \in W_q^{(1)}(\Omega)$ all $a_k^{(i)}(x, \varepsilon_{jl}(u)) \in \mathcal{L}_q(\Omega)$, where

(3.4.6)
$$\varepsilon_{jl}(u) = D_j u^{(l)} + D_l u^{(j)}.$$

Let

(3.4.7)
$$\lambda = \inf_{i,x,\varepsilon_{jl}} \lambda_i \qquad \Lambda = \sup_{i,x,\varepsilon_{jl}} \lambda_i,$$

where $\{\lambda_i\}$ is the spectrum of the matrix (3.4.3). If we substitute the relations (3.4.1) in the equilibrium system we will get a system of the type (3.3.1). Considering only the problem with a fixed boundary (3.3.2) and our assumptions folows that the right-hand side $a_0(x)$ of (3.3.1) (the mass forces) belongs to $\mathcal{L}_q(\Omega)$ with some $q > m$. We shall denote $a_0(x) = f(x)$. Define

(3.4.8)
$$L_0(u) = \Delta u + \nabla(divu).$$

This is the linear operator for the homogeneous orthotropic theory with $\lambda = 0$ and $\mu = 1$. Consider for the problem (3.3.1), (3.3.2) the iterative process

(3.4.9)
$$L_0(u_{n+1}) = L_0(u_n) - \varepsilon L(u_n),$$

where L is determined by (3.3.1), all u_n satisfy (3.3.2) and ε =const > 0. This process is analogous to the method of elastic solution for the media with hardening (see, for example, Il'yshin [1]) and is called also the method of elastic solutions. We can write the equations (3.4.9) with conditions (3.3.2) in the weak form

(3.4.10)
$$\int_\Omega (D'u_{n+1}D'v + divu_{n+1}divv)\,dx = \int_\Omega (D'u_n D'v + divu_n divv)\,dx -$$
$$-\varepsilon \int_\Omega \left[a_k^{(i)}(x; \varepsilon_{jl}(u_n)) D_k v^{(i)} + fv\right] dx.$$

Consider the integral on the left-hand side. Integrating by parts and taking into account (3.3.2) we come to the following equalities

$$\int_\Omega \left(D'u_{n+1}D'v + divu_{n+1}^{(k)} divv\right) dx = \int_\Omega \left[D_j u_{n+1}^{(k)} D_j v^{(k)} + D_j u_{n+1}^{(j)} D_k v^{(k)}\right] dx =$$

$$= \int_\Omega \left[D_j u_{n+1}^{(k)} D_j v^{(k)} - u_{n+1}^{(j)} D_j D_k v^{(k)}\right] dx = \int_\Omega \left[D_j u_{n+1}^{(k)} D_j v^{(k)} + D_k u_{n+1}^{(j)} D_j v^{(k)}\right] dx =$$

$$= \int_\Omega \varepsilon_{jk}(u_{n+1}) D_j v^{(k)} dx = \frac{1}{2}\int_\Omega \varepsilon_{jk}(u_{n+1})\varepsilon_{jk}(v) dx.$$

The analogous set of equations is valid for the first integral on the right-hand side of (3.4.10). Therefore we can write (3.4.10) in the form

$$(3.4.11) \quad \int_\Omega \varepsilon_{ik}(u_{n+1}) D_i v^{(k)} dx = \int_\Omega \left[\varepsilon_{ik}(u_n) - \varepsilon a_k^{(i)}(x; \varepsilon_{jl}(u_n))\right] D_k v^{(i)} dx - \varepsilon \int_\Omega f v dx.$$

Taking into account $a_k^{(i)} = a_i^{(k)}$ and $\varepsilon_{ik} = \varepsilon_{ki}$ then from (3.4.11) it follows that

$$(3.4.12) \quad \int_\Omega \varepsilon_{ik}(u_{n+1})\varepsilon_{ik}(v) dx = \int_\Omega \left[\varepsilon_{ik}(u_n) - \varepsilon a_k^{(i)}(x; \varepsilon_{jl}(u_n))\right] \varepsilon_{ik}(v) dx - 2\varepsilon \int_\Omega f v dx.$$

If $\varepsilon = 2(\Lambda + \lambda)^{-1}$, we subtract the two succesive equalities (3.4.11) or (3.4.12) and apply the finite difference theorem, we get the relation

$$(3.4.13) \qquad \int_\Omega \varepsilon_{ik}(u_{n+1} - u_n)\varepsilon_{ik}(v) dx = \int_\Omega \left(I - \frac{2}{\Lambda + \lambda}\overline{A}\right) E(u_n - u_{n-1}) E(v) dx.$$

Here $\overline{A}$ denotes the matrix (3.4.3) with intermediate arguments and $E(z) = \{\varepsilon_{ik}(z)\} \in R^{m^2}$ for any fixed z.

Lemma 3.4.1. *If the conditions 1)-3) of this section are satisfied and $f \in \mathcal{L}_q$ with $q \geq 2$ then the process (3.4.11) converges in $\overset{\circ}{W}{}_2^{(1)}(\Omega)$ as a geometric progression with the common ratio $K = (\Lambda - \lambda)(\Lambda + \lambda)^{-1}$, starting from $\forall u_0 \in \overset{\circ}{W}{}_2^{(1)}(\Omega)$ and $\varepsilon = 2(\Lambda + \lambda)^{-1}$ to the solution of (3.3.1),(3.3.2).*

Proof. Applying to the right-hand side of (3.4.13) the Hölder inequality we arrive at the relation

$$\left|\int_\Omega \varepsilon_{ik}(u_{n+1} - u_n)\varepsilon_{ik}(v) dx\right| \leq \frac{\Lambda - \lambda}{\Lambda + \lambda}\left(\sum_{i,k=1}^m \int_\Omega |\varepsilon_{ik}(u_n - u_{n-1})|^2 dx\right)^{1/2} \times$$

$$\times \left(\sum_{i,k=1}^m \int_\Omega |\varepsilon_{ik}(v)|^2 dx\right)^{\frac{1}{2}}.$$

Letting $v = u_{n+1} - u_n$ we obtain as in theorem 1.2.2 the following estimate

$$\sum_{i,k=1}^m \int_\Omega |\varepsilon_{ik}(u_{n+1} - u_n)|^2 dx \leq \left(\frac{\Lambda - \lambda}{\Lambda + \lambda}\right)^2 \sum_{i,k=1}^m \int_\Omega |\varepsilon_{ik}(u_n - u_{n-1})|^2 dx.$$

This concludes the proof. □

Lemma 3.4.1 proves the existence of the weak solution without restrictions on dispersion for the spectrum of matrix (3.4.3).
We shall show that if the dispersion of the spectrum A is bounded with some explicit constant then the weak solution of the problem (3.3.1) , (3.3.2) will be Hölder continuous in any strongly interior domain $\Omega' \ll \Omega$.
We consider the iterative method given by one of the following formulas (3.4.10), (3.4.11) or (3.4.12). Let $w_n = u_n - u_{n-1}$. As a test function use the function v, which satisfies in $B_\delta(x_0) \in \Omega$ the equalities

$$(3.4.14) \qquad \Delta v^{(k)} + D_k div v = \sum_{j=1}^{m} D_j[\varepsilon_{jk}(w_{n+1})r^\alpha \zeta],$$

$$(3.4.15) \qquad v|_{\partial B_\delta(x_0)} = 0.$$

and is equal to zero outside $B_\delta(x_0)$.

Lemma 3.4.2. *If $w_{n+1} \in W^{(1)}_{2,\beta+2\alpha}(B_\delta(x_0))$ with $0 \le \beta < m$ then the problem has a solution $v \in W^{(1)}_{2,\beta}$, which satisfies the inequality*

$$(3.4.16) \qquad \int_{B_\delta(x_0)} |D'v|^2 r^\beta dx \le C \int_{B_\delta(x_0)} |D'w_{n+1}|^2 r^{2\alpha+\beta} \zeta^2 dx.$$

Proof. It is obvious that the right-hand side of (3.4.14) can be assumed smooth. The solution is written in the following form

$$(3.4.17) \qquad v^{(l)}(x) = \int_{B_\delta} G_{lk}(x,y) D_j \left[\varepsilon_{jk} r_y^\alpha \zeta\right] dy,$$

in which $\{G_{lk}\}$ is the Green function for the operator L_0 ((3.4.8)) with the boundary condition (3.4.15), $B_\delta = B_\delta(x_0)$ and $\varepsilon_{ik} = \varepsilon_{ik}[w_{n+1}]$. The function w_{n+1} under the sign of ε_{jk} will be omitted during the proof. Since

$$(3.4.18) \qquad G_{lk}(x,y) = S_{lk}(x,y) + R_{lk}(x,y),$$

where $S_{lk}(x,y)$ are elements of the fundamental solution and $R_{lk}(x,y)$ - elements of the so-called regular part of $\{G_{lk}\}$, it is evident that for $|x| < 3\delta/4$ we have

$$(3.4.19) \qquad |D^2 R_{lk}(x,y)| < C.$$

Integrating by parts on the right-hand side of (3.4.17) gives

$$v^{(l)}(x) = -\int_{B_\delta} D_j(G_{lk})\varepsilon_{jk} r_y^\alpha \zeta dy.$$

Therefore, with the help of (3.4.18) we derive

$$(3.4.20) \qquad D_i v^{(l)} = -D_i \int_{B_\delta} D_j[S_{lk}(x,y)]\varepsilon_{jk} r_y^\alpha \zeta dy - D_i \int_{B_\delta} D_j[R_{lk}(x,y)]\varepsilon_{jk} r_y^\alpha \zeta dy.$$

The first term on the right-hand side can be regarded as a derivative of the so-called integral with a weak singularity. Referring to the Stein estimate [1] we can write

$$\sum_{i,l=1}^{m}\int_{B_\delta}|D_i\int_{B_\delta}D_j\left[S_{lk}(x,y)\right]\varepsilon_{jk}r_y^{\alpha}\zeta dy|^2 r_x^{\beta}dx \le C\sum_{i,k=1}^{m}\int_{B_\delta}|\varepsilon_{ik}|^2 r_x^{2\alpha+\beta}\zeta^2 dx.$$

Since $\varepsilon_{ik}^2[w_{n+1}] \le 4|D'w_{n+1}|^2$ we have

$$\sum_{i,l=1}^{m}\int_{B_\delta}|D_i\int_{B_\delta}D_j\left[S_{lk}(x,y)\right]\varepsilon_{jk}r_y^{\alpha}\zeta dy|^2 r_x^{\beta}dx \le C\int_{B_\delta}|D'w_{n+1}|^2 r_x^{2\alpha+\beta}\zeta^2 dx. \tag{3.4.21}$$

Applying (3.4.19) we obtain

$$-D_i\int_{B_\delta}D_j[R_{lk}(x,y)]\varepsilon_{jk}r_y^{\alpha}\zeta dy = \int_{B_\delta}K_{i,j,k,l}(x,y)\varepsilon_{jk}r_y^{\alpha}\zeta dy,$$

where $K_{i,j,k,l}(x,y)$ are smooth functions for $|x| \le 3\delta/4$.
Then

$$|\int_{B_\delta}K_{i,j,k,l}(x,y)\varepsilon_{ik}r_y^{\alpha}\zeta dy| \le C\sum_{j,k=1}^{m}\int_{B_\delta}|\varepsilon_{ik}|r_y^{\alpha}\zeta dy.$$

Multiplying and dividing by $r^{\beta/2}$ under the sign of the integral and applying the Hölder inequality we get

$$|\int_{B_\delta}K_{i,j,k,l}(x,y)\varepsilon_{jk}r_y^{\alpha}\zeta dy| \le C\left(\sum_{j,k=1}^{m}\int_{B_\delta}|\varepsilon_{jk}|^2 r_y^{2\alpha+\beta}\zeta^2 dy\right)^{1/2}\left(\int_{B_\delta}r_y^{-\beta}dy\right)^{1/2}.$$

Since $0 \le \beta < m$, the last integral is finite and we have

$$\int_{B_{3\delta/4}}|\int_{B_\delta}K_{i,j,k,l}(x,y)\varepsilon_{jk}r_y^{\alpha}\zeta dy|^2 r_x^{\beta}\zeta dx \le C\sum_{i,k=1}^{m}\int_{B_\delta}|\varepsilon_{ik}|^2 r_y^{2\alpha+\beta}\zeta^2 dy\int_{B_\delta}r_x^{\beta}dx \le$$
$$\le C\sum_{i,k=1}^{m}\int_{B_\delta}|\varepsilon_{ik}|^2 r_y^{2\alpha+\beta}\zeta^2 dy \le C\int_{B_\delta}|D'w_{n+1}|^2 r_y^{2\alpha+\beta}\zeta^2 dy.$$

Combining the above expression with (3.4.18) and (3.4.21) we reach the end of the proof. □

Lemma 3.4.3. *If $m \le 3$ then for the solution v of the problem* (3.4.14), (3.4.15) *the inequality*

$$\int_{B_\delta}|D'v|^2 r^{-\alpha}dx \le \left[1-\frac{5}{4}\frac{(m-2)^2}{m-1+(m-2)^2}+O(\gamma)+\eta\right]^{-2}\times \tag{3.4.22}$$
$$\times\sum_{j,k=1}^{m}\int_{B_\delta}\varepsilon_{ik}^2(w_{n+1})r^{\alpha}\zeta dx + C\int_{B_{\delta/2,\delta}}|D'v|^2 r^{-2\alpha+\alpha'}dx$$

holds, where $\alpha = 2-m-2\gamma$, $\gamma > 0$ is sufficiently small and α' is arbitrary.

Proof. The right-hand side of (3.4.14) is assumed smooth. Take the spherical system of coordinates with the centrum in x_0 and expand v with the help of the complete orthonormal sequence of spherical functions $\{Y_{s,i}(\theta)\}(s = 0, 1, ..., i = 1, ...k_s)$ to give

(3.4.23) $$v(x) = \sum_{s=0}^{+\infty}\sum_{i=1}^{k_s} v_{s,i}(r)Y_{s,i}(\theta).$$

Expansion of the test-function $z(x)$ for the problem (3.4.14), (3.4.15) will be determined by

(3.4.24) $$z(x) = \sum_{s=0}^{+\infty}\sum_{i=1}^{k_s} z_{s,i}(r)Y_{s,i}(\theta),$$

(3.4.25) $$z_{0,1}(r) = -\int_r^{\delta} v'_{0,1}(\rho)\rho^{-\alpha}d\rho,$$

(3.4.26) $$z_{s,i}(r) = v_{s,i}(r)r^{-\alpha}, \quad s \geq 1.$$

The expansions (3.4.23) and (3.4.24) can be presented in the form

(3.4.27) $$\begin{aligned} v(x) &= v_0(x) + v_1(x), \\ z(x) &= z_0(x) + z_1(x), \end{aligned}$$

where $v_0(x), z_0(x)$ represent the zero-harmonics and $z_1(x) = v_1(x)r^{-\alpha}$. Multiply both sides of (3.4.14) by z. After integrating by parts, we get

(3.4.28) $$\begin{aligned} &\int_{B_\delta} \left[D_jD_jv_0^{(k)} + D_kD_jv_0^{(j)}\right] z_o^{(k)}dx + \int_{B_\delta} \left[D_jD_jv_0^{(k)} + D_kD_jv_0^{(j)}\right] z_1^{(k)}dx + \\ &+\int_{B_\delta} \left[D_jD_jv_1^{(k)} + D_kD_jv_1^{(j)}\right] z_o^{(k)}dx + \int_{B_\delta} \left[D_jD_jv_1^{(k)} + D_kD_jv_1^{(j)}\right] z_1^{(k)}dx = \\ &= -\int_{B_\delta} D_jz^{(k)}\varepsilon_{jk}(w_{n+1})r^\alpha\zeta dx. \end{aligned}$$

Taking into account orthogonality we obtain

$$\int_{B_\delta} D_jD_jv_0^{(k)}z_1^{(k)}dx = \int_{B_\delta} D_jD_jv_1^{(k)}z_0^{(k)}dx = 0$$

and (3.4.28) can be written in the form

(3.4.29) $$\begin{aligned} &\sum_{i=0}^{1}\int_{B_\delta} \left[D_jv_i^{(k)} + D_kv_i^{(j)}\right] D_jz_i^{(j)}dx + \\ &+\int_{B_\delta} (div v_1 div z_0 + div v_0 div z_1)dx = \int_{B_\delta} \varepsilon_{jk}(w_{n+1})D_jz^{(k)}r^\alpha\zeta dx. \end{aligned}$$

Let

$$I_0 = \int\limits_{B_\delta} \left[D_j v_0^{(k)} + D_k v_0^{(j)}\right] D_j z_0^{(k)} dx, \tag{3.4.30}$$

$$I_1 = \int\limits_{B_\delta} \left[D_j v_1^{(k)} + D_k v_1^{(j)}\right] D_j z_1^{(k)} dx, \tag{3.4.31}$$

$$I_2 = \int\limits_{B_\delta} (div v_1 div z_0 + div v_0 div z_1) dx. \tag{3.4.32}$$

Since v_0 is independent of $\theta(Y_{0,1}(\theta)$ =const) then from (3.4.25) follows

$$I_0 = \int\limits_{B_\delta} \left[v_0^{(k)'} cos(x_j, r) + v_0^{(j)'} cos(x_k, r)\right] v_0^{(k)'} cos(x_j, r) r^{-\alpha} dx = \tag{3.4.33}$$
$$= \int\limits_{B_\delta} |v_0'|^2 r^{-\alpha} dx + \sum_{k=1}^{m} \int\limits_{B_\delta} |v_0^{(k)'}|^2 cos^2(x_k, r) r^{-\alpha} dx =$$
$$= \frac{1}{2} \sum_{j,k=1}^{m} \int\limits_{B_\delta} \left[D_j v_0^{(k)} + D_k v_0^{(j)}\right]^2 r^{-\alpha} dx.$$

Hence

$$I_0 = \frac{1}{2} \sum_{j,k=1}^{m} \int\limits_{B_\delta} \varepsilon_{jk}^2(v_0) r^{-\alpha} dx. \tag{3.4.34}$$

Now address I_1 ((3.4.31)). Using (3.4.26) after integration by parts gives

$$I_1 = \int\limits_{B_\delta} \left[D_j v_1^{(k)} + D_k v_1^{(j)}\right] \left[D_j v_1^{(k)} r^{-\alpha} + v_1^{(k)} D_j r^{-\alpha}\right] dx =$$
$$= \frac{1}{2} \sum_{j,k=1}^{m} \int\limits_{B_\delta} \left[D_j v_1^{(k)} + D_k v_1^{(j)}\right]^2 r^{-\alpha} dx + \frac{1}{2} \int\limits_{B_\delta} |D_j v_1|^2 D_j r^{-\alpha} dx +$$
$$+ \int\limits_{B_\delta} D_k v_1^{(j)} v_1^{(k)} D_j r^{-\alpha} dx = \frac{1}{2} \sum_{j,k=1}^{m} \int\limits_{B_\delta} \varepsilon_{jk}^2(v_1) r^{-\alpha} dx - \frac{1}{2} \int\limits_{B_\delta} |v_1|^2 D_j D_j r^{-\alpha} dx -$$
$$- \int\limits_{B_\delta} v_1^{(j)} div v_1 D_j r^{-\alpha} dx - \int\limits_{B_\delta} v_1^{(j)} v_1^{(k)} D_j D_k r^{-\alpha} dx.$$

After elementary calculations we obtain

$$I_1 = \frac{1}{2} \sum_{j,k=1}^{m} \int\limits_{B_\delta} \varepsilon_{jk}^2(v_1) r^{-\alpha} dx + \frac{\alpha(m+\alpha)}{2} \int\limits_{B_\delta} |v_1|^2 r^{-\alpha-2} dx + \tag{3.4.35}$$
$$+ \alpha \int\limits_{B_\delta} v_{1,r} div v_1 r^{-\alpha-1} dx - \alpha(\alpha+2) \int\limits_{B_\delta} v_{1,r}^2 r^{-\alpha-2} dx.$$

Since

$$div z_1 = div(v_1 r^{-\alpha}) = div v_1 r^{-\alpha} - \alpha v_{1,r} r^{-\alpha-1}$$

and

$$div z_0 = \sum_{i=1}^{m} D_i z_0^{(i)} = z_0^{(i)'} cos(x_i, r) = v_0^{(i)'} cos(x_i, r) r^{-\alpha} = div v_0 r^{-\alpha},$$

we have

$$I_2 = 2 \int_{B_\delta} div v_0 div v_1 r^{-\alpha} dx - \alpha \int_{B_\delta} v_{1,r} div v_0 r^{-\alpha-1} dx.$$

Using (3.4.34), (3.4.35) and the last equality, we derive from (3.4.29) the following equality

$$\begin{aligned}
&\frac{1}{2}\sum_{i=0}^{1}\sum_{j,k=1}^{m} \int_{B_\delta} \varepsilon_{jk}^2(v_i) r^{-\alpha} dx + 2\int_{B_\delta} div v_0 div v_1 r^{-\alpha} dx - \\
&-\alpha \int_{B_\delta} v_{1,r} div v_0 r^{-\alpha-1} dx + \frac{\alpha(m-\alpha)}{2} \int_{B_\delta} |v_1|^2 r^{-\alpha-2} dx + \\
&+\alpha \int_{B_\delta} v_{1,r} div v_1 r^{-\alpha-1} dx - \alpha(\alpha+2) \int_{B_\delta} v_{1,r}^2 r^{-\alpha-2} dx = \\
&= \int_{B_\delta} \varepsilon_{jk}(w_{n+1}) D_j z^{(k)} r^{\alpha} \zeta dx.
\end{aligned}$$

Applying now the equality (3.2.10) gives

$$\begin{aligned}
&\int_{B_\delta} \Bigg[\sum_{k=1}^{m} |\nabla v_1^{(k)}|^2 + \left(div v - \frac{\alpha}{2} v_{1,r} r^{-1}\right)^2 - \frac{\alpha^2}{4} v_{1,r}^2 r^{-2} + \\
&+ \frac{\alpha(m-\alpha-2)}{2} |v_1|^2 r^{-2} + \frac{1}{2}\sum_{j,k=1}^{m} \varepsilon_{ik}^2(v_0) - (div v_0)^2\Bigg] r^{-\alpha} dx = \\
&= \int_{B_\delta} \varepsilon_{jk}(w_{n+1}) D_j z^{(k)} r^{\alpha} \zeta dx.
\end{aligned}$$

Since $|v_{1,r}|^2 \le |v_1|^2$ after omitting the nonnegative term we come to the following inequality

(3.4.36)
$$\begin{aligned}
&\int_{B_\delta} \Bigg[\sum_{k=1}^{m} |\nabla v_1^{(k)}|^2 + \frac{\alpha(2m-3\alpha-4)}{4} |v_1|^2 r^{-2} + \\
&+\frac{1}{2}\sum_{j,k=1}^{m} \varepsilon_{jk}^2(v_0) - (div v_0)^2\Bigg] r^{-\alpha} dx \le \int_{B_\delta} \varepsilon_{jk}(w_{n+1}) D_j z^{(k)} r^{\alpha} \zeta dx.
\end{aligned}$$

Obviously

$$(3.4.37)\qquad \frac{1}{2}\sum_{j,k=1}^{m}\int\limits_{B_\delta}\varepsilon_{jk}^2(v_0)r^{-\alpha}dx-\int\limits_{B_\delta}(divv_0)^2r^{-\alpha}dx=\sum_{k=1}^{m}\int\limits_{B_\delta}|\nabla v_0^{(k)}|^2r^{-\alpha}dx.$$

Using (3.4.23) gives

$$(3.4.38)\qquad \sum_{s=1}^{+\infty}\sum_{i=1}^{k_s}\int\limits_0^\delta\Big\{|v'_{1,s,i}|^2+$$
$$+\left[\frac{\alpha(2m-3\alpha-4)}{4}+s(s+m-2)\right]|v_{1,s,i}|^2r^{-2}\Big\}r^{-\alpha+m-1}dr+$$
$$+\sum_{k=1}^{m}\int\limits_{B_\delta}|\nabla v_0^{(k)}|^2r^{-\alpha}dx\le\int\limits_{B_\delta}\varepsilon_{jk}(w_{n+1})D_jz^{(k)}r^\alpha\zeta dx.$$

Considering the first term on the left-hand side

$$\sum_{s=1}^{+\infty}\sum_{i=1}^{k_s}\int\limits_0^\delta|v'_{1,s,i}|^2r^{-\alpha+m-1}dr,$$

from (3.4.15) it follows that $v_{1,s,i}(\delta)=0$. Applying Hardy inequality (2.1.8) with $s=\alpha-m+3=5-2m-2\gamma<1$, we obtain

$$(3.4.39)\qquad \int\limits_0^\delta|v'_{1,s,i}|^2r^{-\alpha+m-1}dr\ge\frac{(\alpha-m+2)^2}{4}\int\limits_0^\delta|v_{1,s,i}|^2r^{-\alpha+m-3}dr.$$

Let $q_s(0<q_s<1)$ and write (3.4.38) in the form

$$\sum_{s=1}^{+\infty}\sum_{i=1}^{k_s}\int\limits_0^\delta\Big\{q_s|v'_{1,s,i}|^2+(1-q_s)|v'_{1,s,i}|^2+\left[\frac{\alpha(2m-3\alpha-4)}{4}+s(s+m-2)\right]|v_{1,s,i}|^2r^{-2}\Big\}\times$$
$$\times r^{\alpha+m-1}dr+\sum_{k=1}^{m}\int\limits_{B_\delta}|\nabla v_0^{(k)}|^2r^{-\alpha}dx\le\int\limits_{B_\delta}\varepsilon_{jk}(w_{n+1})D_jz^{(k)}r^\alpha\zeta dx.$$

Applying (3.4.39) we get

$$\int\limits_0^\delta\sum_{s=1}^{+\infty}\sum_{i=1}^{k_s}\Big\{|v'_{1,s,i}|^2+\left[\frac{\alpha(2m-3\alpha-4)}{4}+s(s+m-2)\right]|v_{1,s,i}|^2r^{-2}\Big\}r^{-\alpha+m-1}dr\ge$$
$$\ge\int\limits_0^\delta\sum_{s=1}^{+\infty}\sum_{i=1}^{k_s}q_s|v'_{1,s,i}|^2r^{-\alpha+m-1}dr+$$
$$+\sum_{s=1}^{+\infty}\sum_{i=1}^{k_s}\int\limits_0^\delta\left[(1-q_s)\frac{(\alpha-m+2)^2}{4}+\frac{\alpha(2m-3\alpha-4)}{4}+s(s+m-2)\right]\times$$
$$\times|v_{1,s,i}|^2r^{-\alpha+m-3}dr.$$

Find q_s from the condition

$$q_s s(s+m-2) = (1-q_s)\frac{(\alpha-m+2)^2}{4} + \frac{\alpha(2m-3\alpha-4)}{4} + s(s+m-2),$$

then from (3.4.38) it follows that

$$\sum_{s=1}^{+\infty}\sum_{i=1}^{k_s} q_s \int_0^\delta \left[|v'_{1,s,i}|^2 + s(s+m-2)|v_{1,s,i}|^2 r^{-2}\right] r^{-\alpha+m-1} dr +$$
$$+\sum_{k=1}^{m}\int_{B_\delta} |\nabla v_0^{(k)}|^2 r^{-\alpha} dx \le \int_{B_\delta} \varepsilon_{jk}(w_{n+1}) D_j z^{(k)} r^\alpha \zeta dx.$$

For small $\gamma > 0 (\alpha = 2 - m - 2\gamma)$ we have

$$q_s \ge q_1 = 1 - \frac{5}{4}\frac{(m-2)^2}{m-1+(m-2)^2} + O(\gamma)$$

and the relations

$$\left[1 - \frac{5}{4}\frac{(m-2)^2}{m-1+(m-2)^2} + O(\gamma)\right] \int_{B_\delta} |D'v|^2 r^{-\alpha} dx \le \int_{B_\delta} \varepsilon_{jk}(w_{n+1}) D_j z^{(k)} r^\alpha \zeta dx,$$

(3.4.40)
$$\left[1 - \frac{5}{4}\frac{(m-2)^2}{m-1+(m-2)^2} + O(\gamma)\right] \int_{B_\delta} |D'v|^2 r^{-\alpha} dx \le$$
$$\le \left[\sum_{j,k=1}^{m}\int_{B_\delta} \varepsilon_{jk}^2(w_{n+1}) r^\alpha \zeta^2 dx\right]^{1/2} \left(\int_{B_\delta} |D'z|^2 r^\alpha dx\right)^{1/2}.$$

From (2.3.6′) and Remark 2.3.3 it follows that

(3.4.41)
$$\int_{B_\delta} |D'z|^2 r^\alpha dx \le [1 + O(\gamma)] \int_{B_\delta} |D'v|^2 r^{-\alpha} dx + C \int_{B_{\delta/2,\delta}} |D'v|^2 r^{-2\alpha+\alpha'} dx.$$

Applying (2.2.15), using (3.4.40) and (3.4.41) we come to (3.4.22). □

Remark 3.4.1. For $m > 3$ the existence of $C^{0,\gamma}(\Omega')$ solutions will be proved in the section 4.7.

Theorem 3.4.1. *Let $a_k^{(i)}(x;\varepsilon_{jl})$ satisfy the conditions (1)-(3) of this section and $u_0 \in W_q^{(1)}(\Omega), f \in \mathcal{L}_q(\Omega)$ with some $q > 2m(m+\alpha)^{-1}, (\alpha = 2 - m - 2\gamma)$ with a positive γ sufficiently small). If the inequality*

$$q = 2\frac{\Lambda - \lambda}{\Lambda + \lambda}\left[1 - \frac{5}{4}\frac{(m-2)^2}{m-1+(m-2)^2}\right]^{-1} < 1$$

holds, then for $m = 2$ and $m = 3$ the problem (3.3.1),(3.3.2) has in any $\Omega' \ll \Omega$ a solution from $H_\alpha(\Omega') \subset C^{0,\gamma}(\Omega')$. The iterative process (3.4.10) converges in H_α as a geometric progression with the common ratio q.

Proof. From (3.4.13) and relation $\zeta \leq 1$ we get

$$\sum_{j,k=1}^{m} \int_{B_\delta} \varepsilon_{j,k}^2(w_{n+1}) r^\alpha \zeta dx \leq \frac{\Lambda - \lambda}{\Lambda + \lambda} \left(\sum_{j,k=1}^{m} \int_{B_\delta} \varepsilon_{j,k}^2(w_{n+1}) r^\alpha dx \right)^{1/2} \times$$

$$\times \left(\sum_{j,k=1}^{m} \int_{B_\delta} \varepsilon_{j,k}^2(v) r^{-\alpha} dx \right)^{1/2} .$$

Since for $m = 2$ and $m = 3$ we have $-2\alpha < m$, then letting $\beta = -2\alpha, \alpha' = 0$ in (3.4.16) we get $\int\limits_{B_\delta} |D'v|^2 r^{-2\alpha} dx \leq C \int\limits_{B_\delta} |D'w_{n+1}|^2 \zeta^2 dx$. Using (3.4.22) we get the relation

$$\sum_{j,k=1}^{m} \int_{B_\delta} \varepsilon_{j,k}^2(w_{n+1}) r^\alpha \zeta dx \leq q = 2\frac{\Lambda - \lambda}{\Lambda + \lambda} \left[1 - \frac{5}{4} \frac{(m-2)^2}{m - 1 + (m-2)^2} \right]^{-1} \times$$

$$\times \left[\sum_{j,k=1}^{m} \int_{B_\delta} \varepsilon_{j,k}^2(w_n) r^\alpha \zeta dx + \|w_n\|^2_{W_2^{(1)}} \right]^{1/2} \left[\sum_{j,k=1}^{m} \int_{B_\delta} \varepsilon_{jk}^2(w_n) r^\alpha \zeta dx + \|w_{n+1}\|^2_{W_2^{(1)}} \right] .$$

According to lemma 3.4.1 we have the inequality $\|w_n\|_{W_2^{(1)}}(\Omega) < C\left[(\Lambda - \lambda)(\Lambda + \lambda)^{-1}\right]^n$. Then, from lemma 2.3.2 it follows that $\|w_n\|_{H_\alpha(\Omega')} < Cq^n$ and the theorem is proved. □

Chapter 4

Differentiability of solutions for second order elliptic systems

4.1 Some auxiliary inequalities

Inequality (2.1.8) brings us to some important estimates which were established by Cordes [1].

Lemma 4.1.1. *Let* $u \in \overset{\circ}{W}{}^{(2)}_{2,\alpha}(B_\delta(x_0), x_0)$, $\alpha = 2 - m - 2\gamma$, $0 < \gamma < 1$, $r = |x - x_0|$ *and the equalities*

(4.1.1) $$u(x_0) = \nabla u\Big|_{x=x_0} = u\Big|_{\partial B_\delta} = \nabla u\Big|_{\partial B_\delta} = 0^*$$

be satisfied.

Then the following inequalities

(4.1.2) $$\int\limits_{B_\delta} |u|^2 r^{\alpha-4} dx \le M_1^2(\gamma) \int\limits_{B_\delta} |\Delta u|^2 r^\alpha dx,$$

(4.1.3) $$\int\limits_{B_\delta} |D'u|^2 r^{\alpha-2} dx \le M_2^2(\gamma) \int\limits_{B_\delta} |\Delta u|^2 r^\alpha dx,$$

(4.1.4) $$\int\limits_{B_\delta} |D'^2 u|^2 r^\alpha dx \le M_3^2(\gamma) \int\limits_{B_\delta} |\Delta u|^2 r^\alpha dx$$

hold, where $B_\delta = B_\delta(x_0)$,

(4.1.5) $$M_1(\gamma) = \begin{cases} \frac{1}{(m+\gamma)\gamma}, & \gamma \le \frac{m+1}{m+\sqrt{(m+1)^2+1}}, \\ \frac{1}{(m+1+\gamma)(1-\gamma)}, & \gamma \ge \frac{m+1}{m+\sqrt{(m+1)^2+1}}, \end{cases}$$

(4.1.6) $$M_2^2(\gamma) = M_1(\gamma)\Big[1 + (m + 2\gamma)(1+\gamma)M_1(\gamma)\Big]$$

and

(4.1.7) $$M_3^2(\gamma) = 1 + \frac{(m - 2 + 2\gamma)\Big\{(1+\gamma)^2 + [2 - (1-\gamma)^2]m\Big\}}{(m+1+\gamma)^2(1-\gamma)^2}.$$

---

*For the functions belonging to $W^{(l)}_{2,\alpha}(\Omega, x_0)$ with $\alpha \in (-m, 2-m)$ there exist the values of u and its derivatives up to the order $l-1$ in x_0.

If $m \geq 4, \alpha \in (3-m, 0]$ and only the equalities $u\big|_{\partial B_\delta} = \nabla u\big|_{\partial B_\delta} = 0$ hold, then the inequality

$$(4.1.4') \qquad \int_{B_\delta} |D'^2 u|^2 r^\alpha dx \leq \int_{B_\delta} |\Delta u|^2 r^\alpha dx$$

is true. If, moreover, $u(x_0) = 0$ then inequality (4.1.4') holds even for $\alpha = 3 - m$, $m \geq 3$.

Proof. It is sufficient to prove the required inequality for smooth functions. Let $\alpha \in (-m, 2-m)$. Taking into account (4.1.1) and integrating by parts, we get

$$(4.1.8) \quad \int_{B_\delta} |D'u|^2 r^{\alpha-2} dx + \int_{B_\delta} u \Delta u r^{\alpha-2} dx - \frac{(\alpha-2)(\alpha+m-4)}{2} \int_{B_\delta} |u|^2 r^{\alpha-4} dx = 0,$$

$$(4.1.9) \quad \int_{B_\delta} |\Delta u|^2 r^\alpha dx - \int_{B_\delta} |D'^2 u|^2 r^\alpha dx - \alpha(\alpha-2) \int_{B_\delta} \left|\frac{\partial u}{\partial r}\right|^2 r^{\alpha-2} dx +$$
$$+\alpha(\alpha+m-3) \int_{B_\delta} |D'u|^2 r^{\alpha-2} dx = 0.$$

Let us extend u by zero outside $B_\delta(x_0)$. Substituting the expression (2.3.1), get

$$(4.1.10) \qquad D = \int_{B_\delta} |\Delta u|^2 r^\alpha dx =$$
$$= \sum_{s=0}^{+\infty} \sum_{k=1}^{k_s} \int_0^\infty \Big\{ |u''_{s,k}|^2 + \big[(m-1)^2 - (m-1)(\alpha+m-2) + 2s(s+m-2)\big] \frac{|u'_{s,k}|^2}{r^2} +$$
$$+ s(s+m-2)\big[s(s+m-2) - (\alpha+m-4)(\alpha-2)\big] \frac{|u_{s,k}|^2}{r^4} \Big\} r^{\alpha+m-1} dr,$$

where $\alpha \in (-m, m)$ (the derivatives over r are denoted by a dash). We have $u_{s,k}(0) = u'_{s,k}(0) = 0$, when $s \geq 0$ *.
Using (2.1.8) we get the following inequalities for $-m < \alpha < 2 - m$

$$(4.1.11) \qquad \int_0^{+\infty} |u''_{s,k}|^2 r^{\alpha+m-1} dr \geq \frac{(\alpha+m-2)^2}{4} \int_0^{+\infty} |u'_{s,k}|^2 r^{\alpha+m-3} dr \geq$$
$$\geq \frac{(\alpha+m-2)^2(\alpha+m-4)^2}{16} \int_0^{+\infty} |u_{s,k}|^2 r^{\alpha+m-5} dr.$$

The right-hand side of (4.1.10) can be reduced by applying the inequalities (4.1.11) to give the estimate

$$(4.1.12) \quad \int_{B_\delta} |\Delta u|^2 r^\alpha dx \geq \min_{s=0,1,\dots} \left[\left(s + \frac{m-2}{2}\right)^2 - \left(\frac{\alpha-2}{2}\right)^2\right]^2 \int_{B_\delta} |u|^2 r^{\alpha-4} dx.$$

*It follows from the fact that $u_{s,k}(r)$ is a Fourier coefficient for $s \geq 1$. The equality $u_{0,1}(0) = 0$ follows from (4.1.1). Analogous considerations can be applied to the functions $u'_{s,k}(r)$.

It is obvious the coefficient before the integral on the right-hand side reaches its minimum value at $s = 1$ or $s = 2$.
When

$$\gamma \leq \frac{m+1}{m+\sqrt{(m+1)^2+1}},$$

the inequality

$$(m+2)^2 - (m+2\gamma)^2 \geq -m^2 + (m+2\gamma)^2$$

holds, and when

$$\gamma \geq \frac{m+1}{m+\sqrt{(m+1)^2+1}},$$

we have

$$(m+2)^2 - (m+2\gamma)^2 \leq -m^2 + (m+2\gamma)^2.$$

Hence we get the estimate

$$\int_{B_\delta} |u|^2 r^{\alpha-4} dx \leq M_1^2(\gamma) \int_{B_\delta} |\Delta u|^2 r^\alpha dx,$$

where $M_1(\gamma)$ is defined by the formula (4.1.5) and the inequality (4.1.2) is established.
Applying the Hölder inequality we get from (4.1.8)

$$\int_{B_\delta} |D'u|^2 r^{\alpha-2} dx \leq \left(\int_{B_\delta} |u|^2 r^{\alpha-4} dx\right)^{1/2} \left(\int_{B_\delta} |\Delta u|^2 r^\alpha dx\right)^{1/2} +$$
$$+(m+2\gamma)(1+\gamma)\int_{B_\delta} |u|^2 r^{\alpha-4} dx.$$

Using (4.1.2) we come to (4.1.3).
It follows from (4.1.9) that

(4.1.13)
$$\int_{B_\delta} |D'^2 u| r^\alpha dx - \int_{B_\delta} |\Delta u|^2 r^\alpha dx =$$
$$= \alpha \sum_{s=0}^{+\infty} \sum_{k=1}^{k_s} \int_0^{+\infty} \left[(m-1)|u'_{s,k}|^2 + (\alpha+m-3)s(s+m-2)\frac{|u_{s,k}|^2}{r^2}\right] r^{\alpha+m-3} dr.$$

It is obvious the terms of the right-hand series for $s < 2$ are negative. For $s = 0$ it is self-evident. To prove this statement for $s = 1$ it is sufficient to use inequalities (4.1.11). Hence

$$\int_{B_\delta} |D'^2 u| r^\alpha dx - \int_{B_\delta} |\Delta u|^2 r^\alpha dx \leq$$
$$\leq \alpha \sum_{s=2}^{+\infty} \sum_{k=1}^{k_s} \int_0^{+\infty} \left[(m-1)|u'_{s,k}|^2 + (\alpha+m-3)s(s+m-2)\frac{|u_{s,k}|^2}{r^2}\right] r^{\alpha+m-3} dr.$$

Applying inequalities (4.1.11), we find

$$\int_{B_\delta} |D^2 u| r^\alpha dx - \int_{B_\delta} |\Delta u|^2 r^\alpha dx \leq \alpha \sum_{s=2}^{+\infty} \sum_{k=1}^{k_s} \int_0^{+\infty} \left[\frac{(\alpha+m-4)^2}{4}(m-1) + \right.$$

$$\left. +(\alpha+m-3)s(s+m-2) \right] |u_{s,k}|^2 r^{\alpha+m-5} dr.$$

Multiplying and dividing each term by

$$\left[\left(s+\frac{m-2}{2}\right)^2 - \left(\frac{\alpha-2}{2}\right)^2 \right]^2$$

and using inequalities (4.1.11), we arrive at the estimate

$$\int_{B_\delta} |D^2 u| r^\alpha dx - \int_{B_\delta} |\Delta u|^2 r^\alpha dx \leq \tag{4.1.14}$$

$$\leq \alpha(\alpha+m-3) \sup_{s\geq 2} \frac{s(s+m-2) + \frac{(\alpha+m-4)^2}{4(\alpha+m-3)}(m-1)}{\left[\left(s+\frac{m-2}{2}\right)^2 - \left(\frac{\alpha-2}{2}\right)^2 \right]^2} \int_{B_\delta} |\Delta u|^2 r^\alpha dx.$$

It is clear that the sup attains at $s = 2$. In fact we can write

$$\frac{s(s+m-2) + \frac{(\alpha+m-4)^2}{4(\alpha+m-3)}(m-1)}{\left[\left(s+\frac{m-2}{2}\right)^2 - \left(\frac{\alpha-2}{2}\right)^2 \right]^2} =$$

$$= \frac{\left(s+\frac{m-2}{2}\right)^2 - \left[\frac{(\alpha+m-4)^2(m-1)}{-4(\alpha+m-3)}(m-1) + \left(\frac{m-2}{2}\right)^2 \right]}{\left[\left(s+\frac{m-2}{2}\right)^2 - \left(\frac{\alpha-2}{2}\right)^2 \right]^2}.$$

For $\alpha = 2 - m - 2\gamma (0 < \gamma < 1)$ we have

$$\frac{(\alpha+m-4)^2(m-1)}{-4(\alpha+m-3)}(m-1) + \left(\frac{m-2}{2}\right)^2 < \left(\frac{\alpha-2}{2}\right)^2.$$

Therefore the sup of the coefficient on the right-hand side of (4.1.14) attains at $s = 2$ and we come to (4.1.4) and (4.1.17). Inequality (4.1.4′) follows from (4.1.13). So, the lemma is completely proved.

□

Remark 4.1.1. *The constant* M_3 *can be also written for* $\alpha = 2 - m - 2\gamma$ $(0 < \gamma < 1)$ *in the form*

$$M_3'(\gamma) = M_3(\alpha) = \alpha(\alpha+m-3) \frac{\left(\frac{m+2}{2}\right)^2 + \frac{(\alpha+m-4)^2}{4(\alpha+m-3)} - \left(\frac{m-2}{2}\right)^2}{\left[\left(\frac{m+2}{2}\right)^2 - \left(\frac{\alpha-2}{2}\right)^2 \right]^2}. \tag{4.1.7′}$$

We have mentioned that along with the space $H_{l,p,\alpha}$ we shall consider the space $H_{l,p,-\alpha}$ consisting of all functions with the finite norm

$$\|u\|_{l,p,-\alpha} = \sup_{\substack{\delta \le \delta_0 \\ x_0 \in \overline{\Omega}}} \left(\int\limits_{\Omega_\delta(x_0)} |D^l u|^p |x - x_0|^{-\alpha} dx \right)^{1/p}. \tag{4.1.15}$$

It can be proved C^∞ is dense in $H_{l,p,-\alpha}$. Analogous to H_α, we denote $H_{1,2,-\alpha}$ by $H_{-\alpha}$, and the norm of $u \in H_{-\alpha}$ by $\|u\|_{-\alpha}$. The following lemma was proved by Chelkak (Chelkak and Koshelev [1], p.28) who used the same method as the one in lemma 4.1.1.

Lemma 4.1.2. *Let $u \in W^{(2)}_{2,-\alpha}$ in the ball $B_\delta(x_0)$, $\alpha \in (-m, 0)$ and let the following conditions*

$$u\Big|_{r=\delta} = \nabla u\Big|_{r=\delta} = 0, \quad r = |x - x_0|, \tag{4.1.16}$$

be satisfied.
Then the estimate

$$\int\limits_{B_\delta} |D'^2 u|^2 r^{-\alpha} dx \le \left[1 - \frac{4\alpha(m-1)}{(\alpha+m)^2}\right] \int\limits_{B_\delta} |\Delta u|^2 r^{-\alpha} dx \tag{4.1.17}$$

holds for $m \ge 4$. This inequality also holds for $\alpha = 2 - m - 2\gamma$, $0 < \gamma < 1$, and $m \ge 2$.

Proof. Initially let $m \ge 4$. Then $-\alpha + m - 5 > -1$. Using equalities (4.1.16), we can represent the functions $u_{s,k}$ along with their derivatives in the following way

$$u_{s,k}(r) = -\int\limits_r^\delta u'_{s,k}(\rho) d\rho,$$

$$u'_{s,k}(r) = -\int\limits_r^\delta u''_{s,k}(\rho) d\rho.$$

Now we can apply relations (4.1.11). Performing certain calculations and applying (4.1.10) and (4.1.13), we get

$$\int\limits_{B_\delta} |\Delta u|^2 r^{-\alpha} dx \ge \sum_{s=0}^{+\infty} \sum_{k=1}^{k_s} \int\limits_0^\delta \left\{ \left[\frac{(\alpha+m)^2}{4} + 2s(s+m-2) \right] |u'_{s,k}|^2 + \right.$$

$$\left. + s(s+m-2)\left[s(s+m-2) + (-\alpha+m-4)(\alpha+2) \right] \frac{|u_{s,k}|^2}{r^2} \right\} r^{-\alpha+m-3} dr \ge$$

$$\ge \sum_{s=0}^{+\infty} \sum_{k=1}^{k_s} \int\limits_0^\delta \left\{ \frac{(\alpha+m)^2}{4} |u'_{s,k}|^2 + s(s+m-2) \left[s(s+m-2) + \right.\right.$$

$$\left.\left. + \frac{(\alpha+m)(-\alpha+m-4)}{2} \right] \frac{|u_{s,k}|^2}{r^2} \right\} r^{-\alpha+m-3} dr \ge$$

$$\ge \min \left\{ \frac{(\alpha+m)^2}{-4\alpha(m-1)}, \min_{s \ge 1} \frac{s(s+m-2) + \frac{1}{2}(\alpha+m)(-\alpha+m-4)}{-\alpha(-\alpha+m-3)} \right\} \times$$

$$\times \int\limits_{B_\delta} (|D'^2 u|^2 - |\Delta u|^2) r^{-\alpha} dx.$$

From this follows the inequality

$$\int_{B_\delta} |\Delta u|^2 r^{-\alpha} dx \geq \frac{(\alpha+m)^2}{-4\alpha(m-1)} \int_{B_\delta} (|D'^2 u|^2 - |\Delta u|^2) r^{-\alpha} dx, \tag{4.1.18}$$

which leads to the relation (4.1.17) for $m \geq 4$. The same proof can be used for the case $\alpha = 2 - m - 2\gamma$ and $m = 3$. Inequality (4.1.18) also holds for $m = 2$ and $\alpha = 2 - m - 2\gamma$, but the proof is slightly different. In this case the representations

$$u_{s,k}(r) = \int_0^r u'_{s,k}(\rho) d\rho, \quad u'_{s,i} = -\int_r^\delta u''_{s,k}(\rho) d\rho \quad (s \geq 1),$$

should be used (remembering that $u_{s,i}(0) = 0$ for $s \geq 1$).

□

Corollary 4.1.1 *If the conditions of the last two lemmas (excepting the conditions at $r = 0$ and $r = \delta$) are satisfied, then inequalities*

$$\int_{B_\delta} |D'^2 u|^2 r^\alpha \zeta^2 dx \leq \left(M_3^2(\gamma) + \eta \right) \int_{B_\delta} |\Delta u|^2 r^\alpha \zeta^2 dx + C \|u\|^2_{W_2^{(2)}(B_\delta)}, \tag{4.1.19}$$

$$\int_{B_\delta} |D'^2 u|^2 r^{-\alpha} \zeta^2 dx \leq \left[1 - \frac{4\alpha(m-1)}{(m+\alpha)^2} \right] \int_{B_\delta} |\Delta u|^2 r^{-\alpha} \zeta^2 dx + \tag{4.1.20}$$

$$+C \int_{B_\delta} |D^2 u|^2 r^\beta dx$$

hold, where $\alpha = 2 - m - 2\gamma$ $(0 < \gamma < 1)$ and β is arbitrary. The inequality (4.1.20) *also holds for $\alpha \in (-m, 0)$ and $m \geq 4$.*

Proof. First we shall prove inequality (4.1.19). Let z be a linear function which satisfies the conditions

$$(u - z)\Big|_{x_0} = \nabla(u - z)\Big|_{x_0} = 0.$$

Consider the function $v = (u - z)\zeta$, where ζ is a smooth cut-off function, which is defined by (2.2.1). Then v satisfies all the conditions of lemma 4.1.1 and

$$\int_{B_\delta} |D'^2 v|^2 r^\alpha dx \leq M_3^2(\gamma) \int_{B_\delta} |\Delta v|^2 r^\alpha dx.$$

It is obvious the second derivatives of v can be expressed by u and its derivatives as follows

$$v_{ik} = u_{ik}\zeta + (u_i - u_i\Big|_{x_0})\zeta_k + (u_k - u_k\Big|_{x_0})\zeta_i + (u - z)\zeta_{ik}.$$

Note that the derivatives of ζ vanish in $B_{\delta/2}(x_0)$. Inserting the derivatives of v and its derivatives into the last inequality and taking into account that $r \geq \delta/2$ on $B_{\delta/2,\delta} \equiv B_\delta \setminus B_{\delta/2}$, we get

$$\int_{B_{\delta/2}} |D'^2 u|^2 r^\alpha \zeta^2 dx \leq M_3^2(\gamma) \int_{B_{\delta/2}} |\Delta u|^2 r^\alpha \zeta^2 dx +$$

$$+ \int_{B_{\delta/2,\delta}} |D^2 u|^2 \chi_1(x) dx + C_1 |\nabla u(x_0)|^2 + C_2 |u(x_0)|^2,$$

where the function χ_1 is smooth on $B_{\delta/2,\delta}$ and the second term on the right-hand side can be estimated by $\|u\|^2_{W_2^{(2)}(B_\delta)}$. The last term on the right-hand side can be estimated by applying (2.1.15) and (2.1.17′). Hence we come to (4.1.19). Inequality (4.1.20) can be proved similarly by means of the function $v = u\zeta$.

□

Lemma 4.1.3. *Let one of the following conditions be satisfied:*

1. $\alpha \in (-m, 2-m), u \in \overset{\circ}{W}{}^{(l)}_{2,\alpha}(B_\delta(x_0), x_0)$ *and vanishes at* x_0 *along with all its derivatives up to the order* $l-1$;

2. $m \geq 3, \alpha = 3 - m, u \in \overset{\circ}{W}{}^{(l)}_{2,\alpha}(B_\delta(x_0), x_0)$ *and vanishes at* x_0 *along with all its derivatives up to the order* $l-2$;

3. $m \geq 4, \alpha \in (3-m, 0), u \in \overset{\circ}{W}{}^{(l)}_{2,\alpha}(B_\delta(x_0), x_0)$;

4. $\alpha \in (-m, 2-m), u \in \overset{\circ}{W}{}^{(l)}_{2,-\alpha}(B_\delta(x_0), x_0)$;

5. $m \geq 4, \alpha \in (-m, 0), u \in \overset{\circ}{W}{}^{(l)}_{2,-\alpha}(B_\delta(x_0), x_0)$.

Then for $l = 2l_1$ *the inequality*

$$\int\limits_{B_\delta(x_0)} |D^l u|^2 r^{\pm\alpha} dx \leq C^{l_1}_{\pm\alpha} \int\limits_{B_\delta(x_0)} |\Delta^{l_1} u|^2 r^{\pm\alpha} dx \tag{4.1.21}$$

holds. For $l = 2l_1 + 1$ *we have*

$$\int\limits_{B_\delta(x_0)} |D^l u|^2 r^{\pm\alpha} dx \leq C^{l_1}_{\pm\alpha} \int\limits_{B_\delta(x_0)} |D'(\Delta^{l_1} u)|^2 r^{\pm\alpha} dx, \tag{4.1.22}$$

where the constants $C_{\pm\alpha}$ *are defined by the following equalities*

$$C^2_\alpha = \begin{cases} 1, & \alpha \in (3-m, 0), \quad m \geq 4, \\ M_3(\gamma), & \alpha = 2 - m - 2\gamma, \quad \gamma \in (0,1) \end{cases} \tag{4.1.23}$$

and

$$C^2_{-\alpha} = 1 - \frac{4\alpha(m-1)}{(\alpha+m)^2}. \tag{4.1.24}$$

Proof. Let $l = 2l_1$ and $B_\delta = B_\delta(x_0)$. Applying lemmas 4.1.1 or 4.1.2, we get the following relations

$$\int\limits_{B_\delta} |\Delta^{l_1} u|^2 r^{\pm\alpha} dx = \int\limits_{B_\delta} |\Delta(\Delta^{l_1-1} u)|^2 r^{\pm\alpha} dx \geq \frac{1}{C^2_{\pm\alpha}} \int\limits_{B_\delta} |D'^2(\Delta^{l_1-1} u)|^2 r^{\pm\alpha} dx =$$

$$= \frac{1}{C^2_{\pm\alpha}} \sum_{i_1,i_2=1}^m \int\limits_{B_\delta} \left|\Delta^{l_1-1} \frac{\partial^2 u}{\partial x_{i_1} \partial x_{i_2}}\right|^2 r^{\pm\alpha} dx \geq \ldots \geq$$

$$\geq \frac{1}{C^{l_1}_{\pm\alpha}} \sum_{i_1,\ldots,i_l=1}^m \int\limits_{B_\delta} \left|\frac{\partial^l u}{\partial x_{i_1} \cdot \ldots \cdot \partial x_{i_l}}\right|^2 r^{\pm\alpha} dx = \frac{1}{C^{l_1}_{\pm\alpha}} \int\limits_{B_\delta} |D'^l u|^2 r^{\pm\alpha} dx.$$

Thus we get inequality (4.1.21) for smooth functions. After the closure procedure we get (4.1.21) for $\forall u \in \overset{\circ}{W}{}^{(l)}_{2,\pm\alpha}$. Similarly, inequality (4.1.22) can be proved.

□

Lemma 4.1.4. *If $\beta \in [0,m)$, $u \in W^{(2)}_{2,\beta}(B_\delta(x_0),x_0)$ and vanishes on ∂B_δ then the following inequality*

$$(4.1.25)\qquad \int\limits_{B_\delta(x_0)} \Delta u u' r^{\beta-1} dx \geq \frac{1}{2}\sum_{s=0}^{+\infty}\sum_{k=1}^{k_s}\int\limits_0^\delta \Big[(m-\beta)|u'_{s,k}|^2 + (\beta+m-4)\times$$
$$\times s(s+m-2)|u_{s,k}|^2 r^{-2}\Big] r^{\beta+m-3} dr.$$

is true. If the boundary condition for u is not satisfied then the relation

$$(4.1.26)\qquad \int\limits_{B_\delta(x_0)} \Delta u u' r^{\beta-1}\zeta dx \geq \frac{1}{2}\sum_{s=0}^{+\infty}\sum_{k=1}^{k_s}\int\limits_0^\delta \Big[(m-\beta)|u'_{s,k}|^2 + (\beta+m-4)\times$$
$$\times s(s+m-2)|u_{s,k}|^2 r^{-2}\Big] r^{\beta+m-3} dr - C(\delta)\int\limits_{B_\delta(x_0)} |D^2 u|^2 r^{\beta'} dx$$

is also true. Here $u_{s,k}(r)$ are Fourier coefficents in the expansion (2.3.1) and β' is arbitrary.

Proof. Let

$$(4.1.27)\qquad S(u) = \int\limits_{B_\delta(x_0)} \Delta u u' r^{\beta-1} dx$$

and

$$(4.1.28)\qquad S_\zeta(u) = \int\limits_{B_\delta(x_0)} \Delta u u' r^{\beta-1}\zeta dx.$$

Substituting expansion (2.3.1) in S we get

$$S = \sum_{s,k}\int\limits_0^\delta u'_{s,k}\left[u''_{s,k} + \frac{m-1}{r}u'_{s,k} - \frac{s(s+m-3)}{r^2}u_{s,k}\right] r^{\beta+m-2} dr$$

and

$$(4.1.29)\qquad S = \frac{1}{2}\sum_{s,k}|u'_{s,k}(\delta)|^2\delta^{\beta+m-2} + \frac{1}{2}\sum_{s,k}\int\limits_0^\delta \Big[(m-\beta)|u'_{s,k}|^2 +$$
$$+(\beta+m-4)s(s+m-2)|u_{s,k}|^2 r^{-2}\Big] r^{\beta+m-3} dr,$$

where $\sum\limits_{s,k}$ denotes $\sum\limits_{s=0}^{+\infty}\sum\limits_{k=r}^{k_s}$.

Therefore, the inequality (4.1.25) is proved. It is clear that

$$S_\zeta(u) = \int\limits_{B_\delta(x_0)} \Delta(u\zeta)(u\zeta)' r^{\beta-1} dx -$$
$$\int\limits_{B_\delta(x_0)} \Big[(2\nabla u\zeta' + u\Delta\zeta)u\zeta' + \Delta u u\zeta\zeta'\Big] r^{\beta-1} dx + \int\limits_{B_\delta(x_0)} \Delta u \cdot u u' r^{\beta-1}(\zeta - \zeta^2) dx$$

Since $\zeta - \zeta^2 = 0$ and $\zeta = 0$ for $|r| < \delta/2$, the last two integral can be estimated by the integral

(4.1.30)
$$\int_{B_\delta(x_0)} |D^2 u|^2 r^{\beta'} dx.$$

Then we have that

$$S_\zeta(u) \geq S(\zeta u) - C \int_{B_\delta(x_0)} |D^2 u|^2 r^{\beta'} dx.$$

Applying (4.1.25) and repeating once again the previous calculations we come to (4.1.26).

□

Lemma 4.1.5. *For $\beta \in [0, m)$ and $u \in W^{(2)}_{2,\beta}(B_\delta(x_0), x_0))$ the inequality*

(4.1.31)
$$\int_{B_\delta(x_0)} \Delta u(\Delta u + \beta r^{-1} u') r^\beta \zeta dx \geq \frac{m-\beta}{m+\beta} \int_{B_\delta(x_0)} |\Delta u + \beta u' r^{-1}|^2 r^\beta \zeta dx -$$
$$-C \int_{B_\delta(x_0)} |D^2 u|^2 r^{\beta'} dx,$$

is true where β' is, as usual, arbitrary.

Proof. Repeating the considerations of the previous lemma we can omit ζ under the signs of (4.1.32), (4.1.33) and assume that $u_{s,k}(\delta) = u'_{s,k}(\delta) = 0$. Using the expansion (2.3.1) and expression (4.1.10) we will get the following formula

(4.1.32)
$$J \equiv \int_{B_\delta(x_0)} |\Delta u + \beta u' r^{-1}|^2 r^\beta dx =$$
$$= \sum_{s,k} \int_0^\delta \Big\{ |u''_{s,k}|^2 + \Big[m - 1 + \beta + 2s(s+m-2)\Big] |u'_{s,k}|^2 r^{-2} +$$
$$+ s(s+m-2)\Big[s(s+m-2) + 2(\beta + m - 4)\Big] |u_{s,k}|^2 r^{-4} \Big\} r^{\beta+m-1} \zeta dr.$$

In the same way we have

(4.1.33)
$$I \equiv \int_{B_\delta(x_0)} \Delta u(\Delta u + \beta r^{-1} u') r^\beta dx =$$
$$= \sum_{s,k} \int_0^\delta \Big\{ |u''_{s,k}|^2 + \Big[m - 1 + \frac{\beta(2-m-\beta)}{2} + 2s(s+m-2)\Big] |u'_{s,k}|^2 r^{-2} +$$
$$+ s(s+m-2)\Big[s(s+m-2) + (\beta+m-4)\frac{4-\beta}{2}\Big] |u_{s,k}|^2 r^{-4} \Big\} r^{\beta+m-1}.$$

Construct an expression $I - \kappa J$ with $1 \geq \kappa > 0$.

Using the expressions (4.1.32) and (4.1.33) we get

$$I - \kappa J = \sum_{s,k} \int_0^\delta \Big\{ |u''_{s,k}|^2(1-\kappa) + \Big[m - 1 - \frac{\beta(\beta+m-2)}{2} + \tag{4.1.34}$$
$$+2s(s+m-2) - \kappa\big(m-1+\beta+2s(s+m-2)\big)\Big] |u'_{s,k}|^2 r^{-2} +$$
$$+s(s+m-2)\Big[(1-\kappa)s(s+m-2)+$$
$$+(\beta+m-4)\Big(\frac{4-\beta}{2} - 2\kappa\Big)\Big] |u_{s,k}|^2 r^{-4} \Big\} r^{\beta+m-1} dr.$$

Taking into account that $u'_{s,k} = 0$ for $r = \delta$ we can write

$$u'_{s,k}(r) = -\int_r^\delta u''_{s,k}(\rho)d\rho.$$

In accordance with the inequality (2.1.8) for $m \geq 2$ we have $s \leq -\beta - m + 3 \leq$ $\leq -\beta + 1 < 1$ and therefore

$$\int_0^\delta |u''_{s,k}|^2 r^{\beta+m-1} dr \geq \frac{(\beta+m-2)^2}{4} \int_0^\delta |u'_{s,k}|^2 r^{\beta+m-3} dr. \tag{4.1.35}$$

Using the inequality from (4.1.34) we get

$$I - \kappa J \geq \sum_{s,k} \int_0^\delta \Big\{ \Big(\frac{m^2 - \beta^2}{4} + 2s(s+m-2) - \tag{4.1.36}$$
$$-\kappa\Big[\frac{(m+\beta)^2}{4} + 2s(s+m-2)\Big]\Big) |u'_{s,k}|^2 +$$
$$+s(s+m-2)\Big[(1-\kappa)s(s+m-2)+$$
$$+(\beta+m-4)\Big(\frac{4-\beta}{2} - 2\kappa\Big)\Big] |u_{s,k}|^2 r^{-2} \Big\} r^{\beta+m-3} dr.$$

The coefficient before $|u'_{s,k}|^2$ under the sign of the integral will be nonnegative if κ satisfies the inequality

$$\kappa \leq \frac{m-\beta}{m+\beta}.$$

When $s > 0$ we have $u_{s,k} = 0$ for $r = 0$. Then

$$u_{s,k}(r) = \int_0^r u'_{s,k}(\rho)d\rho$$

and according to (2.1.8), we get the inequality

$$\int_0^\delta |u'_{s,k}|^2 r^{\beta+m-3} dr \geq \frac{(\beta+m-4)^2}{4} \int_0^\delta |u_{s,k}|^2 r^{\beta+m-5} dr. \tag{4.1.37}$$

Applying this inequality to the right-hand side of (4.1.36) we get the relation

$$I - \kappa J \geq \sum_{s,k} \left\{ (1-\kappa)s^2(s+m-2)^2 + \right. \tag{4.1.38}$$

$$+s(s+m-2)(\beta+m-4)\left(\frac{m}{2} - \kappa\frac{\beta+m}{2}\right) +$$

$$\left. +\frac{(\beta+m-4)^2}{4}\left[\frac{m^2-\beta^2}{4} - \kappa\frac{(m+\beta)^2}{4}\right]\right\} \int_0^\delta |u_{s,k}|^2 r^{\beta+m-5} dr$$

with $s \geq 1$. It can easily be shown that for $\kappa = (m-\beta)(m+\beta)^{-1}$ and for $m > 3$ this expression is nonnegative. Consider the cases $m = 2$ and $m = 3$ for small β. Denote

$$y = \frac{2s(s+m-2)}{\beta+m-4}; \quad s = 1, 2, \ldots \tag{4.1.39}$$

It is obvious that $y \leq 0$ only for $m = 2, m = 3$ and small $\beta > 0$. The quadratic expression before the sign of the integral in (4.1.38) can be written in the form

$$\frac{(\beta+m-4)^2}{4}\left\{(1-\kappa)y^2 + \left[m - \kappa(\beta+m)\right]y + \frac{m^2-\beta^2}{4} - \kappa\frac{(m+\beta)^2}{4}\right\}.$$

The roots of this expression will be as follows

$$y_1 = \frac{-m+\kappa(\beta+m)-\beta}{2(1-\kappa)} = -\frac{m+\beta}{2},$$

$$y_2 = \frac{-m+\kappa(\beta+m)+\beta}{2(1-\kappa)}.$$

Clearly $y_1 < 0$ and $y_1 \leq y_2$. Hence the considered quadratic expression will be nonnegative if $y \leq y_1$. From (4.1.39) follows that

$$y\Big|_{s=1} = \frac{2(m-1)}{\beta+m-4}.$$

It is easy to see that

$$\frac{2(m-1)}{\beta+m-4} \leq -\frac{m+\beta}{2}.$$

In fact this inequality is equivalent to the inequality

$$4(m-1) \geq -(m+\beta)(\beta+m-4),$$

which is true for $m \geq 2$. Therefore

$$y\Big|_{s=1} \leq y_1.$$

As far as $m = 2$ and $m = 3$ with small β is concerned the expression y decreases with $s \to +\infty$ we have that

$$y \leq y\Big|_{s=1} \leq y_1.$$

Therefore $I - \kappa J \geq 0$ and we get (4.1.31) for all $m \geq 2$.

□

Lemma 4.1.6. *If $m = 2$, β is sufficiently small and positive and $u \in W^{(2)}_{2,\beta}(B_\delta(x_0), x_0)$, then*

(4.1.40)
$$\int\limits_{B_\delta(x_0)} \left|\Delta u + \beta u' r^{-1}\right|^2 r^\beta \zeta\, dx \geq (1-\eta) \int\limits_{B_\delta(x_0)} |\Delta u|^2 r^\beta \zeta\, dx -$$
$$-\left[\frac{\beta^2(2-\beta)}{2} + \eta\right] \sum_{k=1}^{2} \int_0^\delta |u_{1,k}|^2 r^{\beta-3} dr - C \int\limits_{B_\delta(x_0)} |D^2 u|^2 r^{\beta'} dx,$$

where $\eta > 0$ and β' are arbitrary.

Proof. Consider the expresion $J - \kappa D$ – where J and D are determined by (4.1.32) and (4.1.10) (under the sign (4.1.10) could be also ζ). Then we get for $0 < \kappa \leq 1$ the following equality

(4.1.41)
$$J - \kappa D = \sum_{s,k} \int_0^\delta \Big\{ |u''_{s,k}|^2 (1-\kappa) + \Big[m - 1 + \beta + 2s(s+m-2) -$$
$$-\kappa\Big((m-1)(1-\beta) + 2s(s+m-2)\Big)\Big] |u'_{s,k}|^2 r^{-2} +$$
$$+ s(s+m-2)\Big[s(s+m-2)(1-\kappa) +$$
$$+ (\beta+m-4)\big[2 + \kappa(\beta-2)\big]\Big] |u_{s,k}|^2 r^{-4} \Big\} r^{\beta+m-1} dr$$

(we suppose that ζ like in lemmas 4.1.4 and 4.1.5 is combined with u).
Consider the expression

(4.1.42)
$$H_\mu = J - \kappa D + \mu \sum_{k=1}^{2} \int_0^\delta |u_{1,k}|^2 r^{\beta+m-5} dr,$$

where $\mu > 0$ is a constant we will choose later.
Write (4.1.42) in the form

(4.1.43)
$$J - \kappa D = H_\mu - \mu \sum_{k=1}^{2} \int_0^\delta |u_{1,k}|^2 r^{\beta+m-5} dr.$$

Applying the estimate (4.1.35) we'll come to the inequality

$$H_\mu \geq \sum_{s,k} \int_0^\delta \Big\{ \Big[\frac{(\beta+m)^2}{4} + 2s(s+m-2) -$$
$$-\kappa\left(\frac{(\beta-m)^2}{4} + 2s(s+m-2)\right)\Big] |u'_{s,k}|^2 +$$
$$+ s(s+m-2)\big[(1-\kappa)s(s+m-2) +$$
$$+ (\beta+m-4)(2+\kappa(\beta-2))\big] |u_{s,k}|^2 r^{-2} \Big\} r^{\beta+m-3} dr +$$
$$+ \mu \sum_{k=1}^{2} \int_0^\delta |u_{1,k}|^2 r^{\beta+m-5} dr.$$

When $\kappa \le (m+\beta)^2(m-\beta)^{-2}$ then the coefficient before $|u'_{s,k}|^2$ under the sign of the intergral on the right-hand side will be nonnegative. Then applying inequality (2.1.8) we shall have

$$\int_0^\delta |u'_{1,k}|^2 r^{\beta+m-3} dr \ge \frac{(\beta+m-4)^2}{4} \int_0^\delta |u_{1,k}|^2 r^{\beta+m-5} dr.$$

Therefore

$$(4.1.44)\qquad \begin{aligned} H_\mu \ge \sum_{s,k} \Big\{ & s^2(s+m-2)^2(1-\kappa) + \\ & + \frac{\beta+m-4}{2} s(s+m-2)\Big[m(1-\kappa)+\beta(1+\kappa)\Big] + \\ & + \frac{(\beta+m-4)^2}{4}\left[\frac{(\beta+m)^2}{4} - \kappa\frac{(\beta-m)^2}{4}\right]\Big\} \int_0^\delta |u_{s,k}|^2 r^{\beta+m-5} dr + \\ & + \mu \sum_{k=1}^{2} \int_0^\delta |u_{1,k}|^2 r^{\beta+m-5} dr. \end{aligned}$$

Consider the quadratic expression under the sign of the first sum. Let

$$y = \frac{2s(s+m-2)}{\beta+m-4}.$$

As in the previous case $y \le 0$. The above mentioned expression will be equal to

$$(4.1.45)\qquad \begin{aligned} \frac{(\beta+m-4)^2}{4} \Big\{ & y^2(1-\kappa) + \big[m(1-\kappa)+\beta(1+\kappa)\big] y + \\ & + \frac{(\beta+m)^2}{4} - \kappa\frac{(\beta-m)^2}{4} \Big\}. \end{aligned}$$

This expression has two roots

$$\begin{aligned} y_1 &= \frac{-\Big[m+\beta-\kappa(m-\beta)\Big] - 2\beta\sqrt{\kappa}}{2(1-\kappa)}, \\ y_2 &= \frac{-\Big[m+\beta-\kappa(m-\beta)\Big] + 2\beta\sqrt{\kappa}}{2(1-\kappa)}. \end{aligned}$$

It is easy to see that $y_1 \le y_2$ and

$$y_1 = -\frac{m}{2} - \beta\frac{1+\sqrt{\kappa}}{2(1-\sqrt{\kappa})}.$$

Thus for $s \ge 2$ and $m = 2$ we have

$$y = \frac{2s^2}{\beta-2} \le \frac{8}{\beta-2} < -1 - \beta\frac{1+\sqrt{\kappa}}{2(1-\sqrt{\kappa})} = y_1$$

for $\kappa = 1 - \eta$ and all β, satisfying the inequality

(4.1.46) $$0 \leq \beta < \beta_0(\eta),$$

where $\beta_0(\eta)$ is sufficiently small.
Therefore for $s \geq 2$ and such β, the expression (4.1.45) is positive.
Let us now take from the right-hand side of (4.1.44) the terms with $\int\limits_0^\delta |u_{1,k}|^2 r^{\beta+m-5}$ for $\kappa = 1$ $(m = 2)$. We'll have

(4.1.47) $$\sum_{k=1}^{2} \int\limits_0^\delta \left[\beta(\beta-2) + \frac{(\beta-2)^2\beta}{2} + \mu\right] |u_{1,k}|^2 r^{\beta+m-5} dr.$$

If

$$\mu = \frac{\beta^2(2-\beta)}{2} + \eta,$$

then the expression (4.1.47) will be nonnegatitve.
Combining this fact with the positiveness of (4.1.45) for $s \geq 2$ under the condition (4.1.46) we come to the inequality $H_\mu \geq 0$ for $\kappa = 1 - \eta$.

□

4.2 Coercivity inequalities with explicit constants in singular weighted spaces.

The problem of regularity for weak solutions of elliptic and parabolic systems is closely related to the so-called coercivity inequalities. We shall explain it now using the example of simplest equations and domains. Consider in R^m $(m \geq 2)$ a ball $B_\delta(x_0)$ with the center x_0 and radius δ. The ball $B_1(0)$ will be denoted B. In this ball an equation

(4.2.1) $$\Delta u = f(x)$$

with a boundary condition

(4.2.2) $$u\Big|_{\partial B} = 0$$

is given. It is well known when the right-hand side of (4.2.1), namely $f(x)$, belongs to $C(B)$, the second derivatives of solution (4.2.1)-(4.2.2) should not be continuous. So, a pair of spaces $C^{(2)}(B)$, $C(B)$ is not suitable for solving problem (4.2.1)-(4.2.2). In principle, it is important to find a space X of functions given on B, such that $f \in X$ and the second derivatives of the weak solutions also belong to X. Of course the derivatives can be regarded not only in their classical sense, but also in the generalized, for example, Sobolev sense. We shall attempt to obtain for the solution of (4.2.1)-(4.2.2) an inequality

$$\left\|D^2 u\right\|_X \leq B_X \|f\|_X,$$

or a slightly more general inequality

$$\left\|D^2 u\right\|_X \leq C_X \|f\|_X + C\|Du\|_Y,$$

where Y can be a weaker space than X.
Such inequalities will be named the inequalities of the coercivity type. It appears that the first inequality of this type was obtained by Bernstein for $X = L_2$. At that time it was obtained for the classical solutions of (4.2.1)-(4.2.2). Moreover, it was shown that $C_{L_2} = 1$. Bernstein got this result for $m = 2$. However later (Ladyzhenskaya [1], Mihlin [1], Cacciopoli[1]) showed that this inequality, containing the same constant, holds true for all $m > 2$. The fact that $C_{L_2} = 1$ is essential for the regularity of weak solutions of elliptic and parabolic systems (it is obvious that the constant can not be smaller than one).
For $X = C^{0,\gamma}(\Omega)$ $(0 < \gamma < 1)$ the coercivity inequalities were first obtained by Korn [1], but the constants $B_{C^{0,\gamma}}$ and $C_{C^{0,\gamma}}$ had an implicit form and until now were neither established, nor estimated by an explicit constant. The same situation can be found in the works of Schauder [1] and Gunther [1].
In 1952 the author [1] obtained the coercivity inequalities for $X = L_p$ $(p > 1)$, but the constants remained implicit and until now could not be estimated (except for $p = 2$). The same situation applied to the linear elliptic systems (Koshelev [2]) (see also Agmon, Duglis, Nirenberg [1]). For $X = L_{2,\alpha}(B)$ $(|\alpha| < m)$, with the help of methods proposed by Kondratjev [1], Stein [1], the coercivity inequality could be obtained without explicitly estimating the constants $B_{L_{2,\alpha}}$ on $C_{L_{2,\alpha}}$. Also it should be remembered that the problem is closely related to the estimates in X of the singular integral operators, but even for $X = L_p$ after applying the Zygmund-Calderon theorem it is impossible to find the explicit bounds.
Nevertheless for $L_{2,\alpha}$ and some essential values of α we have managed to estimate these constants, which are demonstrated in section 2.4 of this book. Now let us show that in some cases the coercivity estimates can be found with explicit constants.
Suppose that $u \in L_{2,\alpha}(B)$, where $L_{2,\alpha}$ is the space of squared integrable functions with a weight $|x|^\alpha$. We assume that $\alpha \in (-m, 0]$, and $|x|$ denotes the distance from the origin. The norm in $L_{2,\alpha}(B)$ as usual is determined by

$$\left(\int_B |u|^2 |x|^\alpha dx\right)^{\frac{1}{2}}.$$

By $W^{(2)}_{2,\alpha}(B)$ we shall denote those functions in the Sobolev space $W^{(2)}_2(B)$ whose second derivatives are square summable with the weight $|x|^\alpha$. As a norm in this space we could take for example the expression

$$\left(\int_B \left|D'^2 u\right|^2 |x|^\alpha dx + \int |u|^2 dx\right)^{\frac{1}{2}}. \tag{4.2.3}$$

One of the aims of this paragraph is to prove for the solution of the problem (4.2.1), (4.2.2) the inequality

$$\int_B \left|D'^2 u\right|^2 r^\alpha dx \le C_\alpha^2 \int_B |f|^2 r^\alpha dx,$$

where C_α has an explicit form. For $\alpha' = m - 2 + 2\gamma$ such an inequality was proved by the author in [11]. First we shall prove some lemmas.

Lemma 4.2.1. *If $u \in W^{(2)}_{2,\alpha}(B)$, then the inequalities*

$$|u(0)|^2 < \eta \int\limits_B |D'u|^2 r^\alpha dx + C_0(\eta) \int\limits_B |u|^2 dx \tag{4.2.4}$$

and

$$\sum_{i=1}^m |u_i(0)|^2 < \eta \int\limits_B |D^n u|^2 r^\alpha dx + C_0(\eta) \int\limits_B |D'u|^2 dx \tag{4.2.5}$$

hold. Here η is as usual an arbitrary positive constant and

$$C_0(\eta) = 2m|S|^{-(m/2\gamma+1)}\gamma^{-\frac{m}{2\gamma}}\eta^{-\frac{m}{2\gamma}}, \tag{4.2.6}$$

where $|S|$ is the surface of the unit sphere in R^m.

Proof. Obviously

$$u(0) = u(x) - \int\limits_0^r \frac{\partial u}{\partial \rho} d\rho.$$

Square both sides of this equality and integrate over the ball $B_\delta(0) = B_\delta$. We get

$$|u(0)|^2 |S| m^{-1} \delta^m \leq 2 \int\limits_{B_\delta} \Big| \int\limits_0^r \frac{\partial u}{\partial \rho} d\rho \Big|^2 dx + 2 \int\limits_B |u|^2 dx.$$

The first term on the right-hand side can be written in the equivalent form to give

$$|u(0)|^2 |S| m^{-1} \delta^m \leq 2 \int\limits_{\partial B_\delta} dS \int\limits_0^\delta \Big| \int\limits_0^r \frac{\partial u}{\partial \rho} \rho^{\frac{\alpha+m-1}{2}} \cdot \rho^{-\frac{\alpha+m-1}{2}} d\rho \Big|^2 r^{m-1} dr + 2 \int\limits_B |u|^2 dx.$$

Applying the Hölder inequality to the inner integral, we obtain the estimate

$$|u(0)|^2 |S| m^{-1} \delta^m \leq \frac{1}{\gamma} \int\limits_{\partial B_\delta} dS \int\limits_0^\delta \int\limits_0^r |\nabla u|^2 \rho^{\alpha+m-1} d\rho r^{m-1+2\gamma} dr + 2 \int\limits_B |u|^2 dx.$$

Applying δ instead of the upper bound of the inner integral we get the following inequality

$$|u(0)|^2 |S| m^{-1} \delta^m \leq \frac{\delta^{m+2\gamma}}{\gamma(m+2\gamma)} \int\limits_{B_\delta} |\nabla u|^2 r^{\alpha+m-1} dr + 2 \int\limits_B |u|^2 dx.$$

Taking into account $m(m+2\gamma)^{-1} < 1$, gives the inequality

$$|u(0)|^2 \leq \frac{\delta^{2\gamma}}{\gamma|S|} \int\limits_{B_\delta} |\nabla u|^2 r^\alpha dx + \frac{2m}{|S|\delta^m} \int\limits_B |u|^2 dx.$$

Using the notation (4.2.6) gives the inequalities (4.2.4) and (4.2.5).

□

Corollary 4.2.1 *Let λ be the smallest absolute value of the eigenvalues for the operator Δ with condition* (4.2.2). *If u satisfies* (4.2.2) *then the inequalities*

(4.2.7)
$$|u(0)|^2 \leq \eta \int_B |D'u|^2 r^\alpha dx + \frac{C_0(\eta)}{\lambda^2} \int_B |\Delta u|^2 dx$$

and

(4.2.8)
$$\sum_i^m |u_i(0)|^2 \leq \eta \int_B |D'^2 u|^2 r^\alpha dx + \frac{C_0(\eta)}{\lambda} \int_B |\Delta u|^2 dx$$

hold.

Proof. In fact both of the second terms on the right-hand side of (4.2.4) and (4.2.5) can be estimated easily by the integral of $|\Delta u|^2$.
Using the condition (4.2.2) and integrating by parts we have

$$\int_B |D'u|^2 dx = -\int_B u\Delta u dx \leq \left(\int_B |u|^2 dx\right)^{\frac{1}{2}} \left(\int_B |\Delta u|^2 dx\right)^{\frac{1}{2}}.$$

Then

$$\int_B |u|^2 dx \leq \frac{1}{\lambda^2} \int_B |\Delta u|^2 dx$$

and from the previous inequality we have

$$\int_B |D'u|^2 dx \leq \frac{1}{\lambda} \int_B |\Delta u|^2 dx.$$

The corollary is proved. □

Lemma 4.2.2. *For $u \in W^{(2)}_{2,\alpha}(B)$, satisfying* (4.2.2), *the equality*

(4.2.9)
$$\int_B u_{ik}u_{ik} r^\alpha dx = \int_B |\Delta u|^2 r^\alpha dx + \alpha \int_B \Big[u_i(x) - u_i(0)\Big] \times$$
$$\times \Big[u_{kk}\cos(x_i, r) - u_{ik}\cos(x_k, r)\Big] r^{\alpha-1} dx - (m-1)\int_{\partial B} |u_r|^2 dS$$

holds.

Proof. Integrating twice by parts we have

$$\int_B u_{ik}u_{ik} dx = \int_B \Big[u_i - u_i(0)\Big]_k \Big[u_i - u_i(0)\Big]_k dx =$$
$$= \int_B |\Delta u|^2 dx + \int_{\partial B} \left\{\Big[u_i - u_i(0)\Big] u_{ik}\cos(x_k, r) - \Big[u_i - u_i(0)\Big] u_{kk}\cos(x_k, r)\right\} dS.$$

Therefore

$$\int_{\partial B} \left\{ \left[u_i - u_i(0) \right] u_{ik} \cos(x_k, r) - \left[u_i - u_i(0) \right] u_{kk} \cos(x_k, r) \right\} dS = \tag{4.2.10}$$

$$= \int_B \left(\left| D'^2 u \right|^2 - \left| \Delta u \right|^2 \right) dx.$$

With the same kind of calculations we come to the identity

$$\int_B u_{ik} u_{ik} r^\alpha dx = \int_B \left| \Delta u \right|^2 r^\alpha dx + \alpha \int_B \left[u_i - u_i(0) \right] u_{kk} r^{\alpha-1} \cos(x_i, r) dx -$$

$$-\alpha \int_B \left[u_i - u_i(0) \right] u_{ik} r^{\alpha-1} \cos(x_k, r) dx + \int_{\partial B} \left\{ \left[u_i - u_i(0) \right] u_{ik} \cos(x_k, r) - \right.$$

$$\left. - \left[u_i - u_i(0) \right] \Delta u \cos(x_i, r) \right\} dS.$$

After applying (4.2.10) gives

$$\int_B u_{ik} u_{ik} r^\alpha dx = \int_B \left| \Delta u \right|^2 r^\alpha dx + \alpha \int_B \left[u_i - u_i(0) \right] \left[u_{kk} \cos(x_i, r) - \right.$$

$$\left. - u_{ik} \cos(x_k, r) \right] r^{\alpha-1} dx + \int_B \left[u_{ik} u_{ik} - |\Delta u|^2 \right] dx.$$

Under condition (4.2.2) we have from (4.2.10) that

$$\int_B \left(\left| D'^2 u \right|^2 - \left| \Delta u \right|^2 \right) dx = -(m-1) \int_{\partial B} \left| u_r \right|^2 dx$$

and satisfy (4.2.9).

□

Consider a function

$$v(x) = u(x) - u(0) - u_i(0) x_i, \tag{4.2.11}$$

which obviously satisfies the conditions

$$v(0) = v_i(0) = 0 \quad \text{and} \quad v_{ik} = u_{ik}.$$

With a complete orthonormal set of spherical functions

$$\left\{ Y_{j,l}(\theta) \right\} \left(j = 0, 1, 2, \dots; l = 1, \dots, k_j, \theta \in S \right)$$

consider the expansion

$$v(x) = \sum_{j=0}^{+\infty} \sum_{l=1}^{k_j} v_{j,l}(r) Y_{j,l}(\theta). \tag{4.2.12}$$

Lemma 4.2.3. *For any $u \in W^{(2)}_{2,\alpha}(B)$, satisfying (4.2.2), the identity*

(4.2.13)
$$\int_B \left|D'^2 u\right|^2 r^\alpha dx = \int_B \left|\Delta u\right|^2 r^\alpha dx - (m-1)\int_{\partial B} \left|u_r\right|^2 dS - \frac{\alpha}{2}\int_{\partial B} \left|\nabla v\right|^2 dS +$$
$$+\frac{\alpha}{2}\int_{\partial B} \left|v_r\right|^2 dS - \frac{\alpha}{2}\sum_{j,l} j(j+m-2)v_{j,l}^2(1) + \alpha\sum_{j,l}\int_0^1 \left[(m-1)\left|v'_{j,l}\right|^2 +\right.$$
$$\left. +(\alpha+m-3)\, j(j+m-2)\left|v_{j,l}\right|^2 r^{-2}\right] r^{\alpha+m-3} dr$$

is true (by $\sum_{j,l} \geq 0$ we understand the summation in the same limits as in (4.2.12)).

Proof. We can write the identity (4.2.9) in the form

$$\int_B \left|D'^2 u\right|^2 r^\alpha dx = \int_B \left|\Delta u\right|^2 r^\alpha dx - (m-1)\int_{\partial B} |u_r|^2 dS +$$
$$+\alpha\int_B v_r \Delta v r^{\alpha-1} dx - \alpha\int_B v_i v_{ir} r^{\alpha-1} dx.$$

Using (4.2.11) integrate by parts in the last term on the right-hand side

$$\int_B v_i v_{ir} r^{\alpha-1} dx = \frac{1}{2}\int_B \left(\left|\nabla v\right|^2\right)_r r^{\alpha-1} dx = \frac{1}{2}\int_{\partial B} dS \int_0^1 \left(\left|\nabla v\right|^2\right)_r r^{\alpha+m-2} dr =$$
$$= \frac{1}{2}\int_{\partial B} dS\left[\left|\nabla v\right|^2 r^{\alpha+m-2}\Big|_0^1 - (\alpha+m-2)\int_0^1 \left|\nabla v\right|^2 r^{\alpha+m-3} dr\right] =$$
$$= \frac{1}{2}\int_{\partial B} \left|\nabla v\right|^2 dS - \frac{\alpha+m-2}{2}\int_B \left|\nabla v\right|^2 r^{\alpha-2} dx.$$

So

(4.2.14)
$$\int_B \left|D'^2 u\right|^2 r^\alpha dx =$$
$$= \int_B \left|\Delta u\right|^2 r^\alpha dx - (m-1)\int_{\partial B} \left|u_r\right|^2 dS + \alpha\int_B v_r \Delta v r^{\alpha-1} dx +$$
$$+\frac{\alpha(\alpha+m-2)}{2}\int_B \left|\nabla v\right|^2 r^{\alpha-2} dx - \frac{\alpha}{2}\int_{\partial B} \left|\nabla v\right|^2 dS.$$

Integrating by parts gives

$$\int_B \left|\nabla v\right|^2 r^{\alpha-2} dx = \int_B v_i v_i r^{\alpha-2} dx = \int_B \left(v v_i r^{\alpha-2}\right)_i dx -$$
$$-\int_B v\Delta v r^{\alpha-2} dx - (\alpha-2)\int_B v v_i r^{\alpha-3} r_i dx =$$
$$= \int_{\partial B} v v_r dS - \int_B v\Delta v r^{\alpha-2} dx - (\alpha-2)\times$$

$$\times \int_{\partial B} dS \int_0^1 v v_r r^{\alpha+m-4} dr = \int_{\partial B} v v_r dS - \frac{\alpha-2}{2}\int_{\partial B} |v|^2 dS -$$
$$-\int_B v\Delta v r^{\alpha-2} dx + \frac{(\alpha-2)(\alpha+m-4)}{2}\int_0^1 |v|^2 r^{\alpha-4} dx.$$

Finally

$$\int_B |\nabla v|^2 r^{\alpha-2} dx = \int_{\partial B}\left(v v_r - \frac{\alpha-2}{2}|v|^2\right) dS - \int_B v\Delta v r^{\alpha-2} dx +$$
$$+ \frac{(\alpha-2)(\alpha+m-4)}{2}\int_B |v|^2 r^{\alpha-4} dx.$$

Substituting in (4.2.14) gives

(4.2.15)
$$\int_B \left|D^2 u\right|^2 r^\alpha dx = \int_B \left|\Delta u\right|^2 r^\alpha dx - (m-1)\int_{\partial B}\left|u_r\right|^2 dS -$$
$$-\frac{\alpha}{2}\int_{\partial B}\left[\left|\nabla v\right|^2 - (\alpha+m-2)\left(v v_r - \frac{\alpha-2}{2}|v|^2\right)\right] dS +$$
$$+\alpha\int_B v_r \Delta v r^{\alpha-1} dr - \frac{\alpha(\alpha+m-2)}{2}\int_B v\Delta v r^{\alpha-2} dx +$$
$$+\frac{\alpha(\alpha-2)(\alpha+m-2)(\alpha+m-4)}{4}\int_B |v|^2 r^{\alpha-4} dx.$$

Now transform the last three terms on the right-hand side of (4.2.15) with the help of the expansion (4.2.12). Then

$$I_1 = \int_B v_r \Delta v r^{\alpha-1} dx =$$
$$= \sum_{j,l}\int_0^1 v'_{j,l}\left[\left(r^{m-1} v'_{j,l}\right)' r^{\alpha-m} - j(j+m-2) v'_{j,l} v_{j,l} r^{\alpha-3}\right] r^{m-1} dr =$$
$$= \sum_{j,l}\left\{ r^{\alpha+m-2} v'_{j,l} v'_{j,l}\Big|_0^1 - \int_0^1\left[v''_{j,l} v'_{j,l} r^{\alpha+m-2} + (\alpha-1)|v'_{j,l}|^2 r^{\alpha+m-3}\right] dr -\right.$$
$$\left. -j(j+m-2)\int_0^1 v'_{j,l} v_{j,l} r^{\alpha+m-4} dr\right\} = \int_{\partial B} |v_r|^2 dS - \sum_{j,l}\left\{\frac{1}{2}\left|v'_{j,l}\right|^2 r^{\alpha+m-2}\Big|_0^1 -\right.$$
$$-\frac{(\alpha+m-2)}{2}\int_0^1 \left|v'_{j,l}\right|^2 r^{\alpha+m-3} dr + j(j+m-2)\int_0^1 v'_{j,l} v_{j,l} r^{\alpha+m-4} dr +$$
$$+(\alpha-1)\int_B \left|v_r\right|^2 r^{\alpha+m-3} dr = \frac{1}{2}\int_{\partial B}\left|v_r\right|^2 dS + \frac{m-\alpha}{2}\int_B \left|v'_r\right|^2 r^{\alpha-2} dx -$$
$$-\sum_{j,l} j(j+m-2)\int_B v'_{j,l} v_{j,l} r^{\alpha+m-4} dr = \frac{1}{2}\int_{\partial B}\left|v_r\right|^2 dS -$$

$$-\frac{1}{2}\sum_{j,l} j(j+m-2)|v_{j,l}(1)|^2 + \frac{\alpha+m-4}{2}\sum_{j,l} j(j+m-2)\int_0^1 |v_{j,l}|^2 r^{\alpha+m-5} dr +$$
$$+\frac{m-\alpha}{2}\int_B \left|v_r\right|^2 r^{\alpha-2} dx.$$

So

$$I_1 = \int_B v_r \Delta v r^{\alpha-1} dx = \tag{4.2.16}$$
$$= \frac{1}{2}\int_{\partial B} \left|v_r\right|^2 dS - \frac{1}{2}\sum_{j,l} j(j+m-2)|v_{j,l}(1)|^2 +$$
$$+\frac{\alpha+m-4}{2}\sum_{j,l} j(j+m-2)\int_0^1 |v_{j,l}|^2 r^{\alpha+m-5} dr + \frac{m-\alpha}{2}\times$$
$$\times \int_B \left|v_r\right|^2 r^{\alpha-2} dx.$$

Now

$$I_2 = \int_B v\Delta v r^{\alpha-2} dx = \sum_{j,l}\int_0^1 v_{j,l}\Big[v''_{j,l} + (m-1)r^{-1}v'_{j,l} -$$
$$-j(j+m-2)r^{-2}v_{j,l}\Big] r^{\alpha+m-3} dr = \sum_{j,l}\Big[\int_0^1 v_{j,l}v''_{j,l} r^{\alpha+m-3} dr + (m-1)\times$$
$$\times \int_0^1 v_{j,l}v'_{j,l} r^{\alpha+m-4} dr - j(j+m-2)\int_0^1 |v_{j,l}|^2 r^{\alpha+m-5} dr\Big] =$$
$$= \sum_{j,l}\Big[v_{j,l}v'_{j,l} r^{\alpha+m-3}\Big|_0^1 - \int_0^1 \left|v'_{j,l}\right|^2 r^{\alpha+m-3} dr -$$
$$-(\alpha+m-3)\int_0^1 v_{j,l}v'_{j,l} r^{\alpha+m-4} dr + \frac{m-1}{2}|v_{j,l}|^2 r^{\alpha+m-4}\Big|_0^1 -$$
$$-(m-1)\frac{\alpha+m-4}{2}\int_0^1 |v_{j,l}|^2 r^{\alpha+m-5} dr - j(j+m-2)\int_0^1 |v_{j,l}|^2 r^{\alpha+m-5} dr\Big].$$

Then

$$I_2 = \int_B v\Delta v r^{\alpha-2} dx = \sum_{j,l}\Big\{v_{j,l}(1)v'_{j,l}(1) - \frac{\alpha-2}{2}|v_{j,l}(1)|^2 + \tag{4.2.17}$$
$$+\Big[\frac{(\alpha+m-4)(\alpha-2)}{2} - j(j+m-2)\Big]\int_0^1 |v_{j,l}|^2 r^{\alpha+m-5} dr -$$
$$-\int_0^1 \left|v'_{j,l}\right|^2 r^{\alpha+m-3} dr\Big\}.$$

Combining (4.2.15), (4.2.16) and (4.2.17) gives (4.2.13) □

Lemma 4.2.4. *If $v \in W^{(2)}_{2,\alpha}(B)$ satisfies (4.2.2) and vanishes with all first derivatives at the center of B, then the identity*

(4.2.18)
$$\int_B |\Delta v|^2 r^\alpha dx =$$
$$= (m-1)\int_{\partial B} |v_r|^2 dS - 2\sum_{j,l} j(j+m-2)v'_{j,l}(1)v_{j,l}(1) +$$
$$+(\alpha-2)\sum_{j,l} j(j+m-2)|v_{j,l}|^2(1) + \sum_{j,l}\int_0^1 \Big\{\left|v''_{j,l}\right|^2 +$$
$$+\Big[(m-1)(1-\alpha)+2j(j+m-2)\Big]\left|v'_{j,l}\right|^2 r^{-2} + j(j+m-2)\times$$
$$\times\Big[j(j+m-2)+(2-\alpha)(\alpha+m-4)\Big]\left|v_{j,l}\right|^2 r^{-4}\Big\} r^{\alpha+m-1}dr$$

is true.

Proof. Using the expansion (4.2.12) we get

$$\int_B |\Delta v|^2 r^\alpha dx =$$
$$= \sum_{j,l}\Big\{\int_0^1 \Big[\left|v''_{j,l}\right|^2 + (m-1)^2 r^{-2}\left|v'_{j,l}\right|^2 + j^2(j+m-2)^2 r^{-4}\left|v_{j,l}\right|^2\Big]\times$$
$$\times r^{\alpha+m-1}dr + 2(m-1)\int_0^1 v''_{j,l}v'_{j,l}r^{\alpha+m-2}dr - 2j(j+m-2)\times$$
$$\times\int_0^1 v''_{j,l}v_{j,l}r^{\alpha+m-3}dr - 2(m-1)j(j+m-2)\int_0^1 v'_{j,l}v_{j,l}r^{\alpha+m-4}dr\Big\} =$$
$$= \sum_{j,l}\Big\{\int_0^1 \Big[\left|v''_{j,l}\right|^2 + (m-1)^2\left|v'_{j,l}\right|^2 r^{-2} + j^2(j+m-2)^2 r^{-4}\times$$
$$\times\left|v_{j,l}\right|^2\Big] r^{\alpha+m-1}dr + 2(m-1)\Big[\frac{|v'|^2}{2}r^{\alpha+m-2}\Big|_0^1 - \frac{\alpha+m-2}{2}\times$$
$$\times\int_0^1 \left|v'_{j,l}\right|^2 r^{\alpha+m-3}dr\Big] - 2j(j+m-2)\Big[v'_{j,l}v_{j,l}r^{\alpha+m-3}\Big|_0^1 - \int_0^1 \left|v'_{j,l}\right|^2\times$$
$$\times r^{\alpha+m-3}dr - (\alpha+m-3)\int_0^1 v'_{j,l}v_{j,l}r^{\alpha+m-4}dr - \frac{2(m-1)j(j+m-2)}{2}\times$$
$$\times\Big[\left|v_{j,l}\right|^2 r^{\alpha+m-4}\Big|_0^1 - (\alpha+m-4)\int_0^1 \left|v_{j,l}\right|^2 r^{\alpha+m-5}dr\Big]\Big\}.$$

Continuing this process we get

$$\int_B |\Delta v|^2 r^\alpha dx =$$

$$= \sum_{j,l} \Bigg\{ \int_0^1 \left[\left|v''_{j,l}\right|^2 + (m-1)^2 \left|v'_{j,l}\right|^2 r^{-2} + j^2(j+m-2)^2 \left|v_{j,l}\right|^2 r^{-4} \right] \times$$

$$\times r^{\alpha+m-1} dr + (m-1)\left|v'_{j,l}(1)\right|^2 - (m-1)(\alpha+m-2) \int_0^1 \left|v'_{j,l}\right|^2 r^{\alpha+m-3} dr -$$

$$-2j(j+m-2) v'_{j,l}(1) v_{j,l}(1) +$$

$$+2j(j+m-2) \int_0^1 \left|v'_{j,l}\right|^2 r^{\alpha+m-3} dr + 2(\alpha+m-3) \times$$

$$\times j(j+m-2) \int_0^1 v'_{j,l} v_{j,l} r^{\alpha+m-4} dr - (m-1) j(j+m-2) \left|v_{j,l}(1)\right|^2 +$$

$$+(m-1)(\alpha+m-4) j(j+m-2) \int_0^1 \left|v_{j,l}\right|^2 r^{\alpha+m-5} dr \Bigg\}.$$

In the same way we get

$$\int_B |\Delta v|^2 r^\alpha dx =$$

$$= \sum_{j,l} \int_0^1 \left[\left|v''_{j,l}\right|^2 + (m-1)^2 \left|v'_{j,l}\right|^2 r^{-2} + j^2(j+m-2)^2 \left|v_{j,l}\right|^2 r^{-4} \right] \times$$

$$\times r^{\alpha+m-1} dr + (m-1)\left|v'_{j,l}(1)\right|^2 - (m-1)(\alpha+m-2) \int_0^1 \left|v'_{j,l}\right|^2 \times$$

$$\times r^{\alpha+m-3} dr - 2j(j+m-2) v'_{j,l}(1) v_{j,l}(1) + 2j(j+m-2) \int_0^1 \left|v'_{j,l}\right|^2 \times$$

$$\times r^{\alpha+m-3} dr + 2(\alpha+m-3) j(j+m-2) \left[\frac{\left|v_{j,l}\right|^2}{2} r^{\alpha+m-4} \Big|_0^1 - \right.$$

$$\left. -\frac{(\alpha+m-4)}{2} \int_0^1 \left|v_{j,l}\right|^2 r^{\alpha+m-5} dr \right] - (m-1) j(j+m-2) \left|v_{j,l}(1)\right|^2 +$$

$$+(m-1)(\alpha+m-4) j(j+m-2) \int_0^1 \left|v_{j,l}\right|^2 r^{\alpha+m-5} dr.$$

which can easily to be shown lead to (4.2.18). □

Lemma 4.2.5. *For any $u \in W^{(2)}_{2,\alpha}(B)$ satisfying* (4.2.2) *the inequality*

$$(4.2.19)\quad \int_B \left|D''^2 u\right|^2 r^\alpha dx \le \left(1+M_\gamma^2\right)\int_B \left|\Delta u\right|^2 r^\alpha dx - (m-1)\int_{\partial B}\left|u_r\right|^2 dS +$$

$$+\frac{(m-2+2\gamma)(m-1)+mM_\gamma^2}{m}\sum_{i=1}^m \left|u_i(0)\right|^2 |S| + \left[(m+1+2\gamma)M_\gamma^2 +\right.$$

$$\left. +\frac{m-2+2\gamma}{2}\right](m-1)\left|u(0)\right|^2 |S|$$

is true, where

$$(4.2.20)\qquad M_\gamma^2 = \frac{(m-2+2\gamma)\left\{(1+\gamma)^2+\left[2-(1-\gamma)^2\right]m\right\}}{(m+1+\gamma)^2(1-\gamma)^2}$$

Proof. From (4.2.18) we have

$$\sum_{j,l}\int_0^1 \left\{\left|v''_{j,l}\right|^2 + \left[(m-1)(1-\alpha)+2j(j+m-2)\right]\left|v'_{j,l}\right|^2 r^{-2} + j(j+m-2)\times\right.$$

$$\left.\times\left[j(j+m-2)+(2-\alpha)(\alpha+m-4)\right]\left|v_{j,l}\right|^2 r^{-4}\right\} r^{\alpha+m-1}dr =$$

$$= \int_B |\Delta v|^2 r^\alpha dx - (m-1)\int_{\partial B}\left|v_r\right|^2 dS + 2\sum_{j,l} j(j+m-2)v'_{j,l}(1)v_{j,l}(1) -$$

$$-(\alpha-2)\sum_{j,l} j(j+m-2)|v_{j,l}(1)|^2.$$

As has been shown in the process of proving (4.1.4) the following inequality is achieved

$$\alpha\sum_{j,l}\int_0^1\left[(m-1)\left|v'_{j,l}\right|^2 + (\alpha+m-3)j(j+m-2)\left|v_{j,l}\right|^2 r^{-2}\right] r^{\alpha+m-3}dr \le$$

$$\le M_\gamma^2\sum_{j,l}\int_0^1\left\{\left|v''_{j,l}\right|^2 + \left[(m-1)(1-\alpha)+2j(j+m-2)\right]\left|v'_{j,l}\right|^2 r^{-2} +\right.$$

$$\left. + j(j+m-2)\left[j(j+m-2)+(2-\alpha)(\alpha+m-4)\right]\left|v_{j,l}\right|^2 r^{-4}\right\} r^{\alpha+m-1}dr.$$

Then from (4.2.13) and (4.2.18) we have

$$\int_B\left(\left|D''^2 u\right| - \left|\Delta u\right|^2\right) r^\alpha dx \le -(m-1)\int_{\partial B}\left|u_r\right|^2 dS + \frac{\alpha}{2}\int_{\partial B}\left|v_r\right|^2 dS -$$

$$-\frac{\alpha}{2}\int_{\partial B}\left|\nabla v\right|^2 dS + M_\gamma^2\left[\int_B\left|\Delta v\right|^2 r^\alpha dx - (m-1)\int_{\partial B}\left|v_r\right|^2 dS +\right.$$

$$+2\sum_{j,l} j(j+m-2)v'_{j,l}(1)v_{j,l}(1) - (\alpha-2)\sum_{j,l} j(j+m-2)|v_{j,l}(1)|^2\Big] -$$

$$-\frac{\alpha}{2}\sum_{j,l} j(j+m-2)\big|v_{j,l}(1)\big|^2.$$

Since u vanishes on ∂B and, according to (4.2.11), v is a linear function on ∂B we have $v_{j,l} = 0$ on ∂B for $j > 1$. Therefore

$$\Big|\sum_{j,l} j(j+m-2)v'_{j,l}(1)v_{j,l}(1)\Big| = (m-1)\Big|\sum_{l=1}^{k_1} v'_{1,l}(1)v_{1,l}(1)\Big| \le$$

$$\le \Big(\sum_{l=1}^{k_1}|v'_{1,l}(1)|^2\Big)^{\frac12}\Big(\sum_{l=1}^{k_1}\big|v_{1,l}(1)\big|^2\Big)^{\frac12}(m-1) \le$$

$$\le \Big(\int_{\partial B}\big|v_r\big|^2 dS\Big)^{\frac12}\Big(\int_{\partial B}|v|^2 dS\Big)^{\frac12}(m-1).$$

So

$$\Big|\sum_{j,l} j(j+m-2)v'_{j,l}(1)v_{j,l}(1)\Big| \le (m-1)\Big(\int_{\partial B}|v_r|^2 dS\Big)^{\frac12}\Big(\int_{\partial B}|v|^2 dS\Big)^{\frac12}. \tag{4.2.21}$$

In the same way we come to the inequality

$$\sum_{j,l} j(j+m-2)\big|v_{j,l}(1)\big|^2 \le (m-1)\int_{\partial B}|v|^2 dS. \tag{4.2.22}$$

With the help of (4.2.21) and (4.2.22) we get

$$\int_B \big|D'^2 u\big|^2 r^\alpha dx \le \tag{4.2.23}$$

$$\le \big(1+M_\gamma^2\big)\int_B |\Delta v|^2 r^\alpha dx - (m-1)\int_{\partial B}\big|u_r\big|^2 dS - \frac{\alpha}{2}\int_{\partial B}|\nabla v|^2 dS +$$

$$+\frac{\alpha}{2}\int_{\partial B}\big|v_r\big|^2 dS + (m-1)M_\gamma^2\Big[-\int_{\partial B}\big|v_r\big|^2 dS + 2\Big(\int_{\partial B}\big|v_r\big|^2 dS\Big)^{\frac12}\times$$

$$\times\Big(\int_{\partial B}\big|v\big|^2 dS\Big)^{\frac12} - (\alpha-2)\int_{\partial B}|v|^2 dS\Big] - \frac{\alpha}{2}(m-1)\int_{\partial B}|v|^2 dS.$$

Now estimating the right-hand side of (4.2.23) obviously (4.2.11) gives

$$\frac{\alpha}{2}\int_{\partial B}\big(|v_r|^2 - |\nabla v|^2\big)dS =$$

$$= \frac{\alpha}{2}\int_{\partial B}\Big\{\Big|u_r - \Big(\sum_{i=1}^m u_i(0)x_i\Big)'_r\Big|^2 - \sum_{i=1}^m\Big|u_i - u_i(0)\Big|^2\Big\}dS =$$

$$= \frac{\alpha}{2}\int_{\partial B}\Big[\big|u_r\big|^2 - 2u_r\sum_{i=1}^m u_i(0)\cos(r,x_i) + \Big|\sum_{i=1}^m u_i(0)\cos(r,x_i)\Big|^2 -$$

$$-\sum_{i=1}^m |u_i|^2 + 2\sum_{i=1}^m u_i u_i(0) - \sum_{i=1}^m |u_i(0)|^2\Big]dS.$$

Taking into account that on ∂B $|\nabla u|^2 = |u_r|^2$ and $u_i = u_r \cos(r, x_i)$ after cancelling some terms we come to the equality

$$\frac{\alpha}{2}\int_{\partial B}\left(\left|v_r\right|^2 - |\nabla v|^2\right)dS = \frac{\alpha}{2}\int_{\partial B}\left\{\sum_{i=1}^{m}|u_i(0)|^2\left[\cos^2(r,x_i)-1\right]+\right.$$

$$\left.+2\sum_{i<k}u_i(0)u_k(0)\cos(r,x_i)\cos(r,x_k)\right\}dS.$$

It can now easily be shown that

$$\frac{\alpha}{2}\int_{\partial B}\left(\left|v_r'\right|^2 - |\nabla v|^2\right)dS = \frac{\alpha(1-m)}{2m}\sum_{i=1}^{m}|u_i(0)|^2|S|. \tag{4.2.24}$$

Applying the inequality

$$2ab < a^2 + b^2$$

to the middle term in quadratic brackets on the right-hand side of (4.2.23) and taking into account that $\alpha - 3 = -(m+1+2\gamma)$, we obtain

$$\int_B \left|D'^2 u\right| r^\alpha dx \le (1+M_\gamma^2)\int_B |\Delta u|^2 r^\alpha dx - (m-1)\int_{\partial B}\left|u_r\right|^2 dS +$$

$$+\frac{(m-2+2\gamma)(m-1)}{2m}\sum_{i=1}^{m}|u_i(0)|^2|S| + (m+1+2\gamma)(m-1)M_\gamma^2\int_{\partial B}|v|^2 dS -$$

$$-\alpha\frac{(m-1)}{2}\int_{\partial B}|v|^2 dS.$$

Since

$$\int_{\partial B}|v|^2 dS = |u(0)|^2|S| + \sum_{i=1}^{m}|u_i(0)|^2|S|m^{-1}$$

we get (4.2.19). □

Lemma 4.2.6. *If $u \in W_{2,\alpha}^{(2)}(B)$ satisfies (4.2.2), then for any $\eta > 0$ the inequality*

$$\int_B |D'^2 u|^2 r^\alpha dx\left\{1 - \eta|S|\left[\frac{(m-2+2\gamma)(m-1)+mM_\gamma^2}{m}+\right.\right. \tag{4.2.25}$$

$$\left.\left.+\frac{2(m+1+2\gamma)M_\gamma^2 + m - 2 + 2\gamma}{(1-\gamma)^2}(m-1)\right]\right\} \le$$

$$\le (1+M_\gamma^2)\int_B |\Delta u|^2 r^\alpha dx + C_0(\eta)|S|\left\{\frac{(m-2+2\gamma)(m-1)+mM_\gamma^2}{m}\times\right.$$

$$\left.\times\int_B |\nabla u|^2 dx + (m-1)\left[(m+1+2\gamma)M_\gamma^2 + (m-2+2\gamma)/2\right]\int_B |u|^2 dx\right\}+$$

$$+(m-1)\left[|S|\frac{(m+1+2\gamma)M_\gamma^2 + (m-2+2\gamma)/2}{1-\gamma}\eta - 1\right]\int_{\partial B}|u_r|^2 dS$$

is true. Here $\alpha = 2 - m - 2\gamma$ $(0 < \gamma < 1), C_0(\eta)$ *and* M_γ^2 *are determined by* (4.2.6) *and* (4.2.20). *The value* $|S|$ *(the area of the unit sphere in* R^m*) is determined by the formula*

$$|S| = 2\pi^{\frac{m}{2}}/\Gamma\left(\frac{m}{2}\right).$$

Proof. From the identity

$$u_i = u_i\Big|_{r=1} + \int_1^r (u_i)'_\rho d\rho$$

follows the inequality

$$\int_B |u_i|^2 r^\alpha dx \le 2\int_B \left|u_i\Big|_{r=1}\right|^2 r^\alpha dx + 2\int_B \left|\int_1^r u_{i\rho} d\rho\right|^2 r^\alpha dx. \tag{4.2.26}$$

Obviously

$$\int_B |u_i\Big|_{r=1}|^2 r^\alpha dx = \frac{1}{2(1-\gamma)}\int_{\partial B} |u_i\Big|_{r=1}|^2 dS.$$

Since u satisfies the condition (4.2.2) we have

$$\sum_{i=1}^m \int_B \left|u_i\Big|_{r=1}\right|^2 r^\alpha dx = \frac{1}{2(1-\gamma)}\int_{\partial B} \left|u'_r\right|^2 dS. \tag{4.2.27}$$

From the Hardy inequality follows

$$\int_B \left|\int_1^r u_{i\rho} d\rho\right|^2 r^\alpha dx \le \frac{1}{(1-\gamma)^2}\int_B \left|u_{ir}\right|^2 r^{\alpha+2} dx.$$

So, taking into account $r \le 1$, we come to the inequality

$$\sum_{i=1}^m \int_B \left|\int_1^r u_{i\rho} d\rho\right|^2 r^\alpha dx \le \frac{1}{(1-\gamma)^2}\int_B \left|D'^2 u\right|^2 r^\alpha dx.$$

Applying (4.2.26) and (4.2.27) we get

$$\int_B \left|D' u\right|^2 r^\alpha dx \le \frac{1}{1-\gamma}\int_{\partial B} \left|u_r\right|^2 dS + \frac{2}{(1-\gamma)^2}\int_B \left|D'^2 u\right|^2 r^\alpha dx. \tag{4.2.28}$$

Now combining (4.2.4), (4.2.5), (4.2.19) and (4.2.28) we come to the inequality (4.2.25). □

Corollary 4.2.2 *We can also apply the inequalities* (4.2.4) *and* (4.2.5). *Then we arrive at the relation*

$$\int_B \left|D'^2 u\right|^2 r^\alpha dx \Big\{1 - \eta|S|\Big[\frac{(m-2+2\gamma)(m-1) + mM_\gamma^2}{m} + \tag{4.2.29}$$

$$+\frac{2(m+1+2\gamma)M_\gamma^2+m-2+2\gamma}{(1-\gamma)^2}(m-1)\Big]\Big\}\leq$$

$$\leq\left(1+M_\gamma^2\right)\int_B|\Delta u|^2r^\alpha dx+\frac{C_0(\eta)|S|}{\lambda}\Big\{\frac{(m-2+2\gamma)(m-1)+mM_\gamma^2}{m}+$$

$$+\frac{(m-1)[(m+1+2\gamma)M_\gamma^2+(m-2+2\gamma)/2]}{\lambda}\Big\}\int_B|\Delta u|_r^2dx+$$

$$+(m-1)\left[|S|\frac{(m+1+2\gamma)M_\gamma^2+(m-2+2\gamma)/2}{1-\gamma}\eta-1\right]\int_{\partial B}|u_r|^2dS.$$

Corollary 4.2.3 *Taking into account* $r\leq 1$, *we can write*

$$\int_B|\Delta u|^2dx\leq\int_B|\Delta u|^2r^\alpha dx$$

and from (4.2.29) *then follows*

(4.2.30) $$\int_B|D'^2u|^2r^\alpha dx\Big\{1-\eta|S|\Big[\frac{(m-2+2\gamma)(m-1)+mM_\gamma^2}{m}+$$

$$+\frac{2(m+1+2\gamma)M_\gamma^2+m-2+2\gamma}{(1-\gamma)^2}(m-1)\Big]\Big\}\leq$$

$$\leq\Big(1+M_\gamma^2+\frac{C_0(\eta)|S|}{\lambda}\Big\{\frac{(m-2+2\gamma)(m-1)+mM_\gamma^2}{m}+$$

$$+\frac{(m-1)[(m+1+2\gamma)M_\gamma^2+(m-2+2\gamma)/2]}{\lambda}\Big\}\Big)\int_B|\Delta u|^2r^\alpha dx+$$

$$+(m-1)\left[|S|\frac{(m+1+2\gamma)M_\gamma^2+(m-2+2\gamma)/2}{1-\gamma}\eta-1\right]\int_{\partial B}|u_r|^2dS.$$

It is easy to see, that if

$$1-\eta|S|\left[\frac{(m-2+2\gamma)(m-1)+mM_\gamma^2}{m}+\frac{2(m+1+2\gamma)M_\gamma^2+m-2+2\gamma}{(1-\gamma)^2}(m-1)\right]$$

is nonnegative, then the expression

(4.2.31) $$|S|\frac{(m+1+2\gamma)M_\gamma^2+(m-2+2\gamma)/2}{1-\gamma}\eta-1$$

is nonpositive. After rescaling x we come to the following

Theorem 4.2.1. *Let* $u\in W_{2,\alpha}^{(2)}(B_R)$ *satisfies the condition* (4.2.2). *Let also the inequality*

(4.2.32) $$E\equiv1-\eta|S|\Big[\frac{(m-2+2\gamma)(m-1)+mM_\gamma^2}{m}+$$

$$+\frac{2(m+1+2\gamma)M_\gamma^2+m-2+2\gamma}{(1-\gamma)^2}(m-1)\Big]>0$$

holds. Then the following estimates

$$(4.2.33)\qquad \int\limits_{B_R} |D'^2u|^2 r^\alpha dx \leq$$

$$\leq \frac{1}{E}\Bigg\{(1+M_\gamma^2)\int\limits_{B_R} |\Delta u|^2 r^\alpha dx + C_0(\eta)|S| \times$$

$$\times\left[\frac{(m-2+2\gamma)(m-1)+mM_\gamma^2}{m} R^{\alpha-2}\int\limits_{B_R} |\nabla u|^2 dx + \right.$$

$$\left. +(m-1)\Big[(m+1+2\gamma)M_\gamma^2 + (m-2+2\gamma)/2\Big] R^{\alpha-4}\int\limits_{B_R} |u|^2 dx\right]\Bigg\},$$

$$(4.2.34)\qquad \int\limits_{B_R} |D'^2u|^2 r^\alpha dx \leq$$

$$\leq \frac{1}{E}\Bigg\{(1+M_\gamma^2)\int\limits_{B_R} |\Delta u|^2 r^\alpha dx + \frac{C_0(\eta)|S|}{\lambda} R^\alpha \times$$

$$\times\left[\frac{(m-2+2\gamma)(m-1)+mM_\gamma^2}{m} + \right.$$

$$\left. +\frac{(m-1)((m+1+2\gamma)M_\gamma^2 + (m-2+2\gamma)/2)}{\lambda}\right]\int\limits_{B_R} |\Delta u|^2 dx\Bigg\}$$

and

$$(4.2.35)\qquad \int\limits_{B_R} |D'^2u|^2 r^\alpha dx \leq$$

$$\leq \frac{1}{E}\Bigg\{1+M_\gamma^2 + \frac{C_0(\eta)|S|}{\lambda}\left[\frac{(m-2+2\gamma)(m-1)+mM_\gamma^2}{m} + \right.$$

$$\left. +\frac{(m-1)\Big((m+1+2\gamma)M_\gamma^2 + (m-2+2\gamma)/2\Big)}{\lambda}\right]\Bigg\}\int\limits_{B_R} |\Delta u|^2 r^\alpha dx$$

are true. □

Let us recall that $\alpha = 2-m-2\gamma$ $(0<\gamma<1)$, M_γ and $C_0(\eta)$ as defined by (4.2.20) and (4.2.6) respectively, λ is the least absolute value of the eigenvalues for the operator Δ in B with condition (4.2.2) and

$$|S| = 2\pi^{\frac{m}{2}}/\Gamma\left(\frac{m}{2}\right).$$

is the area of the unit sphere in R^m.

Now return to inequalities (4.2.4) and (4.2.5) of lemma 4.2.1. Since the power of the integrals on the right-hand side of these inequalities is equal to one they belong to the so-called class of linear inequalities. However in some problems it is important to have the so-called multiplicative inequalities.

Lemma 4.2.7. *If $u \in W^{(2)}_{2,\alpha}(B_R)$ and (4.2.2) takes place, then the inequalities*

(4.2.36)
$$|u(0)|^2 \le C\Big(\int_{B_R} |D'u|^2 r^\alpha dx\Big)^{\frac{m}{m+2\gamma}} \Big(\int_{B_R} |u|^2 dx\Big)^{\frac{2\gamma}{m+2\gamma}},$$
$$\sum_{i=0}^{m} \big|u_i(0)\big|^2 \le C\Big(\int_{B_R} \big|D'^2 u\big|^2 r^\alpha dx\Big)^{\frac{m}{m+2\gamma}} \Big(\int_{B_R} |Du|^2 dx\Big)^{\frac{2\gamma}{m+2\gamma}}$$

are true.

Proof. We will prove only the first of the inequalities (4.2.36). Substituting in (4.2.4) the expression (4.2.6), we get

$$|u(0)|^2 < \eta \int_{B_R} |D'u|^2 r^\alpha dx + C\eta^{-\frac{m}{2\gamma}} \int_{B_R} |u|^2 dx.$$

Now take

$$\eta = \Big(\int_{B_R} |D'u|^2 r^\alpha dx\Big)^{-2\gamma/(m+2\gamma)} \Big(\int_{B_R} |u|^2 dx\Big)^{2\gamma/(m+2\gamma)}$$

and we get (4.2.36) (if $|D'u| = 0$ then $u \equiv 0$ and (4.2.36) is trivial). The second inequality can be proved analogously. □

Remark 4.2.1. *Under the assumption of the lemma the inequality*

(4.2.37)
$$|u(0)|^2 \le C\Big(\int_{B_R} |D'^2 u|^2 r^\alpha dx\Big)^{\frac{m}{m+2\gamma}} \Big(\int_{B_R} |Du|^2 dx\Big)^{\frac{2\gamma}{m+2\gamma}}$$

holds.

In fact

$$\int_{B_R} |D'u|^2 r^\alpha dx \le 2\int_{B_R} |D'u - Du|_0|^2 r^\alpha dx + 2\int_{B_R} |D'u|_0|^2 r^\alpha dx \le$$
$$\le C\Big[\int_{B_R} |D'u - D'u|_0|^2 r^{\alpha-2} dx + |D'u|_0|^2\Big].$$

From the Hardy inequality (2.1.9) and from (4.2.5) it follows that

$$\int_{B_R} |D'u|^2 r^\alpha dx \le C\Big(\int_{B_R} |D'^2 u|^2 r^\alpha dx + \int_{B_R} |Du|^2 dx\Big).$$

Then from (4.2.2) and inequalities, $r \le R$ and $\alpha < 0$ we get

$$\int_{B_R} |D'u|^2 r^\alpha dx \le C\Big(\int_{B_R} |D'^2 u|^2 r^\alpha dx + \int_{B_R} |D^2 u|^2 dx\Big) \le$$
$$\le C\int_{B_R} |D'^2 u|^2 r^\alpha dx.$$

Applying (4.2.36) and (4.2.37) to (4.2.19) gives

Theorem 4.2.2. *Let $u \in W^{(2)}_{2,\alpha}(B_R)$ and satisfy* (4.2.2). *Then the inequality*

(4.2.38)
$$\int_{B_R} |D'^2u|^2 r^\alpha dx \le (1+M_\gamma^2)\int_{B_R} |\Delta u|^2 r^\alpha dx + +C\left(\int_{B_R} |D'^2u|^2 r^\alpha dx\right)^{\frac{m}{m+2\gamma}}\left(\int_{B_R} |Du|^2 dx\right)^{\frac{2\gamma}{m+2\gamma}}$$

holds true.

In fact if we omit on the right-hand side of (4.2.19) the negative term and apply (4.2.36) and (4.2.37) we get

$$\int_{B_R} |D'^2u|^2 r^\alpha dx \le (1+M_\gamma^2)\int_{B_R} |\Delta u|^2 r^\alpha dx + C\left[\left(\int_{B_R} |D'^2u|^2 r^\alpha dx\right)^{\frac{m}{m+2\gamma}} + +\left(\int_{B_R} |D'u|^2 r^\alpha dx\right)^{\frac{m}{m+2\gamma}}\right]\left(\int_{B_R} |Du|^2 dx\right)^{\frac{2\gamma}{m+2\gamma}}.$$

After using the relation

$$\int_{B_R} |u|^2 dx \le \int_{B_R} |D'u|^2 dx$$

we come to the result.

□

Suppose now that condition (4.2.2) is not satisfied; how will estimate (4.2.38) change in this case ?

Theorem 4.2.3. *Let $u \in W^{(2)}_{2,\alpha}(B_R)$. Then the inequality*

(4.2.39)
$$\int_{B_R} |D'^2u|^2 r^\alpha \zeta dx \le (1+M_\gamma^2+\eta)\int_{B_R} |\Delta u|^2 r^\alpha \zeta dx + +C\left\{\left(\int_{B_R} |D'^2u|^2 r^\alpha \zeta dx\right)^{\frac{m}{m+2\gamma}}\left[\int_{B_R} |Du|^2 dx\right]^{\frac{2\gamma}{m+2\gamma}} + +\int_{B_R} |Du|^2 dx\right\}$$

whith ζ defined by (2.2.1) *and η an arbitrary small positive number is true.*

The result follows immediately after substituting the function $u\zeta$ for u in (4.2.38).

4.3 Coercivity inequalities with explicit constants for bounded weights with zeros

In lemma 4.1.2, the estimate (4.1.17) was achieved by Chelkak (Koshelev,Chelkak[1],p.29) using the assumptions (4.1.16), which required the function and its first derivatives

on ∂B_δ to be zero. We want to get the corresponding inequality using the assumption that only the condition (4.2.2) is satisfied.
Take $\beta = m-2+2\gamma$ $(0<\gamma<1/2)$ and suppose that $u \in W^{(2)}_{2,\beta}(B,0)$. It is worth mentioning we can initially assume the function $u(x)$ to be smooth thereby satisfying (4.2.2). Using the expansion (2.3.1) and (4.2.12) derive the following equalities

(4.3.1)
$$\int_B |\Delta u|^2 r^\beta dx = (m-1)\sum_{j,l} |u'_{j,l}(1)|^2 + \sum_{j,l}\int_0^1 \{|u''_{j,l}(r)|^2 +$$
$$+[(m-1)(1-\beta)+2j(j+m-2)]|u'_{j,l}(r)|^2 r^{-2} +$$
$$+j(j+m-2)[j(j+m-2)-(\beta+m-4)(\beta-2)]|u_{j,l}(r)|^2 r^{-4}\} r^{\beta+m-1}dr,$$

(4.3.2)
$$\int_B (|D'^2 u|^2 - |\Delta u|^2) r^\beta dx = -(m-1)\sum_{j,l} |u'_{j,l}(1)|^2 +$$
$$+\beta \sum_{j,l}\int_0^1 [(m-1)|u'_{j,l}(r)|^2 +$$
$$+(\beta+m-3)j(j+m-2)|u_{j,l}(r)|^2 r^{-2}] r^{\beta+m-3}dr\,,$$

where $'$ denotes the derivative with respect to r and the summation for j,l is resolved in the same manner as in (2.3.1) or (4.2.12).
Let us show, for example, how to prove the equality (4.3.2). First, integrate by parts to give

$$\int_B u_{ik}u_{ik} r^\beta dx = \int_B (u_k u_{ik} r^\beta)_i dx - \int_B u_k u_{iik} r^\beta dx -$$
$$-\beta \int_B u_k u_{rk} r^{\beta-1} dx = \int_S u_k u_{rk} dS - \int_S u_r \Delta u dS +$$
$$+\int_B |\Delta u|^2 r^\beta dx + \beta \int_B u_r \Delta u r^{\beta-1} dx - \beta \int_B u_k u_{rk} r^{\beta-1} dx\,.$$

Then, after simple calculations we derive the equality

$$\int_B u_{ik}u_{ik} r^\beta dx = \int_S \left(u_k u_{rk} - u_r\Delta u - \frac{\beta}{2}|\nabla u|^2\right) dS +$$
$$+\int_B |\Delta u|^2 r^\beta dx + \beta\int_B u_r \Delta u r^{\beta-1}dx + \frac{\beta(\beta+m-2)}{2}\int_B |\nabla u|^2 r^{\beta-2}dx\,.$$

Since $\beta>0$ we have $\beta-2>-2\geq -m$ and all the integrals are determined. Applying the boundary condition (4.2.2) we have

$$\left(u_k u_{rk} - u_r\Delta u - \frac{\beta}{2}|\nabla u|^2\right)\Big|_{r=1} = -\left(m-1+\frac{\beta}{2}\right)|u'|^2\Big|_{r=1}.$$

Then

(4.3.3)
$$\int_B \left(|D'^2 u|^2 - |\Delta u|^2\right) r^\beta dx = -\left(m-1+\frac{\beta}{2}\right)\int_S |u'|^2 dS +$$
$$+\beta\int_B u_r\Delta u r^{\beta-1}dx + \frac{\beta(\beta+m-2)}{2}\int_B |\nabla u|^2 r^{\beta-2}dx\,.$$

Consider initially $m = 2$. Then

$$|\nabla u|^2 = |u_r|^2 + r^{-2}|u_\theta|^2.$$

and expanding the integral $\int\limits_B u_r \Delta u r^{\beta-1} dx$ with the help of the expansion (2.3.1) we come to (4.3.2) for this case. For $m \geq 3$ the expansion (2.3.1) should be applied to the right-hand side of the identity

$$\int\limits_B |\nabla u|^2 r^{\beta-2} dx = -\int\limits_B u \Delta u r^{\beta-2} dx + \frac{(\beta-2)(\beta+m-4)}{2} \int\limits_B |u|^2 r^{\beta-4} dx,$$

which we proved earlier for $\int |\nabla u|^2 r^\alpha dx$. After substituting the expression of this integral and the analogous expression for the integral $\int\limits_B u_r \Delta u r^{\beta-1} dx$ in (4.3.3) we also come to (4.3.2) with $m \geq 3$. Rememeber that for $j > 0$ we have the equality $u_{j,l}(0) = 0$.

Theorem 4.3.1. *Let $\beta = m - 2 + 2\gamma$ and $0 < \gamma < 1/2$. For any $u \in W^{(2)}_{2,\beta}(B_\delta)$ satisfying the boundary condition (4.2.2) the inequality*

(4.3.4) $$\int\limits_{B_\delta} |D^2 u|^2 r^\beta dx \leq \Bigg\{1 + \frac{4\beta(m-1)}{(m-\beta)^2} + \frac{4\beta(\beta+m-2)^4(m-1)}{(m-\beta)^2(m+\beta-3)(m-\beta-1)^2\left[m-1+\frac{\beta+m-2}{4}+\frac{(m-\beta)^2}{4\beta}\right]} + O(\gamma)\Bigg\} \times \int\limits_{B_\delta} |\Delta u|^2 r^\beta dx$$

holds.

Proof. It is sufficient to prove (4.3.4) for $\delta = 1$. Obviously

(4.3.5) $$u'_{j,l}(r) - u'_{j,l}(1) = -\int\limits_r^1 u''_{j,l}(\rho) d\rho.$$

For all β $(0 < \beta < m)$ the inequality $s = -\beta - m + 3 < 1$ holds. Then, according to (2.1.8) we have

(4.3.6) $$\int\limits_0^1 |u''_{j,l}|^2 r^{\beta+m-1} dr \geq \frac{(\beta+m-2)^2}{4} \int\limits_0^1 |u'_{j,l}(r) - u'_{j,l}(1)|^2 r^{\beta+m-3} dr.$$

Estimating with the help of (4.3.6) the term with the second derivative on the right-hand side of (4.3.1) we come to the inequality

(4.3.7) $$\int\limits_B |\Delta u|^2 r^\beta dx \geq \left(m - 1 + \frac{\beta+m-2}{4}\right) \sum_{j,l} |u'_{j,l}(1)|^2 - \frac{(\beta+m-2)^2}{2} \sum_{j,l} \int\limits_0^1 u'_{j,l} r^{\beta+m-3} dr u'_{j,l}(1) +$$

$$+\sum_{j,l}\int_0^1\left\{\left[\frac{(m-\beta)^2}{4}+2j(j+m-2)\right]|u'_{j,l}|^2+\right.$$

$$\left.+j(j+m-2)\Big[j(j+m-2)+(\beta+m-4)(2-\beta)\Big]|u^2_{j,l}|r^{-2}\right\}r^{\beta+m-3}dr\,.$$

Since for $m \geq 4$ (all $\beta > 0$) we have $s = -\beta - m + 5 < 1$ then from (2.1.8) follows

$$\int_0^1 |u'_{j,l}|^2 r^{\beta+m-3}dr \geq \frac{(\beta+m-4)^2}{4}\int_0^1 |u_{j,l}|^2 r^{\beta+m-5}dr\,. \tag{4.3.8}$$

The same will occur for $\beta > 1$ and $m = 3$. For $m = 2$ with $0 < \beta < 2$ and $m = 3$ with $0 < \beta < 1$ using the fact that $u_{j,l}(0) = 0$ for $j > 0$ instead of the expression analogous to (4.3.5) we can write

$$u_{j,l}(r) = \int_0^r u'_{j,l}(\rho)d\rho\,.$$

These cases give $s = -\beta-m+5 > 1$ and we can apply (2.1.8). Using the last equality we also come to (4.3.8) (except for $m = 3$ and $\beta = 1$). Estimating the right-hand side of (4.3.7) from below with the help of (4.3.8) we come to

$$\int_B |\Delta u|^2 r^\beta dx \geq \left(m-1+\frac{\beta+m-2}{4}\right)\sum_{j,l}|u'_{j,l}(1)|^2+ \tag{4.3.9}$$

$$+\sum_{j,l}\int_0^1\left\{\frac{(m-\beta)^2}{4}|u'_{j,l}|^2+j(j+m-2)\Big[j(j+m-2)+\right.$$

$$\left.+\frac{(m-\beta)(m+\beta-4)}{2}\Big]|u_{j,l}|^2r^{-2}\right\}r^{\beta+m-3}dr-$$

$$-\frac{(\beta+m-2)^2}{2}\sum_{j,l}\int_0^1 u'_{j,l}r^{\beta+m-3}dru'_{j,l}(1)\,.$$

The middle term of the right-hand side can be estimated in the same way as it was done by Chelkak (lemma 4.1.2)

$$\sum_{j,l}\int_0^1\left\{\frac{(m-\beta)^2}{4}|u'_{j,l}|^2+j(j+m-2)\Big[j(j+m-2)+\right.$$

$$\left.+\frac{(m-\beta)(\beta+m-4)}{2}\Big]|u_{j,l}|^2r^{-2}\right\}r^{\beta+m-3}dr \geq$$

$$\geq \frac{(m-\beta)^2}{4\beta(m-1)}\cdot\beta\sum_{j,l}\int_0^1\Big[(m-1)|u'_{j,l}|^2+$$

$$+(\beta+m-3)j(j+m-2)|u_{j,l}|^2r^{-2}\Big]r^{\beta+m-3}dr\,.$$

Now applying the equality (4.3.2) from (4.3.9) we come to the estimate

$$\int_B |\Delta u|^2 r^\beta dx \geq \left(m-1+\frac{\beta+m-2}{4}\right)\sum_{j,l}|u'_{j,l}(1)|^2 +$$
$$+\frac{(m-\beta)^2}{4\beta(m-1)}\left[\int_B \left(|D'^2 u|^2 - |\Delta u|^2\right) r^\beta dx + (m-1)\sum_{j,l}|u'_{j,l}(1)|^2\right] -$$
$$-\frac{(\beta+m-2)^2}{2}\sum_{j,l}\int_0^1 u'_{j,l} r^{\beta+m-3} dr u'_{j,l}(1).$$

For all $m \geq 2$ (except $m=3$ and $\beta = 1$) follows the inequality

$$\left[1+\frac{4\beta(m-1)}{(m-\beta)^2}\right]\int_B |\Delta u|^2 r^\beta dx \geq \int_B |D'^2 u|^2 r^\beta dx + \tag{4.3.10}$$
$$+\frac{4\beta(m-1)}{(m-\beta)^2}\left[m-1+\frac{\beta+m-2}{4}+\frac{(m-\beta)^2}{4\beta}\right]\sum_{j,l}|u'_{j,l}(1)|^2 -$$
$$-\frac{2\beta(\beta+m-2)^2(m-1)}{(m-\beta)^2}\sum_{j,l}\int_0^1 u'_{j,l}(r) r^{\beta+m-3} dr u'_{j,l}(1).$$

Now consider the integral

$$I = \int_0^1 u'_{j,l}(r) r^{\beta+m-3} dr \tag{4.3.11}$$

for $m \geq 3$. Integrating by parts gives

$$\int_0^1 u'_{j,l}(r) r^{\beta+m-3} dr = -(\beta+m-3)\int_0^1 u_{j,l}(r) r^{\beta+m-4} dr\,.$$

Applying the Hölder inequality gives

$$\left|\int_0^1 u'_{j,l}(r) r^{\beta+m-3} dr\right| \leq \left(\int_0^1 |u_{j,l}(r)|^2 r^{\beta+m-4} dr\right)^{1/2} (\beta+m-3)^{1/2}\,.$$

Using (4.3.10) we obtain

$$\left[1+\frac{4\beta(m-1)}{(m-\beta)^2}\right]\int_B |\Delta u|^2 r^\beta dx \geq \int_B |D'^2 u|^2 r^\beta dx + \frac{4\beta(m-1)}{(m-\beta^2)}\times$$
$$\times\left[m-1+\frac{\beta+m-2}{4}+\frac{(m-\beta)^2}{4\beta}\right]\sum_{j,l}|u'_{j,l}(1)|^2 -$$
$$-\frac{2\beta(\beta+m-2)^2(m-1)(\beta+m-3)^{1/2}}{(m-\beta)^2}\sum_{j,l}\left(\int_B |u_{j,l}(r)|^2 r^{\beta+m-4} dr\right)^{1/2} |u'_{j,l}(1)|.$$

Applying the inequality (2.2.15) gives the relation

$$\left[1+\frac{4\beta(m-1)}{(m-\beta)^2}\right]\int_B |\Delta u|^2 r^\beta dx \geq \int_B |D'^2 u|^2 r^\beta dx +$$

$$+\frac{4\beta(m-1)}{(m-\beta)^2}\left[m-1+\frac{\beta+m-2}{4}+\frac{(m-\beta)^2}{4\beta}\right]\sum_{j,l}|u'_{j,l}(1)|^2-$$
$$-\frac{2\beta(\beta+m-2)^2(\beta+m-3)^{1/2}(m-1)}{(m-\beta)^2}\cdot\eta\sum_{j,l}|u'_{j,l}(1)|^2-$$
$$-\frac{\beta(\beta+m-2)^2(m-1)(\beta+m-3)^{1/2}}{2(m-\beta)^2\eta}\sum_{j,l}\int_0^1|u_{j,l}|^2r^{\beta+m-4}dr$$

with

$$\eta=\frac{2\left[m-1+\frac{\beta+m-2}{4}+\frac{(m-\beta)^2}{4\beta}\right]}{(\beta+m-2)^2(\beta+m-3)^{1/2}}\,.$$

The terms in $\sum|u'_{j,l}(1)|^2$ will be abolished and give

$$\left[1+\frac{4\beta(m-1)}{(m-\beta)^2}\right]\int_B|\Delta u|^2r^\beta dx\geq\int_B|D'^2u|^2r^\beta dx-$$
$$-\frac{\beta(\beta+m-2)^4(m-1)(\beta+m-3)}{4(m-\beta)^2\left[m-1+\frac{\beta+m-2}{4}+\frac{(m-\beta)^2}{4\beta}\right]}\sum_{j,l}\int_0^1|u_{j,l}|^2r^{\beta+m-4}dr\,.$$

Using the equality

$$\sum_{j,l}\int_0^1|u_{j,l}|^2r^{\beta+m-4}dr=\int_B|u|^2r^{\beta-3}dx$$

gives

$$\left[1+\frac{4\beta(m-1)}{(m-\beta)^2}\right]\int_B|\Delta u|^2r^\beta dx\geq \tag{4.3.12}$$
$$\geq\int_B|D'^2u|^2r^\beta dx-\frac{\beta(\beta+m-2)^4(m-1)(\beta+m-3)}{4(m-\beta)^2\left[m-1+\frac{\beta+m-2}{4}+\frac{(m-\beta)^2}{4\beta}\right]}\int_B|u|^2r^{\beta-3}dx\,.$$

Now it is necessary to estimate the integral

$$\int_B|u|^2r^{\beta-3}dx\,.$$

Given earlier

$$-\int_B\Delta uur^{\beta-1}dx=\int_B|\nabla u|^2r^{\beta-1}dx+(\beta-1)\int_B u'ur^{\beta-2}dx\,.$$

Using the condition (4.2.2) and once more integrating by parts the second term in the right-hand side gives

$$-\int_B\Delta uur^{\beta-1}dx=\int_B|\nabla u|^2r^{\beta-1}dx-\frac{(\beta-1)(\beta+m-3)}{2}\int_B|u|^2r^{\beta-3}dx\,.$$

Since

$$u(r) = -\int_r^1 u'(\rho)d\rho$$

and $-\beta - m + 4 < 1$, then from the inequality (2.1.8) we have

$$\int_0^1 |u|^2 r^{\beta+m-4} dr \le \frac{4}{(\beta+m-3)^2} \int_0^1 |u'|^2 r^{\beta+m-2} dr\,.$$

Therefore

$$-\int_B \Delta u u r^{\beta-1} dx \ge \frac{(m+\beta-3)(m-\beta-1)}{4} \int_B |u|^2 r^{\beta-3} dx\,.$$

Since $\beta = m - 2 + 2\gamma$ $(0 < \gamma < 1/2)$ then the right-hand side coefficient will be positive and gives

$$\int_B |u|^2 r^{\beta-3} dx \le -\frac{4}{(m+\beta-3)(m-\beta-1)} \int_B \Delta u u r^{\beta-1} dx\,.$$

From the Hölder inequality follows that

$$\int_B |u|^2 r^{\beta-3} dx \le \frac{16}{(m+\beta-3)^2(m-\beta-1)^2} \int_B |\Delta u|^2 r^{\beta+1} dx\,.$$

Since $r \le 1$ then

$$\int_B |u|^2 r^{\beta-3} dx \le \frac{16}{(m+\beta-3)^2(m-\beta-1)^2} \int_B |\Delta u|^2 r^{\beta} dx\,.$$

Using the estimate (4.3.12) gives the inequality (4.3.4) for $m \ge 3$.
Now consider the case $m = 2$. In the inequality (4.3.9) the last term is estimated in a different way. Evidently

$$\left|\sum_{j,l} \int_0^1 u'_{j,l} r^{\beta+m-3} dr u'_{j,l}(1)\right| \le (\beta+m-2)^{-1/2} \sum_{j,l} \left(\int_0^1 |u_{j,l}|^2 r^{\beta+m-3} dr \right)^{1/2} |u'_{j,l}(1)|\,.$$

Applying (2.2.15) gives

$$\frac{(\beta+m-2)^2}{2} \sum_{j,l} \int_0^1 u'_{j,l} r^{\beta+m-3} dr u'_{j,l}(1) \le \eta \sum_{j,l} \int_0^1 |u'_{j,l}|^2 r^{\beta+m-3} dr + \frac{(\beta+m-2)^3}{16\eta} \sum_{j,l} |u'_{j,l}(1)|^2\,.$$

Then (4.3.9) gives

$$\int_B |\Delta u|^2 r^\beta dx \ge \left[m - 1 + \frac{\beta+m-2}{4} - \frac{(\beta+m-2)^3}{16\eta} \right] \sum_{j,l} |u'_{j,l}(1)|^2 +$$

$$+ \sum_{j,l} \int_0^1 \left\{ \left[\frac{(m-\beta)^2}{4} - \eta \right] |u'_{j,l}|^2 + j(j+m-2) \times \right.$$

$$\left. \times \left[j(j+m-2) + \frac{(m-\beta)(m+\beta-4)}{2} \right] |u_{j,l}|^2 r^{-2} \right\} r^{\beta+m-3} dr$$

with $\eta = O(\beta^2)$. Since

$$\min\left\{\frac{(m-\beta)^2}{4\beta(m-1)}, \min_{j\geq 1}\frac{j(j+m-2)+\frac{(m-\beta)(m+\beta-4)}{2}}{\beta(\beta+m-3)}\right\} = \frac{(m-\beta)^2}{4\beta(m-1)}$$

then for small $\gamma > 0$

$$\int_B |\Delta u|^2 r^\beta dx \geq \left[m-1+\frac{\beta+m-2}{4}-\frac{(\beta+m-2)^3}{16\eta}\right]\sum_{j,l}|u'_{j,l}(1)|^2+$$
$$+\left[\frac{(m-\beta)^2}{4\beta}-O(\beta^2)\right]\beta\sum_{j,l}\int_0^1\Big[(m-1)|u'_{j,l}(r)|^2+$$
$$+(\beta+m-3)j(j+m-2)|u_{j,l}(r)|^2 r^{-2}\Big] r^{\beta+m-3}dr$$

and (4.3.2) gives the inequality

$$\int_B |\Delta u|^2 r^\beta dx \geq \left[m-1+\frac{\beta+m-2}{4}-\frac{(\beta+m-2)^3}{16\eta}\right]\sum_{j,l}|u'_{j,l}(1)|^2+$$
$$+\left[\frac{(m-\beta)^2}{4\beta}-O(\beta^2)\right]\left[\int_B\left(|D'^2u|^2-|\Delta u|^2\right)r^\beta dx+(m-1)\sum_{j,l}|u'_{j,l}(1)|^2\right].$$

With $\eta = 0(\beta^2)$ the first bracket on the right-hand side will be positive and

$$\int_B |\Delta u|^2 r^\beta dx \geq \left[\frac{(m-\beta)^2}{4\beta}-O(\beta^2)\right]\int_B\left(|D'^2u|^2-|\Delta u|^2\right)r^\beta dx$$

completing the proof of the theorem.

□

4.4 Boundedness in $W_2^{(2)}$ for the iterations of the universal process.

Let us now return to the general second order systems (2.3.18) with boundary condition (2.3.20) and consider the problem of Hölder continuity inside Ω for the first derivatives of the solution to problem (2.3.18), (2.3.20). The results of this section were obtained by the author with essential help from the Vishik[1] method.
Later we shall again examine this question by means of a somewhat different method. Suppose conditions 1) – 3) of 1.1 with $s = 2$, are satisfied. In addition, assume $a_i(x,p)$ $(i = 1,\ldots,m)$ are differentiable with respect to x and

(4.4.1) $$\left|\frac{\partial a_i}{\partial x_k}\right| \leq C(1+|p|),$$

is true.

In particular, it follows from these assumptions that, when $u \in W_2^{(2)}(\Omega)$, consequently $L(u) \in \mathcal{L}_2(\Omega)$. Assume g is a trace of some function from $W_2^{(1)}(\Omega)$. Then, consider the iteration process (2.3.22)

$$-\Delta u_{n+1} + u_{n+1} = -\Delta u_n + u_n - \varepsilon L(u_n),$$

where $\varepsilon > 0$ is a constant and $u_n\Big|_{\partial\Omega} = g$, $n = 0, 1, \ldots$. Let $u_0 \in W_2^{(2)}(\omega) \cap W_2^{(1)}(\Omega)$, where ω lies strictly inside Ω. Then all the u_n belong to $W_2^{(2)}(\Omega') \cap W_2^{(1)}(\Omega)$, where Ω' lies strictly inside ω. In fact, take such a sequence of domains Ω_n, $n = 0, 1, \ldots$, $\Omega_0 = \omega$, containing Ω', so that the strong embedding $\Omega_{n+1} \subset \Omega_n$ holds.
As follows from (2.3.22) the function u_1 is the solution of the problem

$$\Delta u_1 - u_1 = \Delta u_0 - u_0 + \varepsilon L(u_0),$$
$$u_1\Big|_{\partial\Omega} = g.$$

Since g is a trace of some function belonging to $W_2^{(1)}(\Omega)$, the solution u_1 will belong to the same space. Addititionally, from the properties of a_i and u_0 it follows that the right-hand side of the differential equation for u_1 also belongs to $\mathcal{L}_2(\omega)$. Therefore, $u_1 \in W_2^{(2)}(\Omega_1) \cap W_2^{(1)}(\Omega)$. Similarly, we get $u_n \in W_2^{(2)}(\Omega_n) \cap W_2^{(1)}(\Omega)$. It remains to note that $\Omega_n \supset \Omega'$, for $\forall n$. From theorem 1.2.1 process (2.3.22) will converge in $W_2^{(1)}(\Omega)$, when inequality $K_\varepsilon < 1$ accurs. Following the paper by Vishik [1], it is possible to show that if ε is defined by the conditions of lemma 1.1.2, the iterative process (2.3.22) will be bounded in common in the norm of $W_2^{(2)}(\Omega')$. So, the following statement is true

Lemma 4.4.1. *Let ω be a strict subdomain of Ω. If the condition* (4.4.1) *is satisfied, $K_\varepsilon < 1$ and $u_0 \in W_2^{(2)}(\omega) \cap W_2^{(1)}(\Omega)$, then in any strict subdomain Ω'' of ω, for all n, the inequality*

(4.4.2) $$\|u_n\|_{W_2^{(2)}(\Omega'')} \leq C$$

is true.

Proof. As we have just seen, the embedding $u_n \in W_2^{(2)}(\Omega') \cap W_2^{(1)}(\Omega)$ takes place when Ω' is an arbitrary, strictly interior subdomain of ω and Ω'' be a strictly interior subdomain of Ω'. Suppose $\psi(x)$ is a smooth function which vanishes at $\partial\Omega'$ and outside Ω', is positive inside Ω' and satisfies the inequality

(4.4.3) $$|\nabla\psi|^2|\psi|^{-1} \leq C.$$

For the one dimensional case we considered such a function with estimates (2.2.26). Taking into account that $K_\varepsilon < 1$, we get from theorem 1.1.2

(4.4.4) $$\|u_n\| \leq C, \quad n = 0, 1, \ldots.$$

Multiply both sides of (2.3.22) by $\Delta v_{n+1}\psi$, where v_{n+1} is smooth in Ω. Then we have

$$\int_{\Omega'} \Delta u_{n+1} \Delta v_{n+1} \psi dx - \int_{\Omega'} u_{n+1} \Delta v_{n+1} \psi dx =$$
$$= \int_{\Omega'} \left[\Delta u_n - u_n + \varepsilon L(u_n)\right] \Delta v_{n+1} \psi dx.$$

Integrating the second term on the left-hand side by parts, we come to

(4.4.5)
$$\int_{\Omega'} \Delta u_{n+1}\Delta v_{n+1}\psi dx + \int_{\Omega'} D'u_{n+1}D'v_{n+1}\psi dx +$$
$$+\int_{\Omega'} u_{n+1}D'v_{n+1}D'\psi dx =$$
$$= \int_{\Omega'} \left[\Delta u_n - \varepsilon \sum_{i=1}^{m} D_i a_i(x; Du_n)\right] \Delta v_{n+1}\psi dx -$$
$$- \int_{\Omega'} [u_n - \varepsilon a_0(x; Du_n)]\, \Delta v_{n+1}\psi dx.^*$$

Consider the first term on the right-hand side and denote it by

$$I_1 = \int_{\Omega'} [\Delta u_n - \varepsilon D_i a_i(x; Du_n)]\, \Delta v_{n+1}\psi dx.$$

Here the summation is carried out over the index i from 1 to m. In the integral I_1 we shall integrate by parts transfering the derivative from Δv_{n+1} which gives

$$I_1 = -\int_{\Omega'} [D_i u_n - \varepsilon a_i(x; Du_n)]\, D_i\Delta v_{n+1}\psi dx -$$
$$- \int_{\Omega'} [D_i u_n - \varepsilon a_i(x; Du_n)]\, \Delta v_{n+1} D_i\psi dx.$$

In the first term take one of the derivatives constructing the operator Δ and transfer it to the first factor by integrating by parts. Using the fact, that ψ vanishes on the boundary gives

(4.4.6)
$$I_1 = \int_{\Omega'} [D_iD_k u_n - \varepsilon D_k a_i(x; Du_n)]\, D_iD_k v_{n+1}\psi dx +$$
$$+ \int_{\Omega'} [D_i u_n - \varepsilon a_i(x; Du_n)]\, D_iD_k v_{n+1} D_k\psi dx -$$
$$- \int_{\Omega'} [D_i u_n - \varepsilon a_i(x; Du_n)]\, \Delta v_{n+1} D_i\psi dx.$$

Differentiating the expression inside the square brackets in the first term, adding and subtracting some terms which correspond to the index $i = 0$, we get for $j = 0, 1, \ldots, m$ and $k = 1, \ldots, m$ (the summation over the repeated indices)

$$\sum_{i=1}^{m} \int_{\Omega'} \left[D_iD_k u_n - \varepsilon \frac{\partial a_i}{\partial p_j^{(l)}} D_jD_k u_n^{(l)} - \varepsilon \frac{\partial a_i}{\partial x_k}\right] D_iD_k v_{n+1}\psi dx =$$
$$= \sum_{i=0}^{m} \int_{\Omega'} \left[D_iD_k u_n - \varepsilon \frac{\partial a_i}{\partial p_j^{(l)}} D_jD_k u_n^{(l)}\right] D_iD_k v_{n+1}\psi dx -$$

*Here we use the sign of $\sum$ because the limits of the summation are vague.

$$-\int_{\Omega'}\left[D_k u_n - \varepsilon\frac{\partial a_0}{\partial p_j^{(l)}}D_j D_k u_n^{(l)}\right] D_k v_{n+1}\psi dx -$$
$$-\varepsilon\sum_{i=1}^{m}\int_{\Omega'}\frac{\partial a_0}{\partial x_k}D_i D_k v_{n+1}\psi dx.$$

Then

$$I_1 = \sum_{i=0}^{m}\int_{\Omega'}\left[D_i D_k u_n - \varepsilon\frac{\partial a_i}{\partial p_j^{(l)}}D_j D_k u_n^{(l)}\right] D_i D_k v_{n+1}\psi dx - \tag{4.4.7}$$
$$-\int_{\Omega'}\left[D_k u_n - \varepsilon\frac{\partial a_0}{\partial p_0^{(l)}}D_k u_n^{(l)}\right] D_k v_{n+1}\psi dx + \varepsilon\sum_{j=1}^{m}\int_{\Omega'}\frac{\partial a_0}{\partial p_j^{(l)}}D_j D_k u_n^{(l)} D_k v_{n+1}\psi dx +$$
$$+\sum_{i=1}^{m}\int_{\Omega'}\left[D_i u_n - \varepsilon a_i(x; Du_n)\right] D_i D_k v_{n+1} D_k\psi dx -$$
$$-\sum_{i=1}^{m}\int_{\Omega'}\left[D_i u_n - \varepsilon a_i(x; Du_n)\right]\Delta v_{n+1} D_i\psi dx - \varepsilon\sum_{i=1}^{m}\int_{\Omega'}\frac{\partial a_i}{\partial x_k}D_i D_k v_{n+1} dx.$$

To estimate the first sum on the right-hand side of (4.4.7). Separate one term which corresponds to the fixed k. According to the definition of the matrix A by (1.1.4), we get

$$\sum_{i=0}^{m}\int_{\Omega'}\left[D_i D_k u_n - \varepsilon\frac{\partial a_i}{\partial p_j^{(l)}}D_j D_k u_n^{(l)}\right] D_i D_k v_{n+1}\psi dx =$$
$$= \int_{\Omega'}(I-\varepsilon A)D(D_k u_n)D(D_k v_{n+1})\psi dx,$$

where Dz is the following vector $(z^{(1)}, \ldots, z^{(N)}, D_1 z^{(1)}, \ldots, D_1 z^{(N)} \ldots, D_m z^{(1)}, \ldots, D_m z^{(N)})$.
Applying lemma 1.1.1 and the Hölder inequality gives the estimate

$$\left|\sum_{i=0}^{m}\int_{\Omega'}\left[D_i D_k u_n - \varepsilon\frac{\partial a_i}{\partial p_j^{(l)}}D_j D_k u_n^{(l)}\right] D_i D_k v_{n+1}\psi dx\right| \le$$
$$\le K_\varepsilon\left[\int_{\Omega'}|D(D_k u_n)|^2\psi dx\right]^{1/2}\left[\int_{\Omega'}|D(D_k v_{n+1})|^2\psi dx\right]^{1/2}.$$

Using the elementary inequality $2ab \le a^2 + b^2$, we have

$$\left|\sum_{i=0}^{m}\int_{\Omega'}\left[D_i D_k u_n - \varepsilon\frac{\partial a_i}{\partial p_j^{(l)}}D_j D_k u_n^{(l)}\right] D_i D_k v_{n+1}\psi dx\right| \le$$
$$\le \frac{1}{2}K_\varepsilon\left(\int_{\Omega'}|D(D_k u_n)|^2\psi dx + \int_{\Omega'}|D(D_k v_{n+1})|^2\psi dx\right).$$

Summing over k from 1 to m gives that the first term on the right-hand side of (4.4.7) which can be estimated in the following way

(4.4.8)
$$\left|\sum_{i=0}^{m}\int_{\Omega'}\left[D_iD_ku_n-\varepsilon\frac{\partial a_i}{\partial p_j^{(l)}}D_jD_ku_n^{(l)}\right]D_iD_kv_{n+1}\psi dx\right|\le$$
$$\le\frac{1}{2}K_\varepsilon\left(\int_{\Omega'}\left|D^2u_n\right|^2\psi dx+\int_{\Omega'}\left|D^2v_{n+1}\right|^2\psi dx\right).$$

Now consider the remaining terms of the right-hand side of (4.4.7). For example, the modulus of the second term

$$-\int_{\Omega'}\left[D_ku_n-\varepsilon\frac{\partial a_0}{\partial p_0^{(l)}}D_ku_n^{(l)}\right]D_kv_{n+1}\psi dx$$

can be estimated by integrals

$$\int_{\Omega'}|Dv_{n+1}|^2\psi dx \quad \text{and} \quad \int_{\Omega'}|Du_n|^2\psi dx.$$

The second term is bounded under inequality (4.4.4).
The expressions under the sign of the remaining integrals are linear with respect to the second order derivatives. Estimate for example, the expression

$$\int_{\Omega'}D_kv_{n+1}\frac{\partial a_0}{\partial p_j^{(l)}}D_jD_ku_n^{(l)}\psi dx.$$

Present this integral in the following form

$$\int_{\Omega'}\psi^{1/2}D_kv_{n+1}\frac{\partial a_0}{\partial p_j^{(l)}}D_jD_ku_n^{(l)}\psi^{1/2}dx.$$

Then apply the inequality (2.2.15) to the integrand and use the boundedness of $\frac{\partial a_0}{\partial p_j^{(l)}}$ to give

(4.4.9)
$$\left|\int_{\Omega'}D_kv_{n+1}\frac{\partial a_0}{\partial p_j^{(l)}}D_jD_ku_n^{(l)}\psi dx\right|\le$$
$$\le\eta\int_{\Omega'}|D'^2u_n|^2\psi dx+C\|v_{n+1}\|^2.$$

Taking into account that η is positive and arbitrarily small and u_n satisfies (4.4.4) we come to the following estimate

$$|I_1|\le\frac{1}{2}(K_\varepsilon+\eta)\left(\int_{\Omega'}|D^2u_n|^2\psi dx+\int_{\Omega'}|D^2v_{n+1}|^2\psi dx\right)+$$
$$+C_1\|v_{n+1}\|^2+C_2.$$

Now return to the identity (4.4.5). It follows from the last inequality that

$$\int_{\Omega'} \Delta u_{n+1} \Delta v_{n+1} \psi dx + \int_{\Omega'} D' u_{n+1} D' v_{n+1} \psi dx \leq$$

$$\leq \frac{1}{2}(K_\varepsilon + \eta) \left(\int_{\Omega'} |D^2 u_n|^2 \psi dx + \int_{\Omega'} |D^2 v_{n+1}|^2 \psi dx \right) +$$

$$+ \left| \int_{\Omega'} u_{n+1} D' v_{n+1} D' \psi dx \right| + \left| \int_{\Omega'} [u_n - \varepsilon a_0(x; Du_n)] \Delta v_{n+1} \psi dx \right| +$$

$$+ C_1 \|v_{n+1}\|^2 + C_2.$$

Transform by parts the first term of the left-hand side integrating in the following way

$$\int_{\Omega'} \Delta u_{n+1} \Delta v_{n+1} \psi dx = \sum_{i,k=1}^{m} \int_{\Omega'} D_i D_i u_{n+1} D_k D_k v_{n+1} \psi dx =$$

$$= - \sum_{i,k=1}^{m} \int_{\Omega'} D_i u_{n+1} D_i D_k D_k v_{n+1} \psi dx - \sum_{i=1}^{m} \int_{\Omega'} D_i u_{n+1} \Delta v_{n+1} D_i \psi dx =$$

$$= \sum_{i,k=1}^{m} \int_{\Omega'} D_i D_k u_{n+1} D_i D_k v_{n+1} \psi dx - \sum_{i=1}^{m} \int_{\Omega'} D_i u_{n+1} \Delta v_{n+1} D_i \psi dx +$$

$$+ \sum_{i,k=1}^{m} \int_{\Omega'} D_i u_{n+1} D_i D_k v_{n+1} D_k \psi dx.$$

Now the last inequality can be introduced in the form

$$\sum_{i,k=1}^{m} \int_{\Omega'} D_i D_k u_{n+1} D_i D_k v_{n+1} \psi dx + \int_{\Omega'} D' u_{n+1} D' v_{n+1} \psi dx \leq$$

$$\leq \frac{1}{2}(K_\varepsilon + \eta) \left(\int_{\Omega'} |D^2 u_n|^2 \psi dx + \int_{\Omega'} |D^2 v_{n+1}|^2 \psi dx \right) +$$

$$+ \left| \int_{\Omega'} u_{n+1} D' v_{n+1} D' \psi dx \right| + \left| \int_{\Omega'} [u_n - \varepsilon a_0(x; Du_n)] \Delta v_{n+1} \psi dx \right| +$$

$$+ \left| \sum_{i=1}^{m} \int_{\Omega'} D_i u_{n+1} \Delta v_{n+1} D_i \psi dx \right| + \left| \sum_{i,k=1}^{m} \int_{\Omega'} D_i u_{n+1} D_i D_k v_{n+1} D_k \psi dx \right| +$$

$$+ C_1 \|v_{n+1}\|^2 + C_2$$

As we mentioned before, v_{n+1} is smooth in Ω. Since all the iterations u_n belong to $W_2^{(2)}(\Omega')$, the iteration u_{n+1} can be approximated by a smooth function v_{n+1}. According to our assumptions $a_0(x; Du_n) \in \mathcal{L}_2$. Hence both sides of the last inequality pass to closure to give

$$(4.4.10) \quad \int_{\Omega'} |D^2 u_{n+1}|^2 \psi dx \leq \frac{1}{2}(K_\varepsilon + \eta) \left(\int_{\Omega'} |D^2 u_n|^2 \psi dx + \int_{\Omega'} |D^2 u_{n+1}|^2 \psi dx \right) +$$

$$+\left|\int_{\Omega'} u_{n+1} D' u_{n+1} D'\psi dx\right| + \int_{\Omega'} |u_{n+1}|^2 \psi dx +$$

$$+\left|\sum_{i=1}^{m}\int_{\Omega'} D_i u_{n+1} \Delta u_{n+1} D_i \psi dx\right| + \left|\sum_{i,k=1}^{m}\int_{\Omega'} D_i u_{n+1} D_i D_k u_{n+1} D_k \psi dx\right| +$$

$$+\left|\int_{\Omega'} [u_n - \varepsilon a_0(x; Du_n)]\, \Delta u_{n+1} \psi dx\right| + C_1\|u_{n+1}\|^2 + C_2.$$

It follows from inequality (4.4.4) the second and the third terms are estimated from the above by a constant. As we have seen when proving (4.4.9), the remaining terms on the right-hand side of the last inequality are estimated from the above by the following items

$$\eta\|u_{n+1}\|^2_{W_2^{(2)}(\Omega')},\quad C\|u_n\|^2_{W_2^{(1)}(\Omega')} \quad \text{and} \quad C\|u_{n+1}\|^2_{W_2^{(1)}(\Omega')}.$$

The last two terms are also estimated from above by a constant. Thus from (4.4.10) we get the inequality

(4.4.11)
$$\int_{\Omega'} |D^2 u_{n+1}|^2 \psi dx \leq$$

$$\leq \frac{1}{2}(K_\varepsilon + \eta)\left(\int_{\Omega'} |D^2 u_{n+1}|^2 \psi dx + \int_{\Omega'} |D^2 u_n|^2 \psi dx\right) + C.$$

From (4.4.11) the relation

$$\int_{\Omega'} |D^2 u_{n+1}|^2 \psi dx \leq \left(\frac{K_\varepsilon + \eta}{2 - K_\varepsilon - \eta}\right)\int_{\Omega'} |D^2 u_n|^2 \psi dx + C$$

follows. Since $K_\varepsilon < 1$ then $\frac{K_\varepsilon+\eta}{2-K_\varepsilon-\eta} < 1$. Therefore we get the estimate

$$\int_{\Omega'} |D^2 u_n|^2 \psi dx \leq C,$$

where C is independent of n.
As long as ψ is positive inside Ω' we get required inequality (4.4.2).

□

Theorem 4.4.1. *If $\partial\Omega \in C^{1,\gamma}$ $(\gamma > 0)$, (4.4.1) is satisfied and $u_n\big|_{\partial\Omega} = g$, where g is a trace of some function from $W_2^{(2)}$ then in the entire domain Ω the inequality*

(4.4.12)
$$\|u_n\|_{W_2^{(2)}(\Omega)} \leq C$$

holds.

Proof. Take any function $u_0 \in W_2^{(2)}(\Omega)$ whose trace on $\partial\Omega$ coincides with g and denote $u = z + u_0$. Then for z we'll have a homogeneous condition $z\big|_{\partial\Omega} = 0$. All conditions of the theorem for the coefficients and domain will remain true.

Now take a small piece of $\partial\Omega$ and straightened it as in section 2.4. Extend u_{n+1} in an antisymmetric way on $x'_m > 0$. Then u_{n+1} will satisfy "almost" a Poisson equation in some domain which is symmetric with respect to $x'_m = 0$. The difference between operator Δ in new coordinates and the operator of the left-hand side of the iterative process (2.3.25) will be a linear operator with small coefficients. Take into account the problem is reduced to the estimates in the interior domain and refer to lemma 4.1.1 .

□

4.5 Hölder continuity for the first derivatives

Let us prove now that under some additional restrictions the weak solution of system (2.3.18) with boundary condition (2.3.20) will have the Hölder derivatives inside Ω. It follows from theorem 1.2.1 that it is sufficient to show $u \in H_{2,2,\alpha}(\Omega')$, where $\alpha = 2 - m - 2\gamma$, $0 < \gamma < 1$ and Ω' is a strictly interior subdomain of Ω. Suppose the norm in $H_{2,2,\alpha}(\Omega')$ is defined by (2.1.5). As was mentioned in section 2.1, this norm is equivalent to the one in (2.1.2). To simplify the considerations, suppose a_i $i = 1,\ldots,m$, are independent of x and a_0 satisfies the inequality

$$|a_0(x; Du)| \leq C|Du| + \varphi(x),$$

where $\varphi \in \mathcal{L}_q(\Omega)$ for some $q > 2m(m+\alpha)^{-1}$.

Theorem 4.5.1. *If the conditions of sections 1.1,3.4 (lemma 4.4.1) and of this section are satisfied and the inequality*

$$K^2\left[M_3^2(1+|\alpha|) + |\alpha||\alpha - 4|M_2^2\right] < 1 \tag{4.5.1}$$

holds for the positive constants defined, respectively, by (4.1.8) *or* (4.1.9), (4.1.7) *and* (4.1.6), *then the weak solution of the problem* (2.3.18), (2.3.20) *has the Hölder continuous derivatives inside* Ω.

Proof. First note that according to lemma 4.4.1 the norms $\|D^2u_n\|_{\mathcal{L}_2}$ are bounded in common in any strictly interior domain. Take $u_0 \in W_q^{(2)}(\Omega'') \cap W_2^{(1)}(\Omega)$ for $q > \frac{2m}{m+\alpha}$. Then from the conditions of the theorem it follows all u_n belong to $W_q^{(2)}(\Omega') \cap W_2^{(1)}(\Omega)$, where Ω' is a strictly interior subdomain of Ω''. Hence all u_n in Ω' are continuously differentiable. This follows from embedding $W_q^{(2)} \subset C^{1,0}$. Take an arbitrary point $x_0 \in \omega$, where ω is a strictly interior subdomain of Ω'. Surround x_0 with the balls $B_{\delta/2} \equiv B_{\delta/2}(x_0)$ and $B_\delta \equiv B_\delta(x_0)$, where δ is so small that $B_\delta(x_0)$ lies inside Ω'. It is clear that for any integer n the integral $\int_{B_\delta} |D^2u_n|^2|x-x_0|^\alpha dx$ will be finite. In fact, it follows from the Hölder inequality that

$$\int_{B_\delta} |D^2u_n|^2|x-x_0|^\alpha dx \leq \left(\int_{B_\delta} |D^2u_n|^q dx\right)^{2/q} \left(\int_{B_\delta} |x-x_0|^{\frac{q\alpha}{q-2}} dx\right)^{(q-2)/q}.$$

Since $q > 2m(m+\alpha)^{-1}$, we have $q\alpha(q-2)^{-1} > -m$ and the second integral on the right-hand side is finite.

Denote as usual $r = |x - x_0|$. Let $\psi(r)$ be a smooth (except in the point x_0) function equal to r^α inside $B_{\delta/2}$ and vanishes in the neighbourhood of ∂B_δ. Equality (2.3.22) can be introduced in the following form

$$\Delta u_{n+1} - u_{n+1} = \sum_{j=1}^{m} D_j(D_j u_n - D_j u_n\big|_0) - u_n -$$
$$-\varepsilon\left\{\sum_{i=1}^{m} D_i\left[a_i(Du_n) - a_i(Du_n)\big|_0\right] - a_0(x; Du_n)\right\},$$

where $\big|_0$ indicates the functions are calculated for $x = x_0$.
Multiply this equality by $\psi(r)\Delta v_{n+1}$, where v_{n+1} is a smooth function in Ω and integrate the derived equality over B_δ. Integrate the principal terms of the left-hand side two times by parts in the exact way we did it in the proof of lemma 4.1.1. Now transfer the derivatives first on v_{n+1} and then on u_n. From the properities of ψ it follows that all the boundary integrals vanish to give the identity

$$\int_{B_\delta} \Delta u_{n+1}\Delta v_{n+1}\psi dx =$$
$$= \sum_{i,k=1}^{m}\int_{B_\delta} D_i^2 u_{n+1} D_k^2 v_{n+1}\psi dx = \sum_{i,k=1}^{m}\int_{B_\delta} D_i(D_i u_{n+1} - D_i u_{n+1}\big|_0) D_k^2 v_{n+1}\psi dx =$$
$$= -\sum_{i,k=1}^{m}\left\{\int_{B_\delta}(D_i u_{n+1} - D_i u_{n+1}\big|_0) D_i D_k^2 v_{n+1}\psi dx + \int_{B_\delta}(D_i u_{n+1} - D_i u_{n+1}\big|_0) D_k^2 v_{n+1} D_i\psi dx\right\} =$$
$$= \sum_{i,k=1}^{m}\left\{\int_{B_\delta} D_i D_k u_{n+1} D_i D_k v_{n+1}\psi dx + \int_{B_\delta}(D_i u_{n+1} - D_i u_{n+1}\big|_0) D_i D_k v_{n+1} D_k\psi dx +\right.$$
$$+\int_{B_\delta} D_i D_k u_{n+1}(D_k v_{n+1} - D_k v_{n+1}\big|_0) D_i\psi dx +$$
$$\left.+\int_{B_\delta}(D_i u_{n+1} - D_i u_{n+1}\big|_0)(D_k v_{n+1} - D_k v_{n+1}\big|_0) D_i D_k\psi dx\right\}.$$

Similarly, transfering the derivatives by integrating by parts on the right-hand side we come to the following equalities

$$\int_{B_\delta}\left[\Delta u_n - u_n - \varepsilon L(u_n)\right]\Delta v_{n+1}\psi dx = \sum_{i,k=1}^{m}\int_{B_\delta}\left\{D_i(D_i u_n - D_i u_n\big|_0) -\right.$$
$$\left.-\varepsilon D_i\left[a_i(Du_n) - a_i(Du_n)\big|_0\right]\right\} D_k(D_k v_{n+1} - D_k v_{n+1}\big|_0)\psi dx -$$
$$-\int_{B_\delta}\left[u_n - \varepsilon a_0(x; Du_n)\right]\Delta v_{n+1}\psi dx =$$
$$= -\sum_{i,k=1}^{m}\int_{B_\delta}\left\{D_i u_n - D_i u_n\big|_0 - \varepsilon\left[a_i(Du_n) - a_i(Du_n)\big|_0\right]\right\} D_i D_k\left(D_k v_{n+1} - D_k v_{n+1}\big|_0\right)\psi dx -$$
$$-\sum_{i,k=1}^{m}\int_{B_\delta}\left\{D_i u_n - D_i u_n\big|_0 - \varepsilon\left[a_i(Du_n) - a_i(Du_n)\big|_0\right]\right\} D_k\left(D_k v_{n+1} - D_k v_{n+1}\big|_0\right) D_i\psi dx -$$

$$-\int_{B_\delta} [u_n - \varepsilon a_0(x_0; Du_n)]\, \Delta v_{n+1} \psi dx =$$

$$= \sum_{i,k=1}^{m} \int_{B_\delta} D_k [D_i u_n - \varepsilon a_i(Du_n)]\, D_i D_k v_{n+1} \psi dx +$$

$$+ \sum_{i,k=1}^{m} \int_{B_\delta} \left\{ D_i u_n - D_i u_n\Big|_0 - \varepsilon \left[a_i(Du_n) - a_i(Du_n)\Big|_0 \right] \right\} D_i D_k v_{n+1} D_k \psi dx +$$

$$+ \sum_{i,k=1}^{m} \int_{B_\delta} D_k [D_i u_n - \varepsilon a_i(Du_n)] \left(D_k v_{n+1} - D_k v_{n+1}\Big|_0 \right) D_i \psi dx +$$

$$+ \sum_{i,k=1}^{m} \int_{B_\delta} \left\{ D_i u_n - D_i u_n\Big|_0 - \varepsilon \left[a_i(Du_n) - a_i(Du_n)\Big|_0 \right] \right\} \left(D_k v_{n+1} - D_k v_{n+1}\Big|_0 \right) D_i D_k \psi dx -$$

$$- \int_{B_\delta} [u_n - \varepsilon a_0(x; Du_n)]\, \Delta v_{n+1} \psi dx.$$

Each term in these equalities is a linear bounded functional with respect to v_{n+1} in $W_q^{(2)}(B_\delta)$. The terms $\int_{B_\delta} D^2 u_{n+1} D^2 v_{n+1} \psi dx$ are bounded because in $B_{\delta/2}$ the function ψ is equal to $|x - x_0|^\alpha$ and it is smooth in $B_{\delta/2,\delta}$. *. Moreover, u_{n+1} and $v_{n+1} \in W_q^{(2)}(B_\delta)$ for $q > 2m(m+\alpha)^{-1}$. Applying the Hölder inequality, we have

$$\int_{B_\delta} D^2 u_{n+1} D^2 v_{n+1} \psi dx C \left(\int_{B_\delta} |D^2 u_{n+1}|^2 |x - x_0|^\alpha dx \right)^{1/2} \left(\int_{B_\delta} |D^2 v_{n+1}|^2 |x - x_0|^\alpha dx \right)^{1/2}$$

to give the relation

$$\int_{B_\delta} D^2 u_{n+1} D^2 v_{n+1} \psi dx \le$$
$$\le C \|u_{n+1}\|_{W_q^{(2)}(B_\delta)} \|v_{n+1}\|_{W_q^{(2)}(B_\delta)},$$

which establishes the boundedness of the functionals under consideration in $W_q^{(2)}$. For the rest of the terms the same statement is similarly established. For example, in the term

$$\int_{B_\delta} (D_i u_n - D_i u_n\Big|_0) D_i D_k v_{n+1} D_k \psi dx$$

note for $x \in B_{\delta/2}$ the value $D_k \psi$ can be estimated by $C|x - x_0|^{\alpha-1}$. Hence,

$$\left| \int_{B_\delta} (D_i u_n - D_i u_n\Big|_0) D_i D_k v_{n+1} D_k \psi dx \right| \le$$
$$\le C \int_{B_\delta} |D_i u_n - D_i u_n\Big|_0|^2 |x - x_0|^{\alpha-2} dx \int_{B_\delta} |D^2 v|^2 |x - x_0|^\alpha dx.$$

The first integral on the right-hand side can be estimated with the help of (2.1.14) to give the required statement.

*From this it follows that for $x \in B_{\delta/2,\delta}$ the inequality $|\psi| \le C|x - x_0|^\alpha$ holds.

Now by examining the closure in $W_q^{(2)}(B_\delta)$ we can state the equalities considered are true for $v_{n+1} = u_{n+1}$ giving the following identities

$$(4.5.2) \qquad \int_{B_\delta} |\Delta u_{n+1}|^2 \psi dx = \int_{B_\delta} |D'^2 u_{n+1}|^2 \psi dx +$$
$$+ \sum_{i,k=1}^m \Big\{ \int_{B_\delta} (D_i u_{n+1} - D_i u_{n+1}\big|_0) D_i D_k u_{n+1} D_k \psi dx +$$
$$+ \int_{B_\delta} D_i D_k u_{n+1} (D_k u_{n+1} - D_k u_{n+1}\big|_0) D_i \psi dx +$$
$$+ \int_{B_\delta} (D_i u_{n+1} - D_i u_{n+1}\big|_0)(D_k u_{n+1} - D_k u_{n+1}\big|_0) D_i D_k \psi dx \Big\},$$

$$(4.5.3) \qquad \int_{B_\delta} [\Delta u_n - u_n - \varepsilon L(u_n)]\, \Delta u_{n+1} \psi dx =$$
$$= \sum_{i,k=1}^m \Big\{ \int_{B_\delta} D_k [D_i u_n - \varepsilon a_i(Du_n)]\, D_i D_k u_{n+1} \psi dx +$$
$$+ \sum_{i,k=1}^m \int_{B_\delta} \Big\{ D_i u_n - D_i u_n\big|_0 - \varepsilon \big[a_i(Du_n) - a_i(Du_n)\big|_0\big] \Big\} D_i D_k u_{n+1} D_k \psi dx +$$
$$+ \sum_{i,k=1}^m \int_{B_\delta} D_k [D_i u_n - \varepsilon a_i(Du_n)]\, (D_k u_{n+1} - D_k u_{n+1}\big|_0) D_i \psi dx +$$
$$+ \sum_{i,k=1}^m \int_{B_\delta} \Big\{ D_i u_n - D_i u_n\big|_0 - \varepsilon \big[a_i(Du_n) - a_i(Du_n)\big|_0\big] \Big\} \times$$
$$\times \Big(D_k u_{n+1} - D_k u_{n+1}\big|_0\Big) D_i D_k \psi dx - \int_{B_\delta} [u_n - \varepsilon a_0(x; Du_n)]\, \Delta u_{n+1} \psi dx$$

and

$$(4.5.4) \qquad \int_{B_\delta} (\Delta u_{n+1} - u_{n+1}) \Delta u_{n+1} \psi dx = \int_{B_\delta} [\Delta u_n - u_n - \varepsilon L(u_n)]\, \Delta u_{n+1} \psi dx.$$

Since ψ is singular, only when $x = x_0$, it is sufficient to estimate the integrals on the right-hand side of (4.5.2) and (4.5.3) in $B_{\delta/2}$. The integrals over $B_{\delta/2,\delta}$ can be estimated by the norms of u_n and u_{n+1} in $W_2^{(2)}$. It is evident that inside $B_{\delta/2}$ the equalities

$$D_i \psi = \alpha r^{\alpha-2} x_i,$$
$$D_i D_j \psi = \alpha(\alpha-2) r^{\alpha-4} x_i x_j + \alpha \delta_{ij} r^{\alpha-2}$$

hold. Hence, after simple calculations it follows from (4.5.4) that

$$\int_{B_{\delta/2}} |\Delta u_{n+1}|^2 \psi dx = \int_{B_{\delta/2}} [\Delta u_n - \varepsilon L(u_n)]\, \Delta u_{n+1} \psi dx + \ldots =$$
$$= \int_{B_{\delta/2}} \Big\{ \Big(D_i D_k u_n - \varepsilon \frac{\partial a_i}{\partial p_s^{(l)}} D_s D_k u_n^{(l)} \Big) D_i D_k u_{n+1} r^\alpha +$$

$$+\alpha\left(D_iD_ku_n-\varepsilon\frac{\partial a_i}{\partial p_s^{(l)}}D_sD_ku_n^{(l)}\right)\left(D_ku_{n+1}-D_ku_{n+1}\Big|_0\right)x_ir^{\alpha-2}+$$

$$+\alpha\left[D_iu_n-D_iu_n\Big|_0-\varepsilon\frac{\overline{\partial a_i}}{\partial p_s^{(l)}}(D_su_n^{(l)}-D_su_n^{(l)}\Big|_0)\right]D_iD_ku_{n+1}x_kr^{\alpha-2}+$$

$$+\alpha(\alpha-2)\left[D_iu_n-D_iu_n\Big|_0-\varepsilon\frac{\overline{\partial a_i}}{\partial p_s^{(l)}}\left(D_su_n^{(l)}-D_su_n^{(l)}\Big|_0\right)\right]\left(D_ku_{n+1}-D_ku_{n+1}\Big|_0\right)x_ix_kr^{\alpha-4}+$$

$$+\alpha\left[D_iu_n-D_iu_n\Big|_0-\varepsilon\frac{\overline{\partial a_i}}{\partial p_s^{(l)}}\left(D_su_n^{(l)}-D_su_n^{(l)}\Big|_0\right)\right]\left(D_ku_{n+1}-\right.$$

$$\left.-D_ku_{n+1}\Big|_0\right)\delta_{ik}r^{\alpha-2}\Big\}dx+\dots,$$

where the bar over the derivatives of a_i indicates their arguments take the intermediate values of u and p, the dots include the weak terms, for example, $\int\limits_{B_\delta}u_{n+1}\Delta u_{n+1}\psi dx$ or the integrals over $B_{\delta/2,\delta}$ and the summation is taken over all indices from 1 to m or from 1 to N. Assume now that the summation over i,s and k is taken from zero to m. Then the added terms will be subordinated to the previous ones. In fact, these terms contain the products of the second derivatives on the first derivatives, the second derivatives on the functions themselves, or the product of the first derivatives between themselves. These terms can be estimated by means of (2.1.17) by the norms of second derivatives with a small coefficients $\eta>0$. Apply now to each of the the terms the Hölder inequality in such a way that all second derivatives have the factor r^α. Then we shall have

(4.5.5)
$$\int\limits_{B_{\delta/2}}|\Delta u_{n+1}|^2r^\alpha dx\le$$

$$\le(K_\varepsilon+\eta)\Bigg\{\left(\int\limits_{B_{\delta/2}}|D^2u_n|^2r^\alpha dx\right)^{1/2}\left(\int\limits_{B_{\delta/2}}|D^2u_{n+1}|^2r^\alpha dx\right)^{1/2}+$$

$$+|\alpha|\left(\int\limits_{B_{\delta/2}}|D^2u_n|^2r^\alpha dx\right)^{1/2}\left(\int\limits_{B_{\delta/2}}\left|Du_{n+1}-Du_{n+1}\Big|_0\right|^2r^{\alpha-2}dx\right)^{1/2}+$$

$$+|\alpha|\left(\int\limits_{B_{\delta/2}}|D^2u_{n+1}|^2r^\alpha dx\right)^{1/2}\left(\int\limits_{B_\delta}\left|Du_n-Du_n\Big|_0\right|^2r^{\alpha-2}dx\right)^{1/2}+$$

$$+|\alpha||\alpha-3|\left(\int\limits_{B_{\delta/2}}\left|Du_n-Du_n\Big|_0\right|^2r^{\alpha-2}dx\right)^{1/2}\left(\int\limits_{B_{\delta/2}}\left|Du_{n+1}-Du_{n+1}\Big|_0\right|^2r^{\alpha-2}dx\right)^{1/2}+C,$$

where

$$\left|Du-Du\Big|_0\right|^2=\left|D'u-D'u\Big|_0\right|^2+\left|u-u_0\Big|_0\right|^2,$$

$$\left|D'u-D'u\Big|_0\right|^2=\sum_{i=1}^m\left|D_iu-D_iu\Big|_0\right|^2$$

and C depends only on the norms of u_n and u_{n+1} in $W_2^{(2)}(\Omega')$. Applying the Hölder inequality to the right-hand side of (4.5.5), we come to

$$\left|\int\limits_{B_{\delta/2}} |\Delta u_{n+1}|^2 r^\alpha dx\right|^2 \leq \tag{4.5.6}$$

$$\leq (K_\varepsilon+\eta)^2 \int\limits_{B_{\delta/2}} \left[(1+|\alpha|)|D^2u_n|^2 + |\alpha||\alpha-4|\left|Du_{n+1} - Du_{n+1}\Big|_0\right|^2 r^{-2}\right] r^\alpha dx \times$$

$$\times \int\limits_{B_{\delta/2}} \left[(1+|\alpha|)|D^2u_{n+1}|^2 + |\alpha|\,|\alpha-4|\left|Du_n - Du_n\Big|_0\right|^2 r^{-2}\right] r^\alpha dx + C.$$

According to (1.1.9′), we have

$$\int\limits_{B_{\delta/2}} \left|Du - Du\Big|_0\right|^2 r^{\alpha-2}dx = \int\limits_{B_{\delta/2}} \left|D'u - D'u\Big|_0\right| r^{\alpha-2}dx + \int\limits_{B_{\delta/2}} \left|u - u\Big|_0\right|^2 r^{\alpha-2}dx \leq$$

$$\leq \int\limits_{B_{\delta/2}} \left|D'u - D'u\Big|_0\right|^2 r^{\alpha-2}dx + C \int\limits_{B_{\delta/2}} |D'u|^2 r^\alpha dx.$$

From inequality (2.1.17′), applied to the first term on the right-hand side, it follows that

$$\int\limits_{B_{\delta/2}} \left|Du - Du\Big|_0\right|^2 r^{\alpha-2}dx \leq$$

$$\leq \int\limits_{B_{\delta/2}} \left|D'u - D'u\Big|_0\right|^2 r^{\alpha-2}dx + \eta \int\limits_{B_{\delta/2}} |D^2u|^2 r^\alpha dx.$$

Then from (4.5.6) we get

$$\left|\int\limits_{B_{\delta/2}} |\Delta u_{n+1}|^2 r^\alpha dx\right|^2 \leq$$

$$\leq (K_\varepsilon+\eta)^2 \left\{ \int\limits_{B_{\delta/2}} \left[(1+|\alpha|)|D^2u_n|^2 + |\alpha||\alpha-4|\left|D'u_n - D'u_n\Big|_0\right|^2 r^{-2}\right] r^\alpha dx + C\right\} \times$$

$$\times \left\{ \int\limits_{B_{\delta/2}} \left[(1+|\alpha|)|D^2u_{n+1}|^2 + |\alpha||\alpha-4|\left|D'u_{n+1} - D'u_{n+1}\Big|_0\right|^2 r^{-2}\right] r^\alpha dx + C\right\}.$$

Note $D_iu - D_iu\Big|_0 = D_i(u-\tilde{u})$, where $\tilde{u}$ is a linear function.
Applying (4.1.3) to the function $(u-\tilde{u})\zeta$, we find

$$\int\limits_{B_{\delta/2}} \left|D'u - D'u\Big|_0\right|^2 r^{\alpha-2}dx \leq$$

$$\leq (M_2^2+\eta) \int\limits_{B_{\delta/2}} |\Delta u|^2 r^\alpha dx + C.$$

Applying (4.1.8), gives the estimate

$$\left| \int_{B_{\delta/2}} |\Delta u_{n+1}|^2 r^\alpha dx \right|^2 \leq (K_\varepsilon + \eta)^2 \left[(1+|\alpha|)M_3^2 + |\alpha||\alpha - 4|M_2^2 \right]^2 \times$$

$$\times \left(\int_{B_{\delta/2}} |\Delta u_n|^2 r^\alpha dx + C \right) \left(\int_{B_{\delta/2}} |\Delta u_{n+1}|^2 r^\alpha dx + C \right).$$

Denote $\sigma_n = \int_{B_{\delta/2}} |\Delta u_n|^2 r^\alpha dx$. After some simple calculations we shall have

$$\sigma_{n+1}^2 \leq (K_\varepsilon + \eta)^2 \left[(1+|\alpha|)M_3^2 + |\alpha||\alpha - 4|M_2^2 \right] \sigma_n \sigma_{n+1} +$$
$$+ C(\sigma_n + \sigma_{n+1}) + C_1.$$

It follows from inequality (2.2.15) that $C\sigma_{n+1} \leq \eta \sigma_{n+1}^2 + C_2$. Hence, the sequence σ_n satisfies the inequality

$$\sigma_{n+1}^2 \leq (K_\varepsilon + \eta)^2 \left[(1+|\alpha|)M_3^2 + |\alpha||\alpha - 4|M_2^2 \right] \sigma_n \sigma_{n+1} + C\sigma_n + C_1.$$

Then, according to condition (4.5.1), by means of lemma 2.3.2 we come to the boundedness of σ_n in common. Therefore the sequence of norms $\int |D^{2} u_n|^2 r^\alpha dx$ is boundeded in common. From this it follows that the sequence of norms $\int_{w_{\delta/2}} |D^{2} u_n|^2 r^\alpha dx$, where $\omega_{\delta/2} = \omega \cap B_{\delta/2}$, is also bounded. As $u_n \to u$ in $W_2^{(1)}$, the weak solution belongs to $H_{2,2,\alpha}(\Omega')$.

□

4.6 Regularity of solutions for degenerated elliptic systems

In a bounded domain $\Omega \subset R^m$ $(m \geq 2)$ we consider a system

(4.6.1) $$L(u) \equiv \sum_{i=1}^{m} D_i a_i(x, Du) - a_0(x, Du) = 0,$$

where u and $a_i(x,p)(i = 0, 1, \ldots, m)$ are N-dimensional vector functions with components $u^{(k)}(x), a_i^{(k)}(x,p)(k = 1, \ldots, N)$,

$$Du = (D_0 u, D_1 u, \ldots, D_m u), \; D_i = \frac{\partial}{\partial x_i}(i = 1, \ldots, m) \text{ and } D_0 = I.$$

About the functions $a_i(x,p)$ we assume they satisfy the conditions 1) and 3) of 1.1. We also suppose that the following assumptions hold true:

(1) the matrix

(4.6.2) $$A = \left\{ \frac{\partial a_i^{(k)}}{\partial p_j^{(l)}} \right\}$$

is symmetric and the eigenvalues of this matrix satisfy the inequalities

$$\frac{\lambda}{1+|p|^s} \le \lambda_j(x,p) \le \frac{\Lambda}{1+|p|^s}, \tag{4.6.3}$$

where $\lambda, \Lambda =$ const > 0 and $0 \le s < 1$;

(2) the inequality

$$\left|\frac{\partial a_i}{\partial x_k}\right| \le C|p| + b \quad (i = 1, \dots, m) \tag{4.6.4}$$

holds, where b is a sufficiently small nonnegative value;

(3) for all $u \in W_q^{(2)}(\Omega)$ with $q > 1$ the result of substitution of u in $L(u)$ belongs to $L_q(\Omega)$.

Consider the solution of (4.6.1) with the boundary condition

$$u|_{\partial\Omega} = 0. \tag{4.6.5}$$

In theorem 1.3.1 it was proved the universal iterative process

$$\Delta u_{n+1} - u_{n+1} = \Delta u_n - u_n - \Lambda^{-1}L(u_n), \quad u_n|_{\partial\Omega} = 0 (n = 0, 1, \dots,) \tag{4.6.6}$$

converges in $W_{2-s}^{(1)}(\Omega)$ to the weak solution u of (4.6.1), (4.6.5) if $u \in W_2^{(1)}(\Omega)$. Consider also the process (4.6.6) with a penalty term,

$$\Delta u_{n+1} - u_{n+1} = \Delta u_n - u_n - \Lambda^{-1}[\delta \Delta u_n + L(u_n)] \quad (\delta \ge 0), \tag{4.6.7}$$

with the same condition (4.6.5).
In the corollary to the theorem 1.4.1 it was also proved a subsequence of the iterations of process (4.6.7) converges weakly to the solution. So, if we want to prove the solution has the Hölder continuous first derivatives, it is enough to show the iterations of(4.6.6) or (4.6.7) satisfy the inequality

$$\int_{\Omega_R} |D^2 u_n|^2 r^\alpha dx \le C, \tag{4.6.8}$$

where $\Omega_R = B_R(x_0) \cap \Omega, \quad x_0 \in \overline{\Omega}, \quad \alpha = 2 - m - 2\gamma (0 < \gamma < 1), \quad r = |x - x_0|$, and C independent of x_0 and n. It is also assumed that R is sufficiently small and fixed.

Lemma 4.6.1. *If the conditions* (1) - (3) *of this section and conditions 1) and 2) of 1.1 are satisfied and* $u_0(x) \in \overset{\circ}{W}{}_2^{(1)}(\Omega)$, *then the relation*

$$\left(\int_\Omega |Du_{n+1}|^2 dx\right)^{1/2} \le \left(1 - \frac{\lambda\Lambda^{-1}}{1 + [\max\{\sup_\Omega |Du_n|, \sup_\Omega |Du_{n+1}|\}]^s}\right)^{1/2} \times \tag{4.6.9}$$

$$\times \left(\int_\Omega |Du_n|^2 dx\right)^{1/2} + \Lambda^{-1}|a|,$$

holds, where

$$(4.6.10) \qquad |a|^2 = \int_\Omega \sum_{i=0}^{m} |a_i(x,0)|^2 dx.$$

Proof. Multiply both sides of the system (4.6.6) by u_{n+1} and integrate once by parts. Then

$$\int_\Omega |Du_{n+1}|^2 dx = \int_\Omega [D_i u_n - \Lambda^{-1} a_i(x, Du_n)] D_i u_{n+1} dx,$$

where summation is taken over the repeated indices from $i = 0$ to $i = m$.
Adding and subtracting $a_i(x,0)$ under the square brackets on the right-hand side, we get

$$\begin{aligned}\int_\Omega |Du_{n+1}|^2 dx &= \int_\Omega [D_i u_n - \Lambda^{-1}(a_i(x, Du_n) - a_i(x,0))] D_i u_{n+1} dx - \\ &\quad -\Lambda^{-1} \int_\Omega a_i(x,0) D_i u_{n+1} dx.\end{aligned}$$

Applying the mean value theorem gives

$$\begin{aligned}\int_\Omega |Du_{n+1}|^2 dx &= \int_\Omega (I - \Lambda^{-1}\overline{A}) Du_n Du_{n+1} dx - \\ &\quad -\Lambda^{-1} \int_\Omega a_i(x,0) D_i u_{n+1} dx,\end{aligned}$$

where $\bar{A}$ denotes the matrix A with intermediate values of variables.
The Hölder inequality gives

$$\left(\int_\Omega |Du_{n+1}|^2 dx\right)^{1/2} \leq \sup_\Omega ||I - \Lambda^{-1}\overline{A}|| \left(\int_\Omega |Du_n|^2 dx\right)^{1/2} + \Lambda^{-1}|a|.$$

It can be easily proved, that

$$(4.6.11) \qquad \sup_\Omega ||I - \Lambda^{-1}\overline{A}|| \leq \sup_{i,\Omega} |1 - \Lambda^{-1}\overline{\lambda}_i|.$$

Using the right-hand side of the inequalities (4.6.3) we get

$$(4.6.12) \qquad ||I - \Lambda^{-1}\overline{A}|| \leq 1 - \frac{\lambda\Lambda^{-1}}{1 + \left[\max\{\sup_\Omega |Du_n|, \sup_\Omega |Du_{n+1}|\}\right]^s}.$$

□

If conditions (2) and (3) of this section are satisfied and u_0 (the initial iteration of (4.6.6) or (4.6.7)) belongs to $W_q^{(2)}(\Omega) \cap \overset{\circ}{W}{}_q^{(1)}(\Omega) (q > 1)$, then all iterations belong to the same space. The iterations can be extented outside the domain Ω to a sufficiently narrow strip preserving the class. This can be made with the help of the well-known procedure which we have used in the previous two sections. First consider a plane piece of the boundary and expand all of the u_n in an antisymmetric way outside Ω.

This gives the same class of $W_q^{(2)}$ for balls $B_R(x_0)$ which do not completely lie in Ω. It is clear that the conditions (1) - (3) do not change, and the values, s, λ and Λ will be the same. This also gives the possibility to consider only the case where $\Omega_R = \Omega \cap B_R(x_0) = B_R(x_0)$ with a fixed small R_0. Let that the cut-off function $\zeta(r)$ satisfies the inequality

$$|\zeta'||\zeta|^{-1/2} < C \tag{4.6.13}$$

and the boundary of Ω belongs to $C^{1,æ}$ ($æ > 0$).

Lemma 4.6.2. *If conditions of the previous lemma are satisfied and* $u_0 \in W_q^{(2)}(\Omega) \cap \overset{\circ}{W}{}_q^{(1)}(\Omega) (q > 2)$ *then the iterations* (4.6.6) *or* (4.6.7) *satisfy the inequality*

$$\int_{\Omega_R} |D^2 u_{n+1}|^2 \zeta dx \le \left[1 - \frac{\lambda\Lambda^{-1}}{1 + [\max\{\sup_\Omega |Du_n|, \sup_\Omega |Du_{n+1}|\}]^s} + \eta\right] \times \tag{4.6.14}$$
$$\times \int_{\Omega_R} |D^2 u_n|^2 \zeta dx + C|a|^2,$$

where C is independent of $x_0 \in \overline{\Omega}, n$, and in the case of (4.6.7) *of* δ.

Proof. According to theorem 4.4.1 (inequality (4.4.12)) we let $\Omega_R = B_R(x_0)$. Multiply (4.6.6) (or (4.6.7)) by $\Delta u_{n+1}\zeta$ and integrate by parts as in the proof of lemma 4.4.1 . In lemma 4.4.1 it was shown that if ζ satisfies (4.6.13) then

$$\int_{B_R} |D^2 u_{n+1}|^2 \zeta dx \le \sum_{k=0}^{m} \int_{B_R} (I - \Lambda^{-1}\overline{A}) D D_k u_n D D_k u_{n+1} \zeta dx + \tag{4.6.15}$$
$$+ C|a| \left(\int_{B_R} |D^2 u_{n+1}|^2 \zeta dx \right)^{1/2}.$$

From this and from (4.6.12) immediately follows (4.6.14).

□

Let $w_k(x)$ satisfy the equation

$$\Delta w_k = \Delta u_{n+1} r^{\alpha_k} \zeta \tag{4.6.16}$$

and the boundary condition

$$w_k|_{\partial B_R} = 0 (k = 1, 2, \dots, M),$$

where M is a positive integer and α_k is monotone decreasing and satisfies the following relations:

$$\begin{cases} \alpha_1 &= -m/2 + \eta, \\ 0 &< \alpha_{k-1} - 2\alpha_k < m, \\ \alpha_k &\notin [2 - m, 3 - m], \alpha_{M-1} > 2 - m, \\ \alpha_M &= \alpha = 2 - m - 2\gamma (0 < \gamma < 1). \end{cases} \tag{4.6.17}$$

According to results of Stein [1] and Kondratjev [1] the inequality

$$\int_{B_R} (|D'^2 w|^2 + |w|^2) r^\beta dx \le C \int_{B_R} |\Delta w|^2 r^\beta dx \tag{4.6.18}$$

holds, if $-m < \beta < m$. Multiply (4.6.6) or (4.6.7) by $\Delta w_k \zeta$ and integrate twice by parts. It is obvious that ζ^2 also satisfies (4.6.13). We then get

$$\begin{aligned} \int_{B_R} |\Delta u_{n+1}|^2 r^{\alpha_k} \zeta^2 dx &= \int_{B_R} \{u_{n,i,j} - \Lambda^{-1}[a_i(x, Du_n)]_j\} w_{k,i,j} \zeta^2 dx + \\ &+ \int_{B_R} [u_{n,i} - \Lambda^{-1} a_i(x, Du_n)] w_{k,ij} (\zeta^2)_j dx - \int_{B_R} [u_{n,i} - \\ &- \Lambda^{-1} a_i(x, Du_n)] w_{k,ij} (\zeta^2)_i dx + \ldots = I_1 + I_2 + + I_3 \end{aligned} \tag{4.6.19}$$

(the unwritten terms contain only the first derivatives of u_n and w_k).
Estimate initially the integral I_1. It is obvious

$$\begin{aligned} I_1 = \int_{B_R} \{u_{n,ij} - \Lambda^{-1}[a_i(x, Du_n)]_j\} w_{k,ij} \zeta^2 dx \le \sup_{\bar\Omega} \|I - \Lambda^{-1}\overline{A}\| \times \\ \times \left(\int_{B_R} |D^2 u_n|^2 r^{\alpha_k} \zeta^2 dx \right)^{\frac{1}{2}} \left(\int_{B_R} |D^2 w_k|^2 r^{-\alpha_k} \zeta^2 dx \right)^{\frac{1}{2}} + \ldots \end{aligned} \tag{4.6.20}$$

Furthermore,

$$\begin{aligned} \int_{B_R} |D^2 w_k|^2 r^{-\alpha_k} \zeta^2 dx &= \sum_{i,j=1}^m \int_{B_R} w_{k,ij}^2 r^{-\alpha_k} \zeta^2 dx = \\ &= \sum_{i,j=1}^m \int_{B_R} [(w_k\zeta)_{ij} - (w_{k,j}\zeta_i + w_{k,i}\zeta_j) - w_k \zeta_{ij}]^2 r^{-\alpha_k} dx \le \\ &\le \int_{B_R} |D^2(w_k\zeta)|^2 r^{-\alpha_k} dx (1+\eta) + C \int_{B_R} (|D' w_k|^2 + |w_k|^2) r^{-\alpha_k} |D^2\zeta|^2 dx. \end{aligned}$$

According to the inequality (4.1.17), we have

$$\int_{B_R} |D'^2(w_k\zeta)|^2 r^{\alpha_k} dx \le \left[1 - \frac{4\alpha_k(m-1)}{(\alpha_k+m)^2}\right] \int_{B_R} |\Delta(w_k\zeta)|^2 r^{-\alpha_k} dx.$$

Then, from (4.6.16) and the fact that $D\zeta \equiv 0$ for $r \le R/2$ it follows that

$$\begin{aligned} \int_{B_R} |D^2 w_k|^2 r^{-\alpha_k} \zeta^2 dx \;&\le\; \left[1 - \frac{4\alpha_k(m-1)}{(\alpha_k+m)^2}\right] \int_{B_R} |\Delta w_k|^2 r^{-\alpha_k} \zeta^2 dx + \\ &+ C \int_{B_R} (|D' w_k|^2 + |w_k|^2) r^{-\alpha_k} |D^2\zeta|^2 dx \le \\ &\le \left[1 - \frac{4\alpha_k(m-1)}{(\alpha_k+m)^2}\right] \int_{B_R} |\Delta u_{n+1}|^2 r^{\alpha_k} \zeta^2 dx + \\ &+ C \int_{B_R} (|D' w_k|^2 + |w_k|^2) r^{-2\alpha_k + \alpha_{k-1}} dx. \end{aligned}$$

If α_k satisfies (4.6.17) we can apply (4.6.18) and obtain the inequality

$$\int_{B_R} |D^2 w_k|^2 r^{-\alpha_k}\zeta^2 dx \leq \left[1 - \frac{4\alpha_k(m-1)}{(\alpha_k+m)^2}\right] \int_{B_R} |\Delta u_{n+1}|^2 r^{\alpha_k}\zeta^2 dx + $$
$$+C\left(\int_{B_R} |\Delta u_{n+1}|^2 r^{\alpha_k - 1}\zeta^2 dx + \int_{B_R} |Du_n|^2 dx\right).$$

Carrying out the same considerations for I_2 and I_3, (4.6.12), (4.6.19) and (4.6.20) yield the relation

$$\int_{B_R} |\Delta u_{n+1}|^2 r^{\alpha_k}\zeta^2 dx \leq \left[1 - \frac{4\alpha_k(m-1)}{(\alpha_k+m)^2} + \eta\right] \times$$
$$\times \left(1 - \frac{\lambda\Lambda^{-1}}{1 + [\max\{\sup_\Omega |Du_n|, \sup_\Omega |Du_{n+1}|\}]^s}\right) \int_{B_R} |D^2 u_n|^2 r^{\alpha_k}\zeta^2 dx +$$
$$+C\left(|a|^2 + \int_{B_R} |\Delta u_{n+1}|^2 r^{\alpha_k - 1}\zeta^2 dx + \int_{B_R} |Du_n|^2 dx\right).$$

For $k = M$ inequality (4.1.17) gives

(4.6.21)
$$\int_{B_R} |D^2 u_{n+1}|^2 r^{\alpha}\zeta^2 dx \leq (1 + M_\gamma^2)\left[1 - \frac{4\alpha(m-1)}{(\alpha+m)^2} + \eta\right] \times$$
$$\times \left(1 - \frac{\lambda\Lambda^{-1}}{1 + [\max\{\sup_\Omega |Du_n|, \sup_\Omega |Du_{n+1}|\}]^s}\right) \int_{B_R} |D^2 u_n|^2 r^{\alpha}\zeta^2 dx +$$
$$+C\left[|a|^2 + \int_{B_R} |Du_n|^2 dx + \int_{B_R} |\Delta u_{n+1}|^2 r^{\alpha_{M-1}}\zeta^2 dx + \right.$$
$$\left. + \left(\int_{B_R} |D^2 u_{n+1}|^2 r^{\alpha}\zeta^2 dx\right)^{\frac{m}{m+2\gamma}} \left(\int_{B_R} |Du_n|^2 dx\right)^{\frac{2\gamma}{m+2\gamma}}\right].$$

For $k < M$ lemma 4.1.1 gives

(4.6.22)
$$\int_{B_R} |D^2 u_{n+1}|^2 r^{\alpha_k}\zeta^2 dx \leq \left[1 - \frac{4\alpha_k(m-1)}{(\alpha_k+m)^2} + \eta\right] \times$$
$$\times \left(1 - \frac{\lambda\Lambda^{-1}}{1 + [\max\{\sup_\Omega |Du_n|, \sup_\Omega |Du_{n+1}|\}]^s}\right) \int_{B_R} |D^2 u_n|^2 r^{\alpha_k}\zeta^2 dx +$$
$$+C\left(\int_{B_R} |Du_n|^2 dx + |a|^2 + \int_{B_R} |\Delta u_{n+1}|^2 r^{\alpha_{k-1}}\zeta^2 dx\right).$$

Theorem 4.6.1. *Suppose the conditions of lemma 4.6.1 are satisfied and the inequalities*

$$(4.6.23)\qquad \begin{cases} \int_\Omega |Du_0|^2 dx & < \ \eta_0^2, \\ \int_\Omega |D^2u_0|^2 r^{\alpha_k} dx & < \ \eta_k^2 \quad (k=1,\dots,M-1), \\ |a|^2 + \sum_{j=1}^{k-1} \eta_j^2 & < \ \varepsilon\eta_k^2 \end{cases}$$

hold true ($\varepsilon, a, \eta_k =$ *const* > 0 *are suficiently small numbers).*
If the relation

$$(4.6.24)\qquad \frac{\Lambda}{\lambda}\,\frac{(1+M_\gamma^2)\left[1-\frac{4\alpha(m-1)}{(\alpha+m)^2}\right]-1}{(1+M_\gamma^2)\left[1-\frac{4\alpha(m-1)}{(\alpha+m)^2}\right]} < 1$$

is satisfied, where M_γ *is determined by* (4.2.20) *then the solution to the problem* (4.6.1), (4.6.5) *belongs to* $C^{1,\gamma}(\overline{\Omega})$ *with* $\gamma = -(\alpha+m-2)/2$ *and the methods (4.6.6) and (4.6.7) converge to the solution.*

Proof. Consider at first the case $m \geq 4$. As we have mentioned earlier it is sufficient to prove inequality (4.6.8). Assume $u \in W_q^{(2)}(\Omega)$ with $q > m(m+\alpha)^{-1}$. Then all u_n arc in $W_q^{(2)}(\Omega)$ and $\forall u_n \in W_{2,\alpha}^{(2)}(\Omega)$ follows.
If we write (4.2.36) for the functions $u_j\zeta$, we get

$$(4.6.25)\quad |u_{j,i}(x_0)|^2 \leq C\left(|a|^2+\sum_{k=0}^{M-1}\eta_k^2\right)^{\frac{2\gamma}{m+2\gamma}}\left[\left(\int_{B_R}|D^2u_0|^2 r^\alpha\zeta^2 dx\right)^{\frac{m}{m+2\gamma}} + \right.$$
$$\left. + \left(|a|^2+\sum_{k=0}^{M-1}\eta_k^2\right)^{\frac{m}{m+2\gamma}}\right] \quad (j=0,1; i=1,\dots,m)\Bigg),$$

where $u_{j,i} = D_i u_j$. In fact, from (4.2.36) and $u = u_j\zeta$ and some calculations we obtain

$$|u_{j,i}(x_0)|^2 \leq C\left(\int_\Omega |Du_j|^2 dx\right)^{\frac{2\gamma}{m+2\gamma}} \times$$
$$\times \left[\left(\int_{B_R}|D^2u_j|^2 r^\alpha\zeta^2 dx\right)^{\frac{m}{m+2\gamma}} + \left(\int_\Omega |Du_j|^2 dx\right)^{\frac{m}{m+2\gamma}}\right] (j=0,1).$$

Now (4.6.25) follows from (4.6.23) for $j = 0$. Applying (4.6.9) and the inequality $(a+b)^2 \leq 2(a^2+b^2)$, we have

$$\int_\Omega |Du_1|^2 dx \leq 2\left(\int_\Omega |Du_0|^2 dx + |a|^2\right) \leq 2(\eta_0^2 + |a|^2\Lambda^{-2}).$$

After using (4.6.21),(4.6.22), (2.2.15) the estimates give the relation

$$\int_{B_R(x_0)} |D^2u_1|^2 r^\alpha\zeta^2 dx \leq C\left[\int_{B_R(x_0)} |D^2u_0|^2 r^\alpha\zeta^2 dx + |a|^2 + \sum_{k=0}^{M-1}\eta_k^2\right].$$

Therefore

$$|u_{1,i}(x_0)|^2 \le C(\eta_0^2+|a|^2)^{\frac{2\gamma}{m+2\gamma}}\left[\left(\int\limits_{B_R(x_0)}|D^2u_0|^2r^\alpha\zeta^2dx\right)^{\frac{m}{m+2\gamma}}+\left(|a|^2+\sum_{k=0}^{M-1}\eta_k^2\right)^{\frac{m}{m+2\gamma}}\right].$$

From (4.6.25) follows that for $j=0,1$

$$\sup_\Omega|Du_j|^2 \le C\left(|a|^2+\sum_{k=0}^{M-1}\eta_k^2\right)^{\frac{2\gamma}{m+2\gamma}}\times$$

$$\times\left[\left(\sup_{x_0\in\overline{\Omega}}\int\limits_{B_R(x_0)}|D^2u_0|^2r^\alpha\zeta^2dx\right)^{\frac{m}{m+2\gamma}}+\left(|a|^2+\sum_{k=0}^{M-1}\eta_k^2\right)^{\frac{m}{m+2\gamma}}\right].$$

Take $|a|^2+\sum_{k=0}^{M-1}\eta_k^2$ small enough that $C(|a|^2+\sum_{k=0}^{M-1}\eta_k^2)^{\frac{2\gamma}{m+2\gamma}}<1$. Then (4.6.21) gives

(4.6.26)
$$\sup_{x_0\in\Omega}\int\limits_{B_R}|D^2u_1|^2r^\alpha\zeta^2dx \le (1+M_\gamma^2)\left[1-\frac{4\alpha(m-1)}{(\alpha+m)^2}+\eta\right]\left\{1-\right.$$

$$\left.-\frac{\lambda\Lambda^{-1}}{1+\left[(\sup\limits_{x_0\in\overline{\Omega}}\int\limits_{B_R(x_0)}|D^2u_0|^2r^\alpha\zeta^2dx)^{\frac{m}{2(m+2\gamma)}}+\left(C|a|^2+\sum\limits_{k=1}^{M-1}\eta_k^2\right)^{\frac{m}{2(m+2\gamma)}}\right]^s}\right\}\times$$

$$\times\sup_{x_0\in\overline{\Omega}}\int\limits_{B_R}|D^2u_0|^2r^\alpha\zeta dx+C\left(|a|^2+\sum_{k=0}^{M-1}\eta_k^2\right).$$

Set

(4.6.27)
$$\begin{cases} X_l &= \sup\limits_{x_0\in\Omega}\int\limits_{B_R}|D^2u_l|^2r^\alpha\zeta^2dx, (l=0,1),\\ Q &= (1+M_\gamma^2)\left[1-\frac{4\alpha(m-1)}{(\alpha+m)^2}+\eta\right],\\ H &= C(|a|^2+\sum\limits_{k=0}^{M-1}\eta_k^2).\end{cases}$$

Inequality (4.6.26) now becomes

$$X_1\le Q\left(1-\frac{\lambda\Lambda^{-1}}{1+X_0^{\frac{ms}{2(m+2\gamma)}}+H^{\frac{ms}{2(m+2\gamma)}}}\right)X_0+H,$$

which can be written in the form

$$X_1 \;\le\; X_0+(Q-1)\left\{\left[1-\frac{Q\lambda\Lambda^{-1}}{(Q-1)(1+X_0^{\frac{ms}{2(m+2\gamma)}}+H^{\frac{ms}{2(m+2\gamma)}})}\right]X_0+\frac{H}{Q-1}\right\}.$$

Let the condition

(4.6.28)
$$Q\lambda\Lambda^{-1}(Q-1)^{-1}>1,$$

holds. Then there exists such a $q_0\in(0,1)$ that $Q\lambda\Lambda^{-1}(Q-q_0)>1$. Let H be small enough such that

$$H\le(1-q_0)\left[Q\lambda\Lambda^{-1}(Q-q_0)^{-1}-1-H^{\frac{ms}{2(m+2\gamma)}}\right]^{\frac{2(m+2\gamma)}{ms}}\equiv P(1-q_0).$$

Then after simple calculations, we get from the inequality $X_0 \leq P$ the relation

$$X_1 \leq P. \tag{4.6.29}$$

We now return to (4.6.22). From (4.6.25) and $C(|a|^2 + \sum_{k=1}^{M-1} \eta_k^2)^{\frac{2\gamma}{m+2\gamma}} < 1$ and get

$$\int_{B_R} |D^2 u_1|^2 r^{\alpha_k} \zeta^2 dx \leq \left[1 - \frac{4\alpha_k(m-1)}{(\alpha_k+m)^2} + \eta\right] \times$$

$$\times \left(1 - \frac{\lambda\Lambda^{-1}}{1 + \left[(\sup_{x_0 \in \Omega} \int_{B_R} |D^2 u_0|^2 r^{\alpha} \zeta^2 dx)^{\frac{m}{2(m+2\gamma)}} + C(|a|^2 + \sum_{j=0}^{M-1} \eta_j^2)^{\frac{m}{2(m+2\gamma)}}\right]^s}\right) \times$$

$$\times \int_{B_R} |D^2 u_0|^2 r^{\alpha_k} \zeta^2 dx + C \left[\int_{B_R} |Du_0|^2 dx + |a|^2 + \int |\Delta u_1|^2 r^{\alpha_{k-1}} \zeta^2 dx\right].$$

Using (4.6.27) and (4.6.29) gives for $k < M$

$$\int_{B_R(x_0)} |D^2 u_1|^2 r^{\alpha_k} \zeta^2 dx \leq \left[1 - \frac{4\alpha_k(m-1)}{(\alpha_k+m)^2} + \eta\right] \left[1 - \frac{4\alpha(m-1)}{(\alpha+m)^2}\right]^{-1} q_0 \times$$

$$\times (1 + M_\gamma^2)^{-1} \int_{B_R(x_0)} |D^2 u_0|^2 r^{\alpha_k} \zeta^2 dx + C(|a|^2 + \sum_{j=0}^{k-1} \eta_j^2).$$

All α_k are negative and decreasing. From the last inequality we obtain

$$\int_{B_R(x_0)} |D^2 u_1|^2 r^{\alpha_k} \zeta^2 dx \leq (1 + M_\gamma^2)^{-1} \eta_k^2 + C(|a|^2 + \sum_{j=0}^{k-1} \eta_j^2).$$

From (4.6.23) it follows that

$$\int_{B_R(x_0)} |D^2 u_1|^2 r^{\alpha_k} \zeta^2 dx < \eta_k^2$$

and therefore for u_1 all conditions of the theorem are satisfied . Thus inequality (4.6.8) and the theorem are proved for $m \leq 4$.
For $m = 2$ and $m = 3$ take $\alpha_1 = -\frac{m}{2} + \eta$ then the condition

$$-\frac{m}{2} + \eta < 2 - m - 2\gamma$$

is satisfied for at least small γ and all consideration are simplified.

□

Remark 4.6.1. *If $\gamma > 0$ is small then the condition* (4.6.24) *gives*

$$\frac{\Lambda}{\lambda} \frac{(1 + \frac{m-2}{m+1})[1 + (m-2)(m-1)] - 1}{(1 + \frac{m-2}{m+1})[1 + (m-2)(m-1)]} < 1$$

For $m = 2$ this inequality does not restrict the dispersion of the spectrum for the matrix of ellipticity.

4.7 Method of elastic solutions; general case.

In section 3.4 we considered the elastic methods convergence in strong norms for small dimensions ($m \leq 3$). We shall go back and prove the analogous result for any dimension $m > 2$. The previous result was based on estimates for the first derivatives of the test functions in $\mathcal{L}_{2\pm,\alpha}$. The following result were obtained with the help of statements from sections 4.1, 4.2 and 4.3. Here we have the estimates for the second derivatives in $\mathcal{L}_{2,\pm\alpha}$. Hence, the essential difference in getting the regularity theorems for the solutions of elasto-plastic problems in three and more dimensional spaces from two-dimensional space consists in the order of derivatives, for which the coercivity estimates are needed.

Lemma 4.7.1. *Let the conditions of lemma 3.4.2 be satisfied and $\alpha \in (-m, 0]$. Then the solution v of the problems (3.4.14) (3.4.15) satisfies the inequality*

$$(4.7.1) \qquad \int\limits_{B_\delta(x_0)} |divv|^2 r^{-\alpha}\zeta dx \leq \frac{1}{2}(C_\alpha + \eta)\left[\sum_{j,k=1}^{m} \int\limits_{B_\delta(x_0)} |\varepsilon_{jk}(w_{n+1})|^2 r^\alpha \zeta^2 dx\right]^{1/2} +$$

$$+C\left[\int\limits_{B_\delta} |D'v|^2 r^{-2\alpha+\alpha'}\zeta dx\right],$$

where C_α is the constant (4.1.23) and $\forall\alpha'$ satisfies the inequality $-m < \alpha' \leq 0$.

Proof. We can assume w_{n+1} and v are smooth functions. Suppose z is the solution of the problem

$$(4.7.2) \qquad \Delta z = divv r^{-\alpha} \quad (x \in B_\delta), z|_{\partial B_\delta} = 0,$$

where v satisfies (3.4.14), (3.4.15). Multiply (3.4.14) by $D_k(\zeta z)$ and integrate over $B_\delta(x_0)$. After integrating by parts we get the following equality

$$(4.7.3) \qquad \int\limits_{B_\delta} \left(\Delta v^{(k)} + D_k divv\right) D_k(\zeta z) dx = -\int\limits_{B_\delta} \varepsilon_{jk}(w_{n+1}) D_j D_k(\zeta z) r^\alpha \zeta dx$$

(omit for a while the index $n+1$). After integrating by parts on the left-hand-side we obtain

$$\int\limits_{B_\delta} \left(\Delta v^{(k)} + D_k divv\right) D_k(\zeta z) dx = -\int\limits_{B_\delta} \Delta\left(D_k v^{(k)}\right)\zeta z dx -$$

$$-\int\limits_{B_\delta} divv\, \Delta\,(\zeta z) dx = -2\int\limits_{B_\delta} divv\, \Delta\,(\zeta z) dx = -2\int\limits_{B_\delta} \zeta divv\, \Delta\, z dx -$$

$$-4\int\limits_{B_\delta} divv \nabla\zeta\nabla z dx - 2\int\limits_{B_\delta} z divv\, \Delta\,\zeta dx.$$

The relation (4.7.3) can now be written in the form

$$\int\limits_{B_\delta} \zeta(divv)^2 r^{-\alpha} dx = \frac{1}{2}\int\limits_{B_\delta} \varepsilon_{jk}(w_{n+1}) D_j D_k(\zeta z) r^\alpha dx -$$

$$-2\int\limits_{B_\delta} divv \nabla\zeta\nabla z dx - \int\limits_{B_\delta} z divv\, \Delta\,\zeta dx.$$

The second and the third terms on the right-hand side have $\nabla\zeta$ and $\Delta\zeta$ under the sign of the integral; therefore, we can additionally include a factor r^β and the integrals will remain bounded. Applying the Hölder inequality to terms of the right-hand side we arrive at

$$\int\limits_{B_\delta} \zeta(divv)^2 r^{-\alpha} dx \leq$$

$$\leq \frac{1}{2}\left[\sum_{j,k=1}^{m} \int\limits_{B_\delta} \zeta^2 \varepsilon_{jk}^2(w_{n+1}) r^\alpha dx\right]^{1/2} \left[\int\limits_{B_\delta} |D'^2(\zeta z)|^2 r^\alpha dx\right]^{1/2} +$$

$$+C\left(\int\limits_{B_{\delta/2,\delta}} |divv|^2 r^\beta dx\right)^{1/2} \left(\int\limits_{B_{\delta/2,\delta}} |Dz|^2 r^\beta dx\right)^{1/2},$$

where $\beta \leq 0$ is arbitrary. The last integrals on the right-hand side can be extended on $B_{q\delta}(x_0)$ with $3/4 < q < 1$ because $\zeta' \equiv 0$ for $r \geq 3\delta/4$. With the help of (4.1.21) we get the following estimate

$$\int\limits_{B_\delta} \zeta(divv)^2 r^{-\alpha} dx \leq$$

$$\leq \frac{1}{2}\left[\sum_{j,k=1}^{m} \int\limits_{B_\delta} \zeta^2 \varepsilon_{jk}^2(w_{n+1}) r^\alpha dx\right]^{1/2} \left[(C_\alpha + \eta)\int\limits_{B_\delta} |\Delta(\zeta z)|^2 r^\alpha dx + C\int\limits_{B_\delta} |D^2(\zeta z)|^2 r^\beta dx\right]^{1/2} +$$

$$+C\left(\int\limits_{B_{\delta/2,\delta}} |divv|^2 r^\beta dx\right)^{1/2} \left(\int\limits_{B_{\delta/2,\delta}} |Dz|^2 r^\beta dx\right)^{1/2},$$

where C_α is denoted by (4.1.23). Here under the sign of the third integral we write an additional factor r^β. This is possible because from $\beta \leq 0$ follows $(r/\delta)^\beta \geq 1$. The factor $\delta^{-\beta}$ is included in C.

The integral $\int\limits_{B_\delta} |\Delta(\zeta z)|^2 r^\alpha dx$ can be written in the following way

$$\int\limits_{B_\delta} |\Delta(\zeta z)|^2 r^\alpha dx = \int\limits_{B_\delta} \zeta^2 |\Delta z|^2 r^\alpha dx + 2\int\limits_{B_{3\delta/4}} \zeta \Delta z(2\nabla\zeta\nabla z + \Delta\zeta z) r^\alpha dx +$$

$$+\int\limits_{B_{3\delta/4}} |2\nabla\zeta\nabla z + \Delta\zeta z|^2 r^\alpha dx.$$

Applying the inequality (2.2.15) we get

$$\int\limits_{B_\delta} |\Delta(\zeta z)|^2 r^\alpha dx \leq (1+\eta)\int\limits_{B_\delta} \zeta^2 |\Delta z|^2 r^\alpha dx + C\int\limits_{B_{3\delta/4}} |D^2 z|^2 r^\beta dx.$$

Since z satisfies the differential equation of the problem (4.7.2) we arrive at

$$\int\limits_{B_\delta} |\Delta(\zeta z)|^2 r^\alpha dx \leq (1+\eta)\int\limits_{B_\delta} \zeta(divv)^2 r^{-\alpha} dx + C\int\limits_{B_{3\delta/4}} |D^2 z|^2 r^\beta dx.$$

From the inequality $|a+b|^{1/2} \leq |a|^{1/2} + |b|^{1/2}$ follows

$$\int_{B_\delta} \zeta(divv)^2 r^{-\alpha} dx \leq \frac{1}{2} \left[\sum_{j,k=1}^{m} \int_{B_\delta} \zeta^2 \varepsilon_{jk}^2(w_{n+1}) r^\alpha dx \right]^{1/2} \times \tag{4.7.4}$$

$$\times \left[(C_\alpha + \eta)^{1/2} \left(\int_{B_\delta} \zeta^2 (divv)^2 r^{-\alpha} dx \right)^{1/2} + C \left(\int_{B_\delta} |D^2 z|^2 r^\beta dx \right)^{1/2} \right] +$$

$$+ C \left(\int_{B_{3\delta/4}} (divv)^2 r^\beta dx \right)^{1/2} \left(\int_{B_{3\delta/4}} |Dz|^2 r^\beta dx \right)^{1/2}.$$

Taking into account the inequality

$$\int_{B_{3\delta/4}} |divv|^2 r^\beta dx \leq C \int_{B_\delta} |divv|^2 r^{-2\alpha+\alpha'} dx$$

holds, then from (4.7.2) (4.2.35) and (4.7.4) follows for $\beta = \alpha' - 2\alpha$

$$\int_{B_\delta} \zeta(divv)^2 r^{-\alpha} dx \leq$$

$$\leq \frac{1}{2} \left[\sum_{j,k=1}^{m} \int_{B_\delta} \zeta^2 \varepsilon_{jk}^2(w_{n+1}) r^\alpha dx \right]^{1/2} \left[(C_\alpha + \eta)^{1/2} \left(\int_{B_\delta} \zeta^2 (divv)^2 r^{-\alpha} dx \right)^{1/2} + \right.$$

$$\left. + C \left(\int_{B_\delta} (divv)^2 r^{-2\alpha+\alpha'} dx \right)^{1/2} \right] + C \int_{B_\delta} |divv|^2 r^{-2\alpha+\alpha'} dx.$$

After elementary calculations we get

$$\left(\int_{B_\delta} \zeta(divv)^2 r^{-\alpha} dx \right)^{1/2} \leq \frac{1}{2} (C_\alpha + \eta)^{1/2} \left[\sum_{j,k=1}^{m} \int_{B_\delta} \zeta^2 \varepsilon_{jk}^2(w_{n+1}) r^\alpha dx \right]^{1/2} +$$

$$+ C \left\{ \left[\int_{B_\delta} \zeta^2 \sum_{j,k=1}^{m} \varepsilon_{jk}^2(w_{n+1}) r^\alpha dx \right]^{1/4} + \left[\int_{B_\delta} (divv)^2 r^{-2\alpha+\alpha'} \right]^{1/4} dx \right\} \times$$

$$\times \left[\int_{B_\delta} (divv)^2 r^{-2\alpha+\alpha'} dx \right]^{1/4}.$$

Applying (2.2.15) we obtain

$$\left[\int_{B_\delta} \zeta^2 \sum_{j,k=1}^{m} \varepsilon_{jk}^2(w_{n+1}) r^\alpha dx \right]^{1/4} \left[\int_{B_\delta} (divv)^2 r^{-2\alpha+\alpha'} dx \right]^{1/4} \leq$$

$$\leq \eta \left[\int_{B_\delta} \zeta^2 \sum_{j,k=1}^{m} \varepsilon_{jk}^2(w_{n+1}) r^\alpha dx \right]^{1/2} + C \left[\int_{B_\delta} (divv)^2 r^{-2\alpha+\alpha'} dx \right]^{1/2}$$

which leads to the follwing relation

$$\left[\int_{B_\delta} \zeta (divv)^2 r^{-\alpha} dx\right]^{1/2} \leq$$

$$\leq \frac{1}{2}(C_\alpha + \eta)^{1/2}\left[\sum_{j,k=1}^{m}\int_{B_\delta} \zeta^2 \varepsilon_{jk}^2(w_{n+1}) r^{\alpha} dx\right]^{1/2} + C\left[\int_{B_\delta} \zeta (divv)^2 r^{-2\alpha+\alpha'} dx\right]^{1/2}.$$

Since $(divv)^2 \leq C|D'v|^2$ we get

$$\left[\int_{B_\delta} \zeta (divv)^2 r^{-\alpha} dx\right]^{1/2} \leq \tag{4.7.5}$$

$$\leq \frac{1}{2}(C_\alpha + \eta)^{1/2}\left[\sum_{j,k=1}^{m}\int_{B_\delta} \zeta^2 \varepsilon_{jk}^2(w_{n+1}) r^{\alpha} dx\right]^{1/2} + C\left[\int_{B_\delta} |D'v|^2 r^{-2\alpha+\alpha'} dx\right]^{1/2}.$$

□

Theorem 4.7.1. *Let $a_k(x; \varepsilon_{jl})$ satisfy the conditions 1)-3) of section 3.4, $u_0 \in \overset{\circ}{W}{}_q^{(1)}(\Omega), f \in \mathcal{L}_q(\Omega)$ with $q > 2m(m+\alpha_0)^{-1}$, $\alpha_0 = 2 - m - 2\gamma (0 < \gamma < 1)$ and $dist\{\Omega', \partial\Omega\} > 0 (\Omega' \in \Omega)$.*
Assume the inequality

$$2^{3/2}\frac{\Lambda - \lambda}{\Lambda + \lambda}\left[1 + \frac{(m-2)^2}{m-1}\right]^{\frac{1}{2}}\left[1 + \frac{\sqrt{m}}{2}\left(1 + \frac{m-2}{m+1}\right)^{1/2}\right] < 1, \tag{4.7.6}$$

where λ and Λ are determined by (3.4.7), holds true. Then for sufficiently small $\gamma > 0$ the weak solution u of the problem (3.3.1),(3.3.2) belongs to $H_\alpha(\Omega') \subset C^{0,\gamma}(\Omega')$. The method of elastic solutions (3.4.10) with $\varepsilon = 2(\Lambda + \lambda)^{-1}$ converges to this solution in $H_\alpha(\Omega')$ as a geometric progression, and the common ratio coincides with the left-hand side of (4.7.6).

Proof. Let $w_n = u_n - u_{n-1}$ and v satisfy (3.4.14). From (3.4.13) follows that

$$\int_{B_\delta} \varepsilon_{ik}(w_{n+1})\varepsilon_{ik}(v) dx \leq \frac{\Lambda - \lambda}{\Lambda + \lambda} \times \tag{4.7.7}$$

$$\times \left[\sum_{i,k=1}^{m}\int_{B_\delta} \varepsilon_{ik}^2(w_n) r^{\alpha} dx\right]^{1/2}\left[\sum_{i,k=1}^{m}\int_{B_\delta} \varepsilon_{jk}^2(v) r^{-\alpha} dx\right]^{1/2}.$$

Multiplying both sides of (3.4.14) by w_{n+1}, integrating by parts and taking into account $v \equiv 0$ outside $B_\delta(x_0)$ we obtain

$$\sum_{i,k=1}^{m}\int_{B_\delta} \zeta \varepsilon_{ik}^2(w_{n+1}) r^{-\alpha} dx = \sum_{j,k=1}^{m}\int_{B_\delta} \varepsilon_{jk}(w_{n+1})\varepsilon_{jk}(v) dx.$$

Then (4.7.7) can be written in the following form

$$\sum_{i,k=1}^{m}\int_{B_\delta}\varepsilon_{ik}^2(w_{n+1})r^\alpha\zeta dx \le \frac{\Lambda-\lambda}{\Lambda+\lambda}\left[\sum_{j,k=1}^{m}\int_{B_\delta}\zeta\varepsilon_{jk}^2(w_n)r^\alpha dx\right]^{1/2}\left[\sum_{j,k=1}^{m}\int_{B_\delta}\varepsilon_{jk}^2(v)r^{-\alpha}dx\right]^{1/2}.$$

Since $\sum_{j,k=1}^{m}\varepsilon_{jk}^2(v)\le 4\sum_{j,k=1}^{m}\left[D_jv^{(k)}\right]^2$, then

$$\sum_{i,k=1}^{m}\int_{B_\delta}\varepsilon_{ik}^2(w_{n+1})r^\alpha\zeta dx \le 2\frac{\Lambda-\lambda}{\Lambda+\lambda}\left[\sum_{j,k=1}^{m}\int_{B_\delta}\varepsilon_{jk}^2(w_n)r^\alpha dx\right]^{1/2}\times \tag{4.7.8}$$

$$\times\left[\int_{B_\delta}|D'v|^2r^{-\alpha}\zeta dx\right]^{1/2}.$$

Because v is the solution of (3.4.14), (3.4.15) it should satisfy the identity

$$\int_{B_\delta}D_jv^{(k)}D_j\chi^{(k)}dx=\int_{B_\delta}\left[\varepsilon_{jk}(w_{n+1})r^\alpha\zeta-divv\delta_{jk}\right]D_j\chi^{(k)}dx,$$

where χ is a finite in B_δ smooth function. From lemma 2.3.3 it follows for $\alpha\in(-m,2-m)\cup(3-m,0]$

$$\int_{B_\delta}|D'v|^2r^{-\alpha}\zeta dx\le\left[\frac{4(m-1)+(-\alpha+m-2)^2}{m^2-\alpha^2}+\eta\right]\times$$

$$\times\sum_{i,k=1}^{m}\int_{B_\delta}\left[\zeta\varepsilon_{jk}(w_{n+1})r^\alpha-divv\delta_{jk}\right]^2r^{-\alpha}dx+C\int_{B_\delta}|Dv|^2r^{-2\alpha+\alpha'}dx.$$

Taking the square root of both sides and applying the Hölder inequality we get

$$\left(\int_{B_\delta}|D'v|^2r^{-\alpha}\zeta dx\right)^{1/2}\le 2^{\frac12}\left[\frac{4(m-1)+(-\alpha+m-2)^2}{m^2-\alpha^2}+\eta\right]^{1/2}\times$$

$$\times\left\{\left[\sum_{i,k=1}^{m}\int_{B_\delta}\zeta^2\varepsilon_{jk}^2(w_{n+1})r^\alpha dx\right]^{1/2}+\sqrt{m}\left(\int_{B_\delta}|divv|^2r^{-\alpha}\zeta dx\right)^{1/2}\right\}+$$

$$+C\left(\int_{B_\delta}|Dv|^2r^{-2\alpha+\alpha'}dx\right)^{1/2}.$$

With the help of (4.7.5) we can enlarge the integral $\int_{B_\delta}|divv|^2r^{-\alpha}dx$ on the right-hand side to give

$$\left(\int_{B_\delta}|D'v|^2r^{-\alpha}dx\right)^{1/2}\le 2^{\frac12}\left[\frac{4(m-1)+(-\alpha+m-2)^2}{m^2-\alpha^2}+\eta\right]^{1/2}\times \tag{4.7.9}$$

$$\times\left[1+\frac{\sqrt{m}}{2}(C_\alpha+\eta)^{1/2}\right]\left[\sum_{i,k=1}^{m}\int_{B_\delta}\zeta^2\varepsilon_{jk}^2(w_{n+1})r^\alpha dx\right]^{1/2}+$$

$$+C\left(\int_{B_\delta}|Dv|^2r^{-2\alpha+\alpha'}dx\right)^{1/2}.$$

Denote

$$\Pi_n^2(\alpha)=\sum_{i,k=1}^{m}\int_{B_\delta}\zeta\varepsilon_{jk}^2(w_n)r^\alpha dx. \tag{4.7.10}$$

Notice that $\int_{B_\delta}\zeta\varepsilon_{jk}^2(w_n)r^\alpha dx=\int_{B_\delta}\varepsilon_{jk}^2(w_n)r^\alpha\zeta dx+\int_{B_\delta}\varepsilon_{jk}^2(w_n)r^\alpha(1-\zeta)dx$ and the second integral on the right-hand side can be estimated by $\int_{B_\delta}\varepsilon_{jk}^2(w_n)dx$. Now the inequality (4.7.8) with the help of (4.7.9) and (4.7.10) gives

$$\Pi_{n+1}^2(\alpha)\le 2^{3/2}\frac{\Lambda-\lambda}{\Lambda+\lambda}\left[\frac{4(m-1)+(-\alpha+m-2)^2}{m^2-\alpha^2}+\eta\right]^{1/2}\times$$

$$\times\left[1+\frac{\sqrt{m}}{2}(C_\alpha+\eta)^{1/2}\right]\Pi_n(\alpha)\Pi_{n+1}(\alpha)+$$

$$+C\Pi_n(\alpha)\left(\int_{B_\delta}|Dv|^2r^{-2\alpha+\alpha'}dx\right)^{1/2}+C\sum_{i,k=1}^{m}\int_{B_\delta}\varepsilon_{jk}^2(w_n)dx.$$

After small transformation we arrive at

$$\Pi_{n+1}^2(\alpha)\le 8\left\{\frac{\Lambda-\lambda}{\Lambda+\lambda}\left[\frac{4(m-1)+(-\alpha+m-2)^2}{m^2-\alpha^2}+\eta\right]^{1/2}\times\right.$$

$$\left.\times\left[1+\frac{\sqrt{m}}{2}(C_\alpha+\eta)^{1/2}\right]\right\}^2\Pi_n(\alpha)^2+$$

$$+C\Pi_n(\alpha)\int_{B_\delta}|Dv|^2r^{-2\alpha+\alpha'}dx+C\sum_{i,k=1}^{m}\int_{B_\delta}\varepsilon_{jk}^2(w_n)dx. \tag{4.7.11}$$

Consider initially the cases $m=2$ and $m=3$. From (3.4.16) we have for $\beta=-2\alpha+\alpha'$

$$\int_{B_\delta(x_0)}|D'v|^2r^{-2\alpha+\alpha'}dx\le C\int_{B_\delta(x_0)}|D'w_{n+1}|^2r^{\alpha'}dx.$$

If we make $\gamma>0$ small and $\alpha'=0$, then with $\alpha=\alpha_0=2-m-2\gamma$ we have $\beta=-2\alpha_0+\alpha'=2(m-2)+4\gamma<m$. For small $\gamma>0$ and (4.1.23) we obtain from (4.7.11) the following estimate

$$\Pi_{n+1}^2(\alpha)\le 8\left\{\frac{\Lambda-\lambda}{\Lambda+\lambda}\left[1+\frac{(m-2)^2}{m-1}+O(\gamma)+\eta\right]^{1/2}\times\right.$$

$$\times\left[1+\frac{\sqrt{m}}{2}\left(1+\frac{m-2}{m+1}+O(\gamma)+\eta\right)^{1/2}\right]\Bigg\}^2 \Pi_n(\alpha_0)^2+$$
$$+C\int_\Omega \sum_{j,k=1}^{m} \varepsilon_{jk}^2(w_{n+1})dx.$$

Lemma 3.4.1 gives us the estimate

$$\sum_{i,k=1}^{m}\int_\Omega \zeta|\varepsilon_{ik}(w_n)|^2 dx < CK^n.$$

It is obvious now that $\Pi(\alpha_0) < Cq^n$, where $0 \le q < 1$, if (4.7.6) is satisfied.
From the Korn inequality (3.2.18) follows the statement of the theorem.
Consider now $m \ge 4$. Remark at first that the coefficient before $\Pi_n^2(\alpha)$ on the rifght-hand side of (4.7.11) is an increasing function of $-\alpha$. So, from (4.7.6) follows that

$$2^{3/2}\frac{\Lambda-\lambda}{\Lambda+\lambda}\left[\frac{4(m-1)+(-\alpha+m-2)^2}{m^2-\alpha^2}+\eta\right]^{1/2}\times$$
$$\times\left[1+\frac{\sqrt{m}}{2}(C_\alpha+\eta)^{1/2}\right]\le q<1.$$

Take $\alpha_1 = -m/2+2-2\gamma$ and $\alpha_1' = 0$. Hence $-2\alpha_1 < m$ and from (3.4.16) for $\beta = -2\alpha_1+\alpha_1' = -2\alpha_1$ we get

$$\int_{B_\delta} |D'v|^2 r^{-2\alpha_1+\alpha_1'}dx \le C\int_{B_\delta} |D'w_{n+1}|^2 dx.$$

Then from (4.7.11) we find

$$\int_{B_\delta} |D'w_n|^2 r^{\alpha_1}dx < CQ^{2n}.$$

Assume $q \le Q$, construct a sequence of α_k, let $\beta = m-2+2\gamma (0<\gamma<1)$ and let $\alpha_1 = \frac{m}{2}+2(1-\gamma)$. Then $2\alpha_1+\beta = 0$ and the solution of our problem belongs to $H_{\alpha_1}(\Omega')$. Construct a sequence of $\alpha_k (k = 0, 1, \ldots)$ such that

$$2\alpha_{k+1}+\beta = \alpha_k.$$

It means that

$$\alpha_{k+1}+m-2+2\gamma = (\alpha_k+m-2+2\gamma)/2.$$

Convergence in $H^2_{\alpha_k}(\Omega')$ follows convergence in $H^2_{\alpha_{k+1}}(\Omega')$. It follows then this sequence converges to $\alpha = 2-m-2\gamma$. Starting with $\alpha_1 = -\frac{m}{2}+2(1-\gamma)$ the method of elastic solutions will converge in all $H^2_{\alpha_k}(\Omega')$ for each k. So it will also converge for some large k with α_k close to $\alpha = 2-m-2\gamma$. □

The condition (4.7.6) is unavoidable. To prove it consider the variant of the De Giorgi[2] example.

$$D_j\left\{\left[\delta_{ij}\delta_{hk}+\left(c\delta_{ih}+d\frac{x_i x_h}{|x|^2}\right)\left(c\delta_{ik}+d\frac{x_j x_k}{|x|^2}\right)\right]\left(D_i u^{(h)}+D_h u^{(i)}\right)\right\}=0,$$

where c and d are defined by the formula (2.5.8). It is evident that for this c and d

$$\frac{\Lambda-\lambda}{\Lambda+\lambda}\left[1+\frac{(m-2)^2}{m-1}\right] \geq 1$$

and there exists a solution $u = \frac{x}{|x|}$, which is discontinious and can initiate for a crack. From the last theorem follows that the inequality

$$2^{3/2}\frac{\Lambda-\lambda}{\Lambda+\lambda}\left[1+\frac{(m-2)^2}{m-1}\right]^{1/2}\left[1+\frac{\sqrt{m}}{2}\left(1+\frac{m-2}{m+1}\right)^{1/2}\right] \geq 1.$$

which gives the necessary condition for discontinuity of solutions. For $m = 2$ amd $m = 3$ the solutions will be continious if, respectively, the following estimates

(4.7.12)
$$\sqrt{2}(2+\sqrt{2})\frac{\Lambda-\lambda}{\Lambda+\lambda} < 1$$

and

(4.7.13)
$$6.818\frac{\Lambda-\lambda}{\Lambda+\lambda} < 1$$

hold true.

Remark 4.7.1. Theorem 3.4.1 can be proved for $3 < m < 7$ if we use at the end of lemma 3.4.3 the final considerations of the theorem 4.7.1.

Chapter 5

Regularity of solutions for parabolic systems with some applications

5.1 Regularity of weak solutions for quasilinear parabolic systems and the universal iterative process

In this section we shall continue to investigate the weak solutions of the second order parabolic systems

$$L_t(u) \equiv \partial_t u - \left[\sum_{i=1}^{m} D_i a_i(t,x;Du) - a_0(t,x,Du)\right] = 0 \tag{5.1.1}$$

which coefficients $a_i(t,x,p)$ $(i=0,\dots,m)$ satisfy in Q the following conditions:

1) for almost all $(t,x) \in Q$ the coefficients $a_i(t,x,p)$ $(i=1,\dots,m)$ are differentiable with respect to p for all finite p;

2) the inequalities (1.1.2) hold in $\overline{Q}$ for $s=2$ and $l=1$, where μ and ν are independent of t;

3) if $q_0(m) > 1$ is a sufficiently large number then for $\forall q$ with $1 < q \le q_0$ and $\forall u \in W_q^{(0,1)}(Q)$ all the coefficients $a_i(t,x;Du)$ $(i=0,1,\dots,m)$ belong to $\mathcal{L}_q(Q)$;

4) the inequality

$$\left|\frac{\partial a_i}{\partial x_j}\right| < C(1+|p|) \quad (i,j=1,\dots,m) \tag{5.1.2}$$

is true in $\overline{Q}$ for all finite p.

From 2) and 3) follows that if $u \in W_q^{1,2}(Q)$ then $L(u) \in \mathcal{L}_q(Q)$ for any $q: 1 < q \le q_0$. Here

$$L(u) \equiv \sum_{i=1}^{m} D_i a_i(t,x;Du) - a_0(t,x;Du). \tag{5.1.3}$$

In the same way as in section 1.1 we can define the matrices $A((1.1.4))$, A^+, A^-, $C((1.1.5))$, the eigenvalues λ_i, σ_i. All these quantities will depend on t. According to assumption 2) the values of $\lambda = \inf_{i,Q,p} \lambda_i > 0$, $\Lambda = \sup_{i,Q,p} \lambda_i < +\infty$ and K_ε, where $\varepsilon = \text{const}$ are independent of t.
Consider some additional functional spaces. Let $x_0, x \in \Omega$ and $r = |x - x_0|$. We shall consider the spaces $W_{2,\alpha}^{k,l}(Q, x_0)$ of functions with the norm

$$(5.1.4) \qquad \|u\|_{2,\alpha}^{k,l} = \left(\int_Q \left[\left|D^l u\right|^2 + \left|\partial_t^k u\right|^2 \right] r^\alpha dx \right)^{1/2} .$$

Analogously to the stationary case in 2.1, denote the space $H_{2,\alpha}^{k,l}(Q)$ with the norm

$$(5.1.5) \qquad \|u\|_{2,\alpha}^{k,l} = \sup_{x_0 \in \Omega} \|u\|_{2,\alpha}^{k,l}.$$

Let the solution of (5.1.1) satisfies the conditions

$$(5.1.6) \qquad u\Big|_{t=0} = u\Big|_{\partial\Omega} = 0.$$

For $l = 1$ we consider the process

$$(5.1.7) \qquad \varepsilon \partial_t u_{n+1} - \Delta u_{n+1} + u_{n+1} = -\Delta u_n + u_n + \varepsilon L(u_n)$$

with conditions (5.1.6) for $\forall u_n$.
Now, we shall be interested in proving the Hölder continuity of the weak solution of (5.1.1), (5.1.6).
From the theorem 1.2.4 it follows that the method (5.1.7) converges to the weak solution in $W_2^{0,1}(\Omega)$. Therefore it converges weakly and we need to estimate the iterations of these process in appropriate norm. For this choose the norm $H_{2,\alpha}^{1,2}(Q')$, $(Q' = (0,T) \times \Omega')$, where $\alpha = 2 - m - 2\gamma$ $(0 < \gamma < 1)$ and $\Omega' \subset\subset \Omega$. We shall prove the Hölder continuity of the weak solution only in a strictly interior subdomain.
It should be taken into account that for elliptic systems the analogous result follow from the estimate of iterations for the space H_α, which norm (2.1.4) includes only the first order derivatives.
For parabolic systems the estimates for the second derivatives of the iterations with respect to space coordinates $x_k (k = 1, \dots, m)$ play an important role, further the estimates for the time derivative follow from the iterative systems (5.1.7).
As in to the elliptic systems it is enough to obtain the estimates in a cylinder $Q_\delta(x_0) = (0,T) \times B_\delta(x_0)$, where $B_\delta(x_0)$ is the ball of radius δ with center x_0. Obviously the estimates in $W_{2,2-m-2\gamma}^{1,2}(Q_\delta(x_0), x_0)$ should be independent of x_0 and $\delta < \delta_0$.
The important embedding statement

$$(5.1.8) \qquad H_{2,2-m-2\gamma}^{1,2}(Q') \subset C^{0,\kappa}(Q')$$

follows from the theorem of Il'in (Besov, Il'in, Nikolskij [1], Chapter 6).
So, we must prove the estimate

$$(5.1.9) \qquad \int_{Q_\delta(x_0)} \left|D^2 u_n\right| |x - x_0|^\alpha \, dx < C,$$

where $Q_\delta(x_0) = (0,T) \times B_\delta(x_0)$, $x_0 \in \Omega' \subset\subset \Omega$, $\delta \leq \delta_0 < \text{dist}\{x_0, \partial\Omega\}$ and $\alpha = 2 - m - 2\gamma$ with any $\gamma : 0 < \gamma < 1$.
It is sometimes useful to apply the inequality

$$|u(x,t)| < C_1 \|u\|_{2,\alpha}^{1,2} + C\|u\|_{\mathcal{L}_2} \tag{5.1.10}$$

with an explicit value of C_1. Denote by $B_\delta(t_0, x_0)$ a ball in $R^{m+1} = R_t \times R_x^m$ with radius δ and center in $M_0(t_0, x_0)$. Let r_t be the distance between $M(t,x)$ and $M_0(t_0, x_0)$ in R^{m+1}. So $r_t^2 = |x - x_0|^2 + |t - t_0|^2$.
From the properties of the boundary $\partial\Omega$ follows the inequality

$$|Q \cap B_\delta(t_0, x_0)| \geq A_Q \delta^{m+1} \ , \tag{5.1.11}$$

where A_Q can be determined by A_Ω from (2.1.11).

Lemma 5.1.1. *If $u \in H_{2,\alpha}^{1,2}(Q)$ for $\alpha = 1 - m - 2\gamma$ $(0 < \gamma < \frac{1}{2})$ then the estimate*

$$\|u\|_{C^{0,\gamma}(\overline{Q})} \leq \frac{2^{\frac{m+3}{2}} + \eta}{A_Q^{1/2}\gamma} \|u\|_{2,\alpha}^{1,2} + C\|u\|_{\mathcal{L}_2(Q)} \tag{5.1.12}$$

holds, where the norm in $H_{2,\alpha}^{1,2}$ on the right-hand side is taken over Q.

Proof. Of course we can start with a smooth function $u(t,x)$. Applying the inequalities (2.1.12) and (2.1.15) we get

$$\|u\|_{C^{0,\gamma}(\overline{Q})} \leq \frac{2^{\frac{m+3}{2}} + \eta}{A_Q^{1/2}\gamma} \left[\sup_{(t_0,x_0)\in Q} \int_{Q_\delta(t_0,x_0)} \left(|\partial_t u|^2 + |Du|^2\right) r_t^\alpha \, dx dt \right]^{1/2} +$$
$$+ C \left(\int_Q |u|^2 dx dt \right)^{1/2} ,$$

where $Q_\delta(t_0, x_0) = Q \cap B_\delta(t_0, x_0)$. Since $\alpha \leq 0$ and $r \leq r_t$ we have $r_t^\alpha \leq r^\alpha$. Then

$$\|u\|_{C^{0,\gamma}(\overline{Q})} \leq \frac{2^{\frac{m+3}{2}} + \eta}{A_Q^{1/2}\gamma} \left[\sup_{(t_0,x_0)\in Q} \int_{Q_\delta(t_0,x_0)} \left(|\partial_t u|^2 + |Du|^2\right) r^\alpha \, dx dt \right]^{1/2} +$$
$$+ C \left(\int_Q |u|^2 dx dt \right)^{1/2} .$$

Applying now (2.1.17′) we complete of the proof.

□

Corollary 5.1.1 *The inequality* (5.1.10) *now has the form*

$$\max_{\overline{Q}} |u(x,t)| \leq \frac{2^{\frac{m+3}{2}} + \eta}{A_Q^{1/2}\gamma} \|u\|_{2,\alpha}^{1,2} + C\|u\|_{\mathcal{L}_2(Q)} \tag{5.1.13}$$

which follows immediately from (2.1.1) *and* (5.1.12). □

In this section we shall get the analogous inequality in $W_2^{1,2}(Q')$.
Let $\zeta_0(x) \geq 0$ be a smooth cut-off function, independnt of t , which is equal to zero in the neighbourhood of $\partial\Omega$, $\zeta_0(x) \equiv 1$ in Ω' and satisfies the condition

$$|\nabla\zeta_0|^2\zeta_0^{-1} \leq C.$$

Theorem 5.1.1. *If the basic assumptions* 1) - 4) *for* $q_0 = 2$ *are satisfied, the initial iteration of* (5.1.7) $u_0(t,x)$ *is smooth, then the inequality*

$$\int\limits_{Q'} \left(|\partial_t u_n|^2 + \left|D^2 u_n\right|^2\right) dxdt \leq C \quad (n = 0, 1, \ldots) \tag{5.1.14}$$

holds. Therefore the weak solution of the problem (5.1.1), (5.1.6) *belongs to* $W_2^{1,2}(Q')$.

Proof. It is obvious that all $u_n \in W_2^{1,2}(Q)$ and evidently satisfy the conditions (5.1.6) and (5.1.7). In fact, the equality (5.1.7) can be written in the form

$$\varepsilon\dot{u}_{n+1} - \Delta u_{n+1} + u_{n+1} = \varepsilon\sum_{i,j=1}^{m} \frac{\partial a_i}{\partial p_j} D_i D_j u_n - \Delta u_n +$$
$$+\varepsilon\sum_{i=1}^{m} \frac{\partial a_i}{\partial u} D_i u_n + u_n + \varepsilon\sum_{i=1}^{m} \frac{\partial a_i}{\partial x_i} - \varepsilon a_0(t,x,Du_n).$$

Assume that $u_n \in W_2^{1,2}(Q)$. Then according to our assumption 1) - 4) the right-hand side belongs to $\mathcal{L}_2(Q)$. Therefore, from the coercivity estimates Ladyzhenskaya, Ural'zeva, Solonnikov [1] it follows that $u_{n+1} \in W_2^{1,2}(Q)$. Then from $u_0 \in W_2^{1,2}(Q)$ we have all $u_n \in W_2^{1,2}(Q)$.
It is easy to see for each $u \in W_2^{1,2}(Q)$ an inequality

$$\int\limits_Q |L_\varepsilon(u)|^2\zeta_0 dxdt \leq (K_\varepsilon^2 + \eta)\int\limits_Q |D^2u|^2\zeta_0 dxdt + C\left[\int\limits_Q |Du|^2 dxdt + 1\right] \tag{5.1.15}$$

holds, in which

$$L_\varepsilon(u) = -\Delta u + u + \varepsilon L(u). \tag{5.1.16}$$

and K satisfies (1.1.6).
Fix t temporarily. It is sufficient to show the inequality (5.1.16) is valid for any smooth function. Applying (2.2.15) and (5.1.17) we get

$$\int\limits_\Omega |L_\varepsilon(u)|^2 \zeta_0 dx \leq (1+\eta)\int\limits_\Omega \left|\sum_{i=1}^{m} D_i\left[D_i u - \varepsilon a_i(t,x,Du)\right]\right|^2 \zeta_0 dx + \tag{5.1.17}$$
$$+C(\eta)\left[\varepsilon^2\int\limits_\Omega |a_0|^2 \zeta_0 dx + \int\limits_\Omega |u|^2\zeta_0 dx\right].$$

From the inequalities (1.1.2) it follows that

$$\int\limits_\Omega |a_0|^2\zeta_0 dx \leq C\left[\int\limits_\Omega |Du|^2 dx + 1\right]. \tag{5.1.18}$$

In fact

$$\int_\Omega |a_0(t,x,Du)|^2\,\zeta_0 dx \le$$

$$\le C\left[\int_\Omega |a_0(t,x,Du)-a_0(t,x,0)|^2\,\zeta_0 dx + \right.$$

$$\left. + \int_\Omega |a_0(t,x,0)|^2\,\zeta_0 dx\right] \le C\left[\int_\Omega \left|\sum_{j=0}^m \overline{\frac{\partial a_0}{\partial p_j}} D_j u\right|^2 \zeta_0 dx + 1\right] \le$$

$$\le C\left[\sum_{j=0}^m \int_\Omega |D_j u|^2 \zeta_0 dx + 1\right] \le C\left[\int_\Omega |Du|^2 dx + 1\right].$$

Here the bar over the derivatives indicates arguments are chosen in some points between zero and $D_j u$.
Let us estimate now the first integral on the right-hand side of (5.1.18). Taking into account $\zeta_0 = 0$ on $\partial\Omega$ and integrating by parts we come to the following relation

$$\text{(5.1.19)} \qquad \int_\Omega |D_i(D_i u - \varepsilon a_i)|^2\,\zeta_0 dx =$$

$$= \int_\Omega D_i(D_i u - \varepsilon a_i) D_k(D_k - \varepsilon a_k)\zeta_0 dx =$$

$$= \int_\Omega D_k(D_i u - \varepsilon a_i) D_i(D_k u - \varepsilon a_k)\zeta_0 dx +$$

$$+ \int_\Omega (D_i u - \varepsilon a_i)\left[D_i(D_k u - \varepsilon a_k) D_k \zeta_0 - D_k(D_k u - \varepsilon a_k) D_i \zeta_0\right] dx.$$

Begin by estimating the first term on the right-hand side. Differentiating gives

$$\text{(5.1.20)} \qquad \int_\Omega D_k(u_i - \varepsilon a_i) D_i(u_k - \varepsilon a_k)\zeta_0 dx =$$

$$= \int_\Omega \left[u_{ik} - \varepsilon \sum_{j=1}^m \frac{\partial a_i}{\partial p_j} u_{jk}\right]\left[u_{ik} - \varepsilon \sum_{h=1}^m \frac{\partial a_k}{\partial p_h} u_{ih}\right]\zeta_0 dx -$$

$$-\varepsilon \int_\Omega \left[\frac{\partial a_i}{\partial p_0} u_k + \frac{\partial a_i}{\partial x_k}\right]\left[u_{ik} - \varepsilon \sum_{h=1}^m \frac{\partial a_k}{\partial p_h} u_{ih} - \varepsilon \frac{\partial a_k}{\partial p_0} u_i - \varepsilon \frac{\partial a_k}{\partial x_i}\right]\zeta_0 dx -$$

$$-\varepsilon \int_\Omega \left[\frac{\partial a_k}{\partial p_0} u_i + \frac{\partial a_k}{\partial x_i}\right]\left[u_{ik} - \varepsilon \sum_{j=1}^m \frac{\partial a_i}{\partial p_j} u_{jk}\right]\zeta_0 dx.$$

The second and the third terms on the right-hand side can be estimated in a simple way. Applying the inequalities (5.1.2) and (2.2.15), and taking into account the boundedness of the derivatives of a_i with respect to p_j, we'll find

$$\text{(5.1.21)} \qquad \left| -\varepsilon \int_\Omega \left[\frac{\partial a_i}{\partial p_0} u_k + \frac{\partial a_i}{\partial x_k}\right]\left[u_{ik} - \varepsilon \sum_{h=1}^m \frac{\partial a_k}{\partial p_h} u_{ih} - \varepsilon \frac{\partial a_k}{\partial p_0} u_i - \varepsilon \frac{\partial a_k}{\partial x_i}\right]\zeta_0 dx - \right.$$

$$-\varepsilon\int_\Omega\left[\frac{\partial a_k}{\partial p_0}u_i+\frac{\partial a_k}{\partial x_i}\right]\left[u_{ik}-\varepsilon\sum_{j=1}^m\frac{\partial a_i}{\partial p_j}u_{jk}\right]\zeta_0 dx\Bigg|\leq$$

$$\leq\eta\int_\Omega|D^2u|^2\zeta_0 dx+C(\eta)\left[\int_\Omega|Du|^2dx+1\right].$$

To estimate the first integral on the right-hand side in (5.1.21) we note an expression of the type

$$u_{ik}-\varepsilon\sum_{j=1}^m\frac{\partial a_i}{\partial p_j}u_{jk}$$

can be regarded as the projections of the vector $(I-\varepsilon\tilde{A})D'(D_ku)$, where the matrix $\tilde{A}$ is obtained from A by omitting the lines and rows whose indices are equal to zero. Hence, after applying the Hölder inequality, we get

$$\int_\Omega\left[u_{ik}-\varepsilon\sum_{j=1}^m\frac{\partial a_i}{\partial p_j}u_{jk}\right]\left[u_{ik}-\varepsilon\sum_{h=1}^m\frac{\partial a_k}{\partial p_h}u_{ih}\right]\zeta_0 dx\leq$$

$$\leq\left[\sum_{i,k=1}^m\int_\Omega\left|u_{ik}-\varepsilon\sum_{j=1}^m\frac{\partial a_i}{\partial p_j}u_{jk}\right|^2\zeta_0 dx\right]^{1/2}\left[\sum_{i,k=1}^m\int_\Omega\left|u_{ik}-\varepsilon\sum_{h=1}^m\frac{\partial a_k}{\partial p_h}u_{ih}\right|^2\zeta_0 dx\right]^{1/2}\leq$$

$$\leq\sup_{x,p}\left\|I-\varepsilon\tilde{A}\right\|^2\int_\Omega\left|D'^2u\right|^2\zeta_0 dx.$$

It is obvious that

$$\|I-\varepsilon\tilde{A}\|\leq\|I-\varepsilon A\|$$

because the operator A, denoted by matrix A on $R^{(m+1)N}$ is an extension of the operator $\tilde{A}$ denoted by the matrix $\tilde{A}$ on $R^{(m+1)N}$. Thus from (1.1.6) we have $\|I-\varepsilon\tilde{A}\|\leq K_\varepsilon$ and the first term on the right-hand side of (5.1.21) is estimated as follows

$$\int_\Omega\left[u_{ik}-\varepsilon\sum_{j=1}^m\frac{\partial a_i}{\partial p_j}u_{jk}\right]\left[u_{ik}-\varepsilon\sum_{h=1}^m\frac{\partial a_k}{\partial p_h}u_{ih}\right]\zeta_0 dx\leq$$

$$\leq K_\varepsilon^2\int_\Omega|D^2u|^2\zeta_0 dx.$$

Taking into account the inequality (5.1.22) and the relation (5.1.21) gives the following inequality

(5.1.22) $$\int_Q D_k(u_i-\varepsilon a_i)D_i(u_k-\varepsilon a_k)\zeta_0 dxdt\leq$$

$$\leq(K_\varepsilon^2+\eta)\int_Q|D^2u|^2\zeta_0 dxdt+C(\eta)\left[\int_Q|Du|^2dxdt+1\right].$$

Now consider the second term on the right-hand side of (5.1.20). From the inequalities (5.1.14) and (2.2.15) we get

$$\int_Q(D_iu-\varepsilon a_i)\left[D_i(D_ku-\varepsilon a_k)D_k\zeta_0-D_k(D_ku-\varepsilon a_k)D_i\zeta_0\right]dxdt\leq$$

$$\leq \eta \int_Q |D^2u|^2\zeta_0 dxdt + C(\eta)\left[\int_\Omega |Du|^2 dxdt + 1\right].$$

Hence this relation together with inequalities (5.1.18), (5.1.19) and (5.1.23) after integration over $(0,T)$ brings us to (5.1.16).
Then from (5.1.7) follows the inequality

$$\int_Q |\varepsilon \dot u_{n+1} - \Delta u_{n+1}|^2 \zeta_0 dxdt \leq$$

$$\leq (K_\varepsilon^2 + \eta)\int_Q |D^2u_n|^2\zeta_0 dxdt + C(\eta)\left[\int_Q |Du_n|^2 dxdt + 1\right] + \int_\Omega |u_{n+1}|^2\zeta_0 dxdt.$$

As long as the process (5.1.7) converges in $W_2^{(0,1)}(Q)$, the second term of the right-hand side is bounded with a constant, independent of n. Hence,

$$\int_Q |\varepsilon \dot u_{n+1} - \Delta u_{n+1}|^2 \zeta_0 dxdt \leq (K_\varepsilon^2 + \eta)\int_Q |D^2u_n|^2\zeta_0 dxdt + C. \tag{5.1.23}$$

Now consider the left-hand side of this inequality. Integrating by parts gives

$$\int_Q |\varepsilon \dot u_{n+1} - \Delta u_{n+1}|^2 \zeta_0 dxdt =$$

$$= \varepsilon^2 \int_Q |\dot u_{n+1}|^2\zeta_0 dxdt + 2\varepsilon \int_Q \nabla \dot u_{n+1}\nabla u_{n+1}\zeta_0 dxdt +$$

$$+2\varepsilon \int_Q \dot u_{n+1}\nabla u_{n+1}\nabla\zeta_0 dxdt + \int_Q |\Delta u_{n+1}|^2\zeta_0 dxdt.$$

After integrating over $(0,T)$ and using the initial condition (5.1.6) we have

$$\int_Q |\varepsilon \dot u_{n+1} - \Delta u_{n+1}|^2 \zeta_0 dxdt =$$

$$= \varepsilon^2 \int_Q |\dot u_{n+1}|^2\zeta_0 dxdt + \varepsilon \int_\Omega |\nabla u_{n+1}|^2\zeta_0 dx\Big|_{t=T} +$$

$$+ \int_Q |\Delta u_{n+1}|^2\zeta_0 dxdt + 2\varepsilon \int_Q \dot u_{n+1}\nabla u_{n+1}\nabla\zeta_0 dxdt \quad .$$

Applying (2.2.15), we get the estimate

$$\left|\int_Q \dot u_{n+1}\nabla u_{n+1}\nabla\zeta_0 dxdt\right| \leq \eta \int_Q |\dot u_{n+1}|^2\zeta_0 dxdt + C(\eta)\int_Q |Du_n|^2 dxdt.$$

Further, applying this inequality to (5.1.24) and omitting the positive term we get

$$(\varepsilon^2 - \eta)\int_Q |\dot u_{n+1}|^2 \zeta_0 dxdt + \int_Q |\Delta u_{n+1}|^2\zeta_0 dxdt \leq (K_\varepsilon^2 + \eta)\int_Q |D^2u_n|^2 \zeta_0 dxdt + C \quad .$$

We can also omit the first term on the left-hand side expression and obtain the following relation

$$\int_Q |\Delta u_{n+1}|^2 \zeta_0 dxdt \leq (K_\varepsilon^2 + \eta) \int_Q \left|D^2 u_n\right|^2 \zeta_0 dxdt + C \ .$$

From the well known inequality

$$\int_\Omega \left|D^2 u\right|^2 \zeta_0 dx \leq (1+\eta) \int_\Omega |\Delta u|^2 \zeta_0 dx + C \int_\Omega |Du|^2 dx$$

(for example see lemma 4.1.3) it follows that

$$\int_Q \left|D^2 u_{n+1}\right|^2 \zeta_0 dxdt \leq (K_\varepsilon^2 + \eta) \int_Q \left|D^2 u_n\right|^2 \zeta_0 dxdt + C \ .$$

Taking ε as prescribed by (1.2.12) and (1.2.13) here instead of K_ε, we get the lesser quantity of $K < 1$. Then all the integrals $\int_Q |D^2 u_n|^2 \zeta_0 dxdt$ will be bounded. The time deivative is estimated from (5.1.7) and we get (5.1.15).

□

5.2 Hölder continuity of weak solutions for quasilinear parabolic systems.

In this section we shall prove the Hölder continuity of weak solution for a parabolic system in a cylinder $Q' = [0,T) \times \Omega'$, where Ω' is a strictly interior subdomain of Ω and T is any positive number. As we have seen in the previous section it is sufficient to obtain the estimate (5.1.9). For this reason we consider inside the cylinder $Q_\delta = Q_\delta(x_0) = (0,T) \times B_\delta(x_0)$ $(B_\delta(x_0) \subset \Omega)$ a parabolic equation

(5.2.1) $$\varepsilon \dot{w} + \Delta w = f$$

with boundary conditions

(5.2.2) $$w\Big|_{\partial B_\delta} = w\Big|_{t=T} = 0 \ .$$

Denote $W^{0,0}_{2,\alpha} = L_{2,\alpha}$ and suppose that the right-hand side of (5.2.1) belongs to $L_{2,\beta}(Q_\delta)$, which means that the integral

$$\int_{Q_\delta} |f|^2 r^\beta dxdt$$

is finite.

Lemma 5.2.1. *Let β' be an arbitrary number, $\beta + m - 4 > 0$ and $0 \leq \beta < m$. Then the weak solution of the problem* (5.2.1), (5.2.2) *satisfies the following estimate*

(5.2.3) $$\int_{Q_\delta} |\Delta w|^2 r^\beta \zeta dxdt \leq \frac{m}{m-\beta} \int_{Q_\delta} |f|^2 r^\beta \zeta dxdt + C(\delta) \int_{Q_\delta} \left|D^2 w\right| r^{\beta'} dxdt.$$

Proof. Certainly, we can assume f and w are smooth. Multiply the system (5.2.1) by $\Delta w r^{\beta}\zeta(r)$, then integration over Q_δ gives

$$\varepsilon \int_{Q_\delta} \dot{w}\Delta w r^{\beta}\zeta dxdt + \int_{Q_\delta} |\Delta w|^2 r^{\beta}\zeta dxdt = \int_{Q_\delta} f\Delta w r^{\beta}\zeta dxdt \ .$$

In the first integral on the left-hand side integrate once by parts with respect to x. Then

$$-\varepsilon \int_{Q_\delta} \nabla\dot{w}\nabla w r^{\beta}\zeta dxdt - \varepsilon\beta \int_{Q_\delta} \dot{w}\nabla w\nabla r r^{\beta-1}\zeta dxdt -$$
$$-\varepsilon \int_{Q_\delta} \dot{w} r^{\beta}\nabla w\nabla\zeta dxdt + \int_{Q_\delta} |\Delta w|^2 r^{\beta}\zeta dxdt = \int_{Q_\delta} f\Delta w r^{\beta} dxdt \ .$$

The first term on the left-hand side is integrated with respect to t. Taking into account the conditions (5.2.2) and substituting $\varepsilon\dot{w}$ from the system (5.2.1) we come to

(5.2.4)
$$\frac{\varepsilon}{2}\int_{B_\delta} |\nabla w|^2 r^{\beta}\zeta dx\Big|_{t=0} + \beta \int_{Q_\delta} \Delta w w' r^{\beta-1}\zeta dxdt + \int_{Q_\delta} |\Delta w|^2 r^{\beta}\zeta dxdt =$$
$$= \int_{Q_\delta} f\Delta w r^{\beta}\zeta dxdt + \beta \int_{Q_\delta} f w' r^{\beta-1}\zeta dxdt -$$
$$- \int_{Q_\delta} \Delta w\nabla w r^{\beta}\nabla\zeta dxdt + \int_{Q_\delta} f\nabla w r^{\beta}\nabla\zeta dxdt.$$

Here and everywhere later we denote with a dash the derivative with respect to r. Initially we shall estimate the integrals which have under their sign the derivatives of ζ. Let us take into account on the set of r where $\nabla\zeta \neq 0$ we have the inequalities $\frac{\delta}{2} \le r \le \frac{3\delta}{4}$. Then, for example,

$$\left|\int_{Q_\delta} \Delta w\nabla w r^{\beta}\nabla\zeta dxdt\right| \le C\left|\int_{Q_\delta} |\Delta w||Dw| r^{\beta'} dxdt\right| \ .$$

Applying the Hölder inequality we get the estimate

(5.2.5)
$$\left|\int_{Q_\delta} \Delta w\nabla w r^{\beta}\nabla\zeta dxdt\right| \le C\left|\int_{Q_\delta} \left|D^2 w\right|^2 r^{\beta'} dxdt\right| \ .$$

In the same manner we shall estimate similar integrals and occasionally consider them unessential terms.

Now estimate the integral

(5.2.6)
$$S = \int_{Q_\delta} \Delta w w' r^{\beta-1} dx\zeta dt$$

Expand the function w in a series over the complete orthonormal set of spherical harmonics $\{Y_{s,k}(\Theta)\}$ $(s = 0,1,\ldots; k = 1,\ldots,k_s)$, where Θ is a point on a unit sphere in R^m

(5.2.7)
$$w(x,t) = \sum_{s=0}^{+\infty}\sum_{k=1}^{k_s} w_{s,k}(t,r)Y_{s,k}(\Theta) \ .$$

Taking into account the inequality (4.1.26) after integration from zero to T we get

$$S \geq \frac{1}{2}\sum_{s,k}\int_0^T\int_0^\delta \Big[(m-\beta)\left|w'_{s,k}\right|^2 + \tag{5.2.8}$$
$$+(\beta+m-4)s(s+m-2)\left|w_{s,k}\right|^2 r^{-2}\Big] r^{\beta+m-3}\zeta dr dt + \ldots,$$

where the nonwritten term can be estimated as in (5.2.5) by $\int_{Q_\delta} |D^2 w|^2 r^{\beta'} dx dt$ Substituting this expression in (5.2.4) and omitting the positive term

$$\frac{\varepsilon}{2}\int_{B_\delta} |\nabla w|^2 r^\beta \zeta dx\Big|_{t=0}$$

on left-hand side, we get the inequality

$$\int_{Q_\delta} |\Delta w|^2 r^\beta \zeta dx dt + \frac{\beta}{2}\sum_{s,k}\int_0^T\int_0^\delta \Big[(m-\beta)\left|w'_{s,k}\right|^2 + \tag{5.2.9}$$
$$+(\beta+m-4)s(s+m-2)\left|w_{s,k}\right|^2 r^{-2}\Big] r^{\beta+m-3}\zeta dr dt \leq$$
$$\leq \int_{Q_\delta} f(\Delta w + \beta w' r^{-1}) r^\beta \zeta dx dt + \ldots,$$

where the non-written terms contain expressions as $\int_{Q_\delta} |D^2 w|^2 r^{\beta'} dx dt$. We will now consider the right-hand side of (5.2.9). Applying the inequality (2.2.15) and taking into account that β is positive gives the following estimate

$$\left|\int_{Q_\delta} f(\Delta w + \beta w' r^{-1}) r^\beta \zeta dx dt\right| \leq \left|\int_{Q_\delta} f \Delta w r^\beta \zeta dx dt\right| +$$
$$+\eta\beta\int_{Q_\delta} |w'|^2 r^{\beta-2}\zeta dx dt + \frac{\beta}{4\eta}\int_{Q_\delta} |f|^2 r^\beta \zeta dx dt.$$

From the expansion (5.2.7) follows

$$\int_{Q_\delta} |w'|^2 r^{\beta-2}\zeta dx dt = \sum_{s,k}\int_0^T\int_0^\delta \left|w'_{s,k}\right|^2 r^{\beta+m-3}\zeta dr dt.$$

Now the previous inequality can be written in the form

$$\left|\int_{Q_\delta} f(\Delta w + \beta w' r^{-1}) r^\beta \zeta dx dt\right| \leq \left|\int_{Q_\delta} f \Delta w r^\beta \zeta dx dt\right| +$$
$$+\eta\beta\sum_{s,k}\int_0^T\int_0^\delta \left|w'_{s,k}\right|^2 r^{\beta+m-3}\zeta dr dt + \frac{\beta}{4\eta}\int_{Q_\delta} |f|^2 r^\beta \zeta dx dt .$$

Therefore the estimate (5.2.9) leads to the following relation

$$\int_{Q_\delta} |\Delta w|^2 r^\beta \zeta dx dt + \frac{\beta}{2} \sum_{s,k} \int_0^T \int_0^\delta \Big[(m-\beta) \left|w'_{s,k}\right|^2 +$$
$$+ (\beta + m - 4) s(s+m-2) \left|w_{s,k}\right|^2 r^{-2} \Big] r^{\beta+m-3} \zeta dr dt \le$$
$$\le \left| \int_{Q_\delta} f \Delta w r^\beta \zeta dx dt \right| + \eta\beta \sum_{s,k} \int_0^T \int_0^\delta \left|w'_{s,k}\right|^2 r^{\beta+m-3} \zeta dr dt +$$
$$+ \frac{\beta}{4\eta} \int_{Q_\delta} |f|^2 r^\beta \zeta dx dt + \dots \; .$$

As long as $\beta + m - 4 > 0$, the second term in the quadratic brackets on the left-hand side is nonnegative and we can omit it . Take $\eta = (m-\beta)/2$. The sums on both sides will be cancelled and give the estimate

$$\int_{Q_\delta} |\Delta w|^2 r^\beta \zeta dx dt \le \left| \int_{Q_\delta} f \Delta w r^\beta \zeta dx dt \right| + \frac{\beta}{2(m-\beta)} \int_{Q_\delta} |f|^2 r^\beta \zeta dx dt + \dots \; .$$

Apply the inequality (2.2.15) with $\eta = \frac{1}{2}$ to the right-hand side to give the inequality (5.2.3).

□

Lemma 5.2.2. *If* $0 \le \beta < m$ *and* $\beta + m - 4 > 0$*, then for the weak solution of the problem* (5.2.1), (5.2.2) *the inequality*

$$\int_{Q_\delta} |\Delta w|^2 r^\beta dx dt \le \frac{m}{m-\beta} \int_{Q_\delta} |f|^2 r^\beta dx dt \tag{5.2.10}$$

is true.

Proof. The proof is not substantially different from that of lemma 5.2.1. In fact, let us multiply the equalities (5.2.1) by $\Delta w r^\beta$. After integrating by parts and using the boundary conditions (5.2.2) we get the equality

$$\frac{\varepsilon}{2} \int_{B_\delta} |\nabla w|^2 r^\beta dx \Big|_{t=0} + \beta \int_{Q_\delta} \Delta w w' r^{\beta-1} dx dt + \int_{Q_\delta} |\Delta w|^2 r^\beta dx dt =$$
$$= \int_{Q_\delta} f(\Delta w + \beta w' r^{-1}) r^\beta dx dt, \tag{5.2.11}$$

which is similar to the equality (5.2.4). If we apply the inequality (4.1.25) and repeat the end of the proof for the previous lemma we come to (5.2.10).

□

Remark 5.2.1. *The condition* $\beta + m - 4 > 0$ *can be invalid only for* $m = 2$ *and* $m = 3$ *with* $\beta < 1$.

Theorem 5.2.1. *Let the basic assumptions* 1) - 4) *of section* 5.1 *be satisfied for* $q_0 > \frac{2m}{m+\alpha}$ *and let* $\alpha = 2 - m - 2\gamma$ $(0 < \gamma < 1)$. *If for* $m \geq 3$ *the relation*

$$K^2 \left[1 - \frac{4\alpha(m-1)}{(\alpha+m)^2}\right] \left(1 + M_\gamma^2\right) \frac{m}{\alpha+m} < 1 \tag{5.2.12}$$

where K *is denoted by* (1.1.8), (1.1.9) *and* M_γ *by* (4.2.20) *is true, then the weak solution of the problem* (5.1.1), (5.1.6) *will be Hölder continuous in* $Q = [0,T) \times \Omega'$ *where* $\Omega' \subset\subset \Omega$.

Proof. Take $u_0 \in W_q^{1,2}(Q)$ with $q < \frac{2m}{m+\alpha}$ and some fixed sufficiently small $\delta_0 > 0$. Then according the assumptions, $\forall u_n \in W_q^{1,2}(Q) \subset W_{2,\beta}^{1,2}(Q)$.
Take from (5.2.1)

$$f = -\Delta u_{n+1} r^{-\beta_0} \tag{5.2.13}$$

and denote the solution of the problem (5.2.1), (5.2.2) by w_0. Take

$$\beta_0 = \frac{m}{2} - \eta_1 = -\alpha_0 \ .$$

Thus, from (5.2.3) and inequality (4.1.20) with $\beta = \beta_0 = -\alpha_0$ and $\beta' = 2\beta_0 = -2\alpha_0$ follows

$$\int_{Q_\delta} \left|D'^2 w_0\right|^2 r^{\beta_0} \zeta dx dt \leq \tag{5.2.14}$$
$$\leq \frac{m}{m-\beta_0}\left[1 + \frac{4\beta_0(m-1)}{(m-\beta_0)^2} + \eta\right] \int_{Q_\delta} |\Delta u_{n+1}|^2 r^{-\beta_0} \zeta^2 dx dt +$$
$$+ C \int_{Q_\delta} \left|D^2 w_0\right|^2 r^{2\beta_0} dx dt \ .$$

We have mentioned before the well known inequality

$$\int_{B_\delta} \left|D'^2 w\right|^2 r^\kappa dx \leq C \int_{B_\delta} |\Delta w|^2 r^\kappa dx, \tag{5.2.15}$$

which holds for $-m < \kappa < m$ and w, satisfying the homogeneous condition $w = 0$ on ∂B_δ. Integrating from zero to T we arrive at

$$\int_{Q_\delta} \left|D'^2 w\right|^2 r^\kappa dx dt \leq C \int_{Q} |\Delta w|^2 r^\kappa dx dt \ . \tag{5.2.16}$$

Applying this inequality to $\kappa = 2\beta_0$ and w_0 and taking into account (5.2.10) we obtain the following estimate

$$\int_{Q_\delta} \left|D'^2 w_0\right|^2 r^{2\beta_0} dx dt \leq C \int_{Q_\delta} |\Delta u_{n+1}|^2 dx dt$$

and (5.1.15) gives

$$\int_{Q_\delta} \left|D^2 w_0\right|^2 r^{2\beta_0} dx dt \leq C. \tag{5.2.17}$$

From the estimate (5.2.14) we obtain

(5.2.18)
$$\int_{Q_\delta} \left|D'^2 w_0\right|^2 r^{\beta_0} \zeta dxdt \leq$$
$$\leq \frac{m}{m-\beta_0}\left[1+\frac{4\beta_0(m-1)}{(m-\beta_0)^2}+\eta\right]\int_{Q_\delta} |\Delta u_{n+1}|^2 r^{-\beta_0}\zeta dxdt + C \ .$$

Multiply (5.1.7) by $\Delta w_0 \zeta$ and integrate the result over Q_δ. Integrating by parts once by t and twice by x we get the following equality

$$-\int_{Q_\delta} \Delta u_{n+1}(\varepsilon \dot{w}_0 + \Delta w_0)\zeta dxdt = \int_{Q_\delta} L_\varepsilon(u_n)\Delta w_0 \zeta dxdt +$$
$$+2\varepsilon \int_{Q_\delta} \nabla u_{n+1} \nabla \zeta \dot{w}_0 dxdt + \varepsilon \int_{Q_\delta} \dot{w}_0 u_{n+1} \Delta\zeta dxdt - \int_{Q_\delta} u_{n+1} \Delta w_0 \zeta dxdt,$$

where $L_\varepsilon(u)$ is denoted by (5.1.17).
Taking into account (5.2.1) and (5.2.13) we get

$$\int_{Q_\delta} |\Delta u_{n+1}|^2 r^{-\beta_0} \zeta dxdt = \int_{Q_\delta} L_\varepsilon(u_n)\Delta w_0 \zeta dxdt +$$
$$+2\varepsilon \int_{Q_\delta} \nabla u_{n+1} \nabla \zeta \dot{w}_0 dxdt + \varepsilon \int_{Q_\delta} u_{n+1} \Delta\zeta \dot{w}_0 dxdt - \int_{Q_\delta} u_{n+1} \Delta w_0 \zeta dxdt.$$

Now estimate the last two integrals on the right-hand side. Takinig into account the derivatives of ζ are contained under the signs of the integrals and $\dot{w}_0$ satisfies (5.2.1), we get

$$\left|\int_{Q_\delta} \nabla u_{n+1} \nabla \zeta \dot{w}_0 dxdt\right| \leq \int_{Q_\delta} |\nabla u_{n+1}| \ |\nabla \zeta| \left[|\Delta w_0| + |\Delta u_{n+1}| r^{-\beta_0}\right] dxdt \leq$$
$$\leq C\left[\int_{Q_\delta} \left|D^2 u_{n+1}\right|^2 dxdt + \int_{Q_\delta} |\Delta w_0|^2 r^{2\beta_0} dxdt\right].$$

Then from the inequalities (5.1.15) and (5.2.17) follows

(5.2.19)
$$\left|\int_{Q_\delta} \nabla u_{n+1} \nabla \zeta \dot{w}_0 dxdt\right| < C \ .$$

The second integral can be estimated in the same manner to give

(5.2.20)
$$\int_{Q_\delta} |\Delta u_{n+1}|^2 r^{-\beta_0} \zeta dxdt \leq \int_{Q_\delta} L_\varepsilon(u_n)\Delta w_0 \zeta dxdt + \left|\int_{Q_\delta} u_{n+1} \Delta w_0 \zeta dxdt\right| + C.$$

Now consider the integral

$$\int_{Q_\delta} L_\varepsilon(u_n)\Delta w_0 \zeta dx dt \ .$$

After integrating by parts twice with respect to x, taking into account that ζ^2 vanishes on ∂B_δ and omitting temporarily the indices we get

$$\int_{Q_\delta} L_\varepsilon(u)\Delta w\zeta dxdt = -\int_{Q_\delta}(u_{ii} - u + \varepsilon a_0 - \varepsilon D_i a_i) w_{kk}\zeta dxdt =$$

$$= \int_{Q_\delta}(u_i - \varepsilon a_i) w_{ikk}\zeta dxdt + \int_{Q_\delta}(u_i - \varepsilon a_i) w_{kk}\zeta_i dxdt + \int_{Q_\delta}(u - \varepsilon a_0)\Delta w\zeta dxdt$$

$$= -\int_{Q_\delta}[u_{ik} - \varepsilon D_k a_i]\, w_{ik}\zeta dxdt + \int_{Q_\delta}(u_i - \varepsilon a_i) w_{kk}\zeta_i dxdt -$$

$$-\int_{Q_\delta}(u_i - \varepsilon a_i) w_{ik}\zeta_k dxdt + \int_{Q_\delta}(u - \varepsilon a_0)\Delta w\zeta dxdt.$$

The integrals which contain under their signs the derivatives of ζ can be estimated as in (5.2.19). The inequalities (2.1.17′) and (5.1.15) give

$$\left|\int_{Q_\delta} u_{n+1}\zeta dxdt\right| \le \left(\eta\int_{Q_\delta}|D'^2 u_{n+1}|^2 r^{-\beta_0}\zeta dxdt + C\right)^{1/2}\left(\int_{Q_\delta}|D'^2 w_0|^2 r^{\beta_0}\zeta dxdt\right)^{1/2} .$$

Similarily can be estimated the following integral

$$\int_{Q_\delta}(u_n - \varepsilon a_0)\Delta w_0\zeta dxdt.$$

Hence, we get the following estimate

$$\left|\int_{Q_\delta} L_\varepsilon(u)\Delta w\zeta dxdt\right| \le \left|\int_{Q_\delta}(u_{ik} - \varepsilon D_k a_i) w_{ik}\zeta dxdt\right| +$$

$$+\left(\eta\int_{Q_\delta}|D'^2 u|^2 r^{-\beta_0} dxdt + C\right)^{1/2}\left(\int_{Q_\delta}|D'^2 w|^2 r^{\beta_0} dxdt\right)^{1/2} + C.$$

Multiplying and dividing by $r^{\beta_0/2}$ under the sign of the right-hand side integral and applying the Hölder inequality we get

$$\left|\int_{Q_\delta} L_\varepsilon(u)\Delta w\zeta dxdt\right| \le \left[\sum_{i,k=1}\int_{Q_\delta}|u_{ik} - \varepsilon D_k a_i|^2 r^{-\beta_0}\zeta dxdt\right]^{1/2} \times$$

$$\times\left[\int_{Q_\delta}\left|D'^2 w\right|^2 r^{\beta_0}\zeta dxdt\right]^{1/2} + \left(\eta\int_{Q_\delta}|D'^2 u|^2 r^{-\beta_0} dxdt + C\right)^{1/2}\left(\int_{Q_\delta}|D'^2 w|^2 r^{\beta_0} dxdt\right)^{1/2} + C.$$

Thus, applying the analogous estimate, which we used in the proof of (5.1.23) we get the inequality

$$\left|\int_{Q_\delta} L_\varepsilon(u_n)\Delta w_0 \zeta dxdt\right| \leq (K_\varepsilon+\eta)\left\{\left[\int_{Q_\delta} \left|D'^2 u_n\right|^2 r^{-\beta_0}\zeta dxdt\right]^{1/2} + C\right\} \times$$
$$\times \left[\int_{Q_\delta} \left|D'^2 w_0\right|^2 r^{\beta_0}\zeta dxdt\right]^{1/2} + C .$$

Now applying (5.2.18), we have the following estimate

$$\left|\int_{Q_\delta} L_\varepsilon(u_n)\Delta w_0 \zeta dxdt\right| \leq (K_\varepsilon+\eta)\left[1+\frac{4\beta_0(m-1)}{(m-\beta_0)^2}\right]^{1/2} \times$$
$$\times \left[\frac{m}{m-\beta_0}\right]^{1/2}\left[\left(\int_{Q_\delta} \left|D'^2 u_n\right|^2 r^{-\beta_0}\zeta dxdt\right)^{1/2} + C\right]\left[\left(\int_{Q_\delta} |\Delta u_{n+1}|^2 r^{-\beta_0}\zeta dxdt\right)^{1/2} + C\right] .$$

Using the inequality (5.2.20) and taking into account inequality (4.1.19) , we shall have

(5.2.21) $$\int_{Q_\delta} |\Delta u_{n+1}|^2 r^{-\beta_0}\zeta dxdt \leq$$
$$\leq (K_\varepsilon+\eta)\left[1+\frac{4\beta_0(m-1)}{(m-\beta_0)^2}\right]^{1/2}\left[\frac{m}{m-\beta_0}\right]^{1/2}\left(1+M_\gamma^2\right)^{1/2} \times$$
$$\times \left[\left(\int_{Q_\delta} |\Delta u_n|^2 r^{-\beta_0}\zeta dxdt\right)^{1/2} + C\right]\left[\left(\int_{Q_\delta} |\Delta u_{n+1}|^2 r^{-\beta_0}\zeta dxdt\right)^{1/2} + C\right] ,$$

where M_γ^2 is denoted by (4.2.20) with $\gamma = \frac{\beta_0+2-m}{2}$. This corresponds to $\alpha = -\beta_0 = 2-m-2\gamma$.
Take ε such as prescribed by (1.2.12) and (1.2.13). Then $K_\varepsilon = K$. So if $\beta_0 \in [0, m-3) \cup (m-2, m-2+2\gamma)$ we get from (5.2.12) that

$$(K_\varepsilon+\eta)\left[1+\frac{4\beta_0(m-1)}{(m-\beta_0)^2}\right]^{1/2}\left[\frac{m}{m-\beta_0}\right]^{1/2}\left(1+M_\gamma^2\right)^{1/2} < 1$$

and from (5.2.21) follows

$$\int_{Q_\delta} |\Delta u_{n+1}|^2 r^{-\beta_0}\zeta dxdt < C .$$

Hence,

$$\int_{Q_\delta} \left|D'^2 u_{n+1}\right|^2 r^{-\beta_0}\zeta dxdt < C$$

and (2.1.17′) brings us to the relation $\int\limits_{Q_\delta} |D^2u_{n+1}|^2 \zeta dxdt < C$. Take $m = 3$. Then $\beta_0 = 3/2 - \eta_1$ and we can take $\eta_1 = 1/2 - 2\gamma_1$ with a suitable small $\gamma_1 > 0$ and the theorem for $m = 3$ and small γ_1 is proved.

Now let us take $f = -\Delta u_{n+1} r^{-\beta_1}$ in equality (5.2.1), where $2\beta_1 - \beta_0 = m - \eta_1$. Denote the solution of (5.2.1), (5.2.2) by $w_1(t,x)$. From the inequality (5.2.3) (with $\beta = \beta_1$ and $\beta' = 2\beta_1 - \beta_0$) analogous to (5.2.14) we get

$$\int\limits_{Q_\delta} \left|D'^2 w_1\right|^2 r^{\beta_1}\zeta dxdt \le \frac{m}{m-\beta_1}\left[1 + \frac{4\beta_1(m-1)}{(m-\beta_1)^2} + \eta\right] \int\limits_{Q_\delta} |\Delta u_{n+1}|^2 r^{\beta_1}\zeta dxdt +$$

$$+C\int\limits_{Q_\delta} \left|D^2 w_1\right|^2 r^{2\beta_1-\beta_0} dxdt \ .$$

From the inequality (5.2.10) with $\beta = 2\beta_1 - \beta_0$ we will find that

$$\int\limits_{Q_\delta} |\Delta w_1|^2 r^{2\beta_1-\beta_0} dxdt \le C \int\limits_{Q_\delta} |\Delta u_{n+1}|^2 r^{-\beta_0} dxdt$$

which leads to

$$\int\limits_{Q_\delta} \left|D^2 w_1\right|^2 r^{2\beta_1-\beta_0} dxdt \le C \ .$$

Therefore,

$$\int\limits_{Q_\delta} \left|D'^2 w_1\right|^2 r^{\beta_1}\zeta dxdt \le$$

$$\le \left[1 + \frac{4\beta_1(m-1)}{(m-\beta_1)^2} + \eta\right] \frac{m}{m-\beta_1} \int\limits_{Q_\delta} |\Delta u_{n+1}|^2 r^{-\beta_1}\zeta dxdt + C \ .$$

Repeating all the arguments, we have used for w_0, we come to the conclusion that if $m - 2 < \beta_1 < m$ with a suitable γ we will have the inequality

$$\int\limits_{Q_\delta} \left|D^2 u_{n+1}\right|^2 r^{-\beta_1}\zeta dxdt < C$$

for such β_1. Take $m = 4$ and $\eta_1 = 1 - 2\gamma$, the inequalities $m - 2 < 3m/4 - \eta_1 < m$ will be satisfied for $\gamma \in (0,1)$ and the theorem is proved for $m = 4$.

For $m > 4$ this process can be continued. In fact, take in (5.2.1)

$$f = -\Delta u_{n+1} r^{-\beta_k}$$

and denote the solution of (5.2.1), (5.2.2) by $w_k(t,x)$.

Let $\{u_n\}$ satisfy the inequality

$$\int\limits_{Q_\delta} \left|D^2 u_n\right|^2 r^{-\beta_{k-1}}\zeta dxdt \le C \ . \tag{5.2.22}$$

Choose $\{\beta_k\}$, $k = 0, 1, \ldots, k_0$, such that

$$0 < 2\beta_k - \beta_{k-1} < m, \quad \beta_k \notin [m-3, m-2], \quad \beta_0 > 1, \quad \beta_{k_0} = m - 2 + 2\gamma \ .$$

If (5.2.22) holds then for the arguments which we have used for $m = 4$ it follows that

(5.2.23)
$$\int_{Q_\delta} \left|D^2 u_n\right|^2 r^{-\beta_k} \zeta dx dt \le C \ .$$

From (5.2.23) with $\beta_k = \beta_{k_0}$ follows (5.2.19) and theorem is completely proved.

□

Now consider the case $m = 2$.

Theorem 5.2.2. * *Let the assumptions* 1) – 4) *of the section* 5.1 *be satisfied. If the inequality*

(5.2.24)
$$2K^2 < 1$$

holds true, where K is defined by (1.1.8), (1.1.9), *then the weak solution of the problem* (5.1.1), (5.1.16) *is Hölder continuous in $Q' = (0,T) \times \Omega'$, where $\Omega' \subset\subset \Omega$.*

Proof. It is clear that we must estimate the sum

(5.2.25)
$$\sum_{s=0}^{+\infty} \sum_{k=1}^{k_s} s(s+m-2) \int_0^T \int_0^\delta |w_{s,k}|^2 r^{\beta+m-5} \zeta dr dt$$

on the right-hand side of inequality (5.2.8). This sum appears also on the left-hand sides in (5.2.4) and (5.2.11). In fact, we must estimate this sum starting only with $s = 1$. Omitting the first positive term on the left-hand side and the nonessential terms on the right-hand side of (5.2.4) we get

(5.2.26)
$$I \equiv \int_{Q_\delta} \Delta w(\Delta w + \beta r^{-1} w') r^\beta \zeta dx dt \le$$
$$\le \int_{Q_\delta} f(\Delta w + \beta r^{-1} w') r^\beta \zeta dx dt + \ldots \ .$$

Because of the presence of the factor ζunder the integrals signs we can assume that w and w' vanishes for $r = \delta$, and cancel the factor ζ. In fact, changing ζ for ζ^2 we move ζ under the signs of the derivatives. Then we have some additional members, which can be estimated as in (5.2.5).
Using the expansion (5.2.7) and the expression (4.1.10) we'll get the following formula

(5.2.27)
$$J = \int_{Q_\delta} \left|\Delta w + \beta w' r^{-1}\right|^2 r^\beta dx dt =$$
$$= \sum_{s,k} \int_0^T \int_0^\delta \Big\{ \left|w''_{s,k}\right|^2 + [m - 1 + \beta + 2s(s+m-2)] \left|w'_{s,k}\right|^2 r^{-2} +$$
$$+ s(s+m-2)\left[s(s+m-2) + 2(\beta + m - 4)\right] |w_{s,k}|^2 r^{-4} \Big\} r^{\beta+m-1} dr dt \ .$$

*This theorem was proved with the help of Chelkak.

It is easy to see that the equality

$$(5.2.28)\qquad I=\sum_{s,k}\int_0^T\int_0^\delta\Big\{\left|w''_{s,k}\right|^2+\Big[m-1+\frac{\beta(2-m-\beta)}{2}+$$

$$+2s(s+m-2)\Big]\left|w'_{s,k}\right|^2r^{-2}+$$

$$+s(s+m-2)\left[s(s+m-2)+(\beta+m-4)\frac{4-\beta}{2}\right]|w_{s,k}|^2r^{-4}\Big\}r^{\beta+m-1}drdt+\dots$$

is true (I is the left-hand side in (5.2.26)). If we integrate the inequality (4.1.31) from zero to T (u like w is now supposed to be a function in both t and x) we get

$$(5.2.29)\qquad I\ge\frac{m-\beta}{m+\beta}J-C\int_{Q_\delta}\left|D^2w\right|^2r^{\beta'}dxdt,$$

where β' is arbitrary.
From (4.1.40) for $m=2$ and $0\le\beta<m$ follows (after integrating by t)

$$(5.2.30)\qquad J\ge(1-\eta)\int_{Q_\delta}|\Delta w|^2r^\beta\zeta dxdt-\left[\frac{\beta^2(2-\beta)}{2}+\eta\right]\times$$

$$\times\sum_{k=1}^2\int_0^T\int_0^\delta|w_{1,k}|^2r^{\beta-3}drdt-C\int_{Q_\delta}\left|D^2w\right|^2r^{\beta'}dxdt\ .$$

From this inequality with the help of (5.2.27) and (5.2.29) for $m=2$ we get

$$(5.2.31)\qquad (1-\eta)\int_{Q_\delta}|\Delta w|^2r^\beta dxdt\le\le\left[\frac{2+\beta}{2-\beta}+\eta\right]^2\int_{Q_\delta}|f|^2r^\beta\zeta dxdt+$$

$$+\left[\frac{\beta^2(2-\beta)}{2}+\eta\right]\sum_{k=1}^2\int_0^T\int_0^\delta|w_{1,k}|^2r^{\beta-3}drdt+\dots\ .$$

It remains now to estimate the second term on the right-hand side. For this purpose we multiply the equation (5.2.1) by $w_{1,1}r^{\beta-2}\zeta(r)$ and integrate over Q_δ.
Integrating with respect to t, we get the equality

$$-\frac{\varepsilon}{2}\int_{B_\delta}|w_{1,1}|^2r^{\beta-2}\zeta dx\Big|_{t=0}+\int_0^T\int_0^\delta\left[\frac{1}{r}(rw'_{1,1})'-\frac{1}{r^2}w_{1,1}\right]w_{1,1}r^{\beta-1}\zeta drdt=$$

$$=\int_0^T\int_0^\delta f_{1,1}w_{1,1}r^{\beta-1}\zeta drdt,$$

where $f_{1,1}$ is the Fourier coefficients of f.
Integrating by parts after trivial calculations we get the equality

$$(5.2.32) \qquad \frac{\varepsilon}{2}\int_{B_\delta} |w_{1,1}|^2 r^{\beta-2}\zeta dx\Big|_{t=0} +$$

$$+\int_0^T\int_0^\delta \left\{ |w'_{1,1}|^2 + \left[1-\frac{(2-\beta)^2}{2}\right] |w_{1,1}|^2 r^{-2} \right\} r^{\beta-1}\zeta dr dt =$$

$$= -\int_0^T\int_0^\delta f_{1,1} w_{1,1} r^{\beta-1}\zeta dr dt + \dots .$$

All the integrals on the left-hand side are finite because $w_{1,1} = 0$ for $r = 0$. From the Hardy inequality (2.1.8) we have

$$\int_0^\delta |w'_{1,1}|^2 r^{\beta-1}\zeta dr \geq \frac{(2-\beta)^2}{4}\int_0^\delta |w_{1,1}|^2 r^{\beta-3}\zeta dr + \dots .$$

Therefore, from (5.2.32) we get the estimate

$$\left[1-\frac{(2-\beta)^2}{4}\right]\int_0^T\int_0^\delta |w_{1,1}|^2 r^{\beta-3}\zeta dr dt \leq -\int_0^T\int_0^\delta f_{1,1} w_{1,1} r^{\beta-1}\zeta dr dt \dots \leq + \dots$$

$$\leq \left[\int_0^T\int_0^\delta |f_{1,1}|^2 r^{\beta+1}\zeta dr dt\right]^{1/2} \left[\int_0^T\int_0^\delta |w_{1,1}|^2 r^{\beta-3}\zeta dr dt\right]^{1/2} + \dots .$$

Taking into account the relation

$$1-\frac{(2-\beta)^2}{4} = \beta\left[1-\frac{\beta}{4}\right] > 0,$$

we get, with the help of the Hölder inequality from (5.2.32), the following estimate

$$\int_0^T\int_0^\delta |w_{1,1}|^2 r^{\beta-3}\zeta dr dt \leq \left[\frac{4}{(4-\beta)\beta}\right]^2 \int_0^T\int_0^\delta |f_{1,1}|^2 r^{\beta+1}\zeta dr dt + \dots .$$

The analogous inequality holds for $w_{1,2}(r)$ and we have

$$\sum_{k=1}^2 \int_0^T\int_0^\delta |w_{1,k}|^2 r^{\beta-3}\zeta dr dt \leq \frac{16}{(4-\beta)^2\beta^2}\int_{Q_\delta} |f|^2 r^\beta \zeta dx dt + \dots .$$

Therefore with help of (5.2.31) we get the inequality

$$\int_{Q_\delta} |\Delta w|^2 r^\beta \zeta dx dt \leq \left[\left(\frac{2+\beta}{2-\beta}\right)^2 + \frac{8(2-\beta)}{(4-\beta)^2} + \eta\right]\int_{Q_\delta} |f|^2 r^\beta \zeta dx dt + \dots$$

or simply

$$\int_{Q_\delta} |\Delta w|^2 r^\beta \zeta dx dt \leq [2 + O(\beta) + \eta]\int_{Q_\delta} |f|^2 r^\beta \zeta dx dt + \dots .$$

So the inequality, analogous to (5.2.3) for $m = 2$ is established.
Here instead of $m(m-\beta)^{-1}$ we have another constant $2 + O(\beta) + \eta$.
Now let us find the constant corresponding to the coefficient before the integral on the right-hand side of (5.2.10) (of course the equality (5.2.11) is also true for $m = 2$). From (5.2.11) follows the estimate

$$\int_{Q_\delta} |\Delta w|^2 r^\beta dx dt + \frac{\beta(m-\beta)}{2} \sum_{s,k} \int_0^T \int_0^\delta \left|w'_{s,k}\right|^2 r^{\beta+m-3} dr dt \leq$$

$$\leq \int_{Q_\delta} f(\Delta w + \beta w' r^{-1}) r^\beta dx dt - \frac{\beta(\beta+m-4)}{2} \sum_{s,k} s(s+m-2) \times$$

$$\times \int_0^T \int_0^\delta |w_{s,k}|^2 r^{\beta+m-5} dr dt \ .$$

So, all that remains is to estimate the sum

$$\sum_{s \geq 1, k} s(s+m-2) \int_0^T \int_0^\delta |w_{s,k}|^2 r^{\beta+m-5} dr dt \ .$$

For this purpose we multiply (5.2.1) by $w_{s,k} r^{\beta-2}$ and integrate over Q_δ using the conditions (5.2.2). After integrating by parts and using the equality $w_{s,k}(0) = 0$ for $s \geq 1$, we get the relation

$$\frac{\varepsilon}{2} \int_0^\delta |w_{s,k}|^2 r^{\beta+m-3} dr \Big|_{t=0} +$$

$$+ \left[s(s+m-2) - \frac{(\beta+m-4)^2}{2}\right] \int_0^T \int_0^\delta |w_{s,k}|^2 r^{\beta+m-5} dr dt +$$

$$+ \int_0^T \int_0^\delta \left|w'_{s,k}\right|^2 r^{\beta+m-3} dr dt = - \int_0^T \int_0^\delta f_{s,k} w_{s,k} r^{\beta+m-3} dr dt \ .$$

Therefore for $m = 2$ we have

$$\int_0^T \int_0^\delta \left|w'_{s,k}\right|^2 r^{\beta-1} dr dt + \left[s^2 - \frac{(\beta-2)^2}{2}\right] \int_0^T \int_0^\delta |w_{s,k}|^2 r^{\beta-3} dr dt \leq$$

$$\leq - \int_0^T \int_0^\delta f_{s,k} w_{s,k} r^{\beta-1} dr dt \ .$$

Using Hardy inequality (2.1.8) gives the estimate

$$\frac{(\beta-2)^2}{4} \int_0^T \int_0^\delta |w_{s,k}|^2 r^{\beta-3} dr dt + \left[s^2 - \frac{(\beta-2)^2}{2}\right] \int_0^T \int_0^\delta |w_{s,k}|^2 r^{\beta-3} dr dt \leq$$

$$\leq - \int_0^T \int_0^\delta f_{s,k} w_{s,k} r^{\beta-1} dr dt \ .$$

From this follows

$$\left[\left(1-\frac{\beta}{4}\right)\beta+s^2-1\right]\int_0^T\int_0^\delta |w_{s,k}|^2 r^{\beta-3}drdt \leq -\int_0^T\int_0^\delta f_{s,k}w_{s,k}r^{\beta-1}drdt$$

and therefore

$$\sum_{s\geq 1,k} s^2\int_0^T\int_0^\delta |w_{s,k}|^2 r^{\beta-3}drdt \leq C\int_{Q_\delta}|f|^2r^\beta dxdt \ .$$

So, the sum is estimated and as in lemma 5.2.2 we have

$$\int_{Q_\delta}|\Delta w|^2r^\beta dxdt \leq C\int_{Q_\delta}|f|^2r^\beta dxdt \ .$$

The constant is not essential.
Repeating the proof of theorem 5.2.2 and using the inequalities (4.1.19) and (4.1.20) we reach the end of the proof.

□

For small $\gamma > 0$ theorems 5.2.1 and 5.2.2 give

Theorem 5.2.3. *Let conditions of theorems* 5.2.1 *and* 5.2.2 *be satisfied for small positive* γ.
If

(5.2.33) $$K^2\left(1+\frac{m-2}{m+2}\right)[1+(m-2)(m-1)]\frac{m}{2} < 1 \text{ for } m\geq 3$$

and

(5.2.34) $$2K^2 < 1 \text{ for } m=2$$

then the weak solution of the problem (5.1.1), (5.1.6) *is Hölder continuous in* $Q' = (0,T)\times\Omega'$, *where* $\Omega' \subset\subset \Omega$.

5.3 Some coercivity inequalities with explicit constants

Now consider the cylinder $Q_R = (0,T)\times B_R$ ($Q_1 = Q$) and the boundary conditions

(5.3.1) $$u|_{\partial B_R} = u|_{t=0} = 0$$

for a function $u(t,x)$ given in Q. Denote $\beta = -\alpha$ and omit temporarily the index R. For $m > 2$ the inequality

(5.3.2) $$\int_Q |\Delta u|^2 r^\beta dxdt \leq \frac{m}{m-\beta}\int_Q |\varepsilon\dot{u}-\Delta u|^2 r^\beta dxdt$$

was essentialy proved in lemma 5.2.2 (inequality (5.2.10)). The difference between (5.2.10) and (5.3.2) lies in the right-hand side operators $\varepsilon\dot{w}+\Delta w$ and $\varepsilon\dot{u}-\Delta u$. It is evident after changing t for $T-t$ and substituting conditions (5.2.2) by (5.3.1) we get the equivalent inequalities.

Lemma 5.3.1. *Let* $m = 2\ \beta = 2\gamma\ (0 < \gamma < 1)$. *Assume* u *satisfies* (5.3.1) *and* $u \in L_2\left\{(0,T); W^{(2)}_{2,\beta}(B)\right\}$. *Then the inequality*

$$\int_Q |\Delta u|^2 r^\beta dxdt \le \left[1 + \frac{\beta}{2-\beta} + \frac{2-\beta}{(1-\frac{\beta}{4})^2\beta}\right] \int_Q |\varepsilon \dot{u} - \Delta u|^2 r^\beta dxdt \tag{5.3.3}$$

is true where ε *is an arbitrary nonnegative constant.*

Proof. Denote

$$\varepsilon \dot{u} - \Delta u = f, \tag{5.3.4}$$

multiply this equality by $\Delta u r^\beta$ and integrate by parts on the left–hand side. According to lemma 5.2.2 we get

$$\frac{\varepsilon}{2}\int_B |\nabla u|^2 r^\beta dx\,|_{t=T} + \beta \int_Q \Delta u u' r^{\beta-1} dxdt + \int_B |\Delta u|^2 r^\beta dxdt = \int_Q f(\Delta u + \beta u' r^{-1}) r^\beta dxdt.$$

After using for $u(t,x)$ the expansion (2.3.1), analogous to (5.2.7), according to the same lemma 5.2.2, get the inequality

$$\int_Q |\Delta u|^2 r^\beta dxdt + \frac{\beta(2-\beta)}{2}\sum_{s,k}\int_0^T\int_0^1 |u'_{s,k}|^2 r^{\beta-1} drdt \le \tag{5.3.5}$$

$$\le \int_Q f(\Delta u + \beta u' r^{-1}) r^\beta dxdt + \frac{\beta(2-\beta)}{2}\sum_{s,k} s^2 \int_0^T\int_0^1 |u_{s,k}|^2 r^{\beta-3} dr.$$

Now we must estimate the right–hand side of (5.3.5). Multiply (5.3.4) by $u_{s,k} r^{\beta-2} (s \ge 1)$. Integrating by parts gives the inequality

$$\left[(1-\frac{\beta}{4})\beta + s^2 - 1\right]\int_0^T\int_0^1 |u_{s,k}|^2 r^{\beta-3} drdt \le \int_0^T\int_0^1 f_{s,k} u_{s,k} r^{\beta-1} drdt.$$

It is obvious

$$\sum_{s\ge 1,k} s^2 \int_0^T\int_0^1 |u_{s,k}|^2 r^{\beta-3} drdt \le \frac{1}{(1-\frac{\beta}{4})\beta}\sum_{s\ge 1,k}\left[(1-\frac{\beta}{4})\beta + s^2 - 1\right]\int_0^T\int_0^1 |u_{s,k}|^2 r^{\beta-3} drdt.$$

Thus we have

$$\sum_{s\ge 1,k} s^2 \int_0^T\int_0^1 |u_{s,k}|^2 r^{\beta-3} drdt \le \frac{1}{(1-\frac{\beta}{4})\beta}\left|\int_0^T\int_0^1 f_{s,k} u_{s,k} r^{\beta-1} drdt\right|.$$

After applying the Hölder inequality we get

$$\sum_{s\ge 1,k} s^2 \int_0^T\int_0^1 |u_{s,k}|^2 r^{\beta-3} drdt \le \frac{1}{(1-\frac{\beta}{4})^2\beta^2}\int_Q |f|^2 r^\beta dxdt.$$

From (5.3.5) we get the inequality

$$\int_Q |\Delta u|^2 r^\beta dx dt + \frac{\beta(2-\beta)}{2} \sum_{s,k} \int_0^T \int_0^1 \left|u'_{s,k}\right|^2 r^{\beta-1} dr dt \leq \tag{5.3.6}$$
$$\leq \frac{2-\beta}{2(1-\frac{\beta}{4})^2\beta} \int_Q |f|^2 r^\beta dx dt + \int_Q f\left(\Delta u + \beta u' r^{-1}\right) r^\beta dx dt.$$

Applying inequality (2.2.15) we get

$$\int_Q f\left(\Delta u + \beta u' r^{-1}\right) r^\beta dx dt \leq \left|\int_Q f \Delta u r^\beta dx dt\right| + \eta\beta \int_Q |u'|^2 r^{\beta-2} dx dt + \frac{\beta}{4\eta} \int_Q |f|^2 r^\beta dx dt.$$

From the expansion (2.3.1) we obtain

$$\int_Q |u'|^2 r^{\beta-2} dx dt = \sum_{s,k} \int_0^T \int_0^1 \left|u'_{s,k}\right|^2 r^{\beta-1} dr dt$$

and we can write

$$\left|\int_Q f(\Delta u + \beta u' r^{-1}) r^\beta dx dt\right| \leq \left|\int_Q f \Delta u r^\beta dx dt\right| +$$

$$+\eta\beta \sum_{s,k} \int_0^T \int_0^1 \left|u'_{s,k}\right|^2 r^{\beta-1} dr dt + \frac{\beta}{4\eta} \int_Q |f|^2 r^\beta dx dt.$$

So, from (5.3.6) we derive

$$\int_Q |\Delta u|^2 r^\beta dx dt + \frac{\beta(2-\beta)}{2} \sum_{s,k} \int_0^T \int_0^1 |u'_{s,k}|^2 r^{\beta-1} dr dt \leq \left|\int_Q f \Delta u r^\beta dx dt\right| +$$
$$+\eta\beta \sum_{s,k} \int_0^T \int_0^1 \left|u'_{s,k}\right|^2 r^{\beta-1} dr dt + \frac{\beta}{4\eta} \int_Q |f|^2 r^\beta dx dt + \frac{2-\beta}{2(1-\frac{\beta}{4})^2\beta} \int_Q |f|^2 r^\beta dx dt.$$

Taking $\eta = \frac{2-\beta}{2}$ and applying the inequality

$$\left|\int_Q f \Delta u r^\beta dx dt\right| \leq \frac{1}{2} \int_Q |f|^2 r^\beta dx dt + \frac{1}{2} \int_Q |\Delta u|^2 r^\beta dx dt,$$

we get (5.3.3).

□

Set

$$A^2_{\alpha,m} = \begin{cases} 1 - \frac{\alpha}{2+\alpha} - \frac{2+\alpha}{(1+\frac{\alpha}{4})^2\alpha}, & m = 2, \\ \frac{m}{m+\alpha}, & m > 2. \end{cases} \tag{5.3.7}$$

Theorem 5.3.1. *Suppose $u \in \left\{W^{1,2}_{2,\alpha}(Q_R)\right\}$, satisfies the boundary conditions (5.3.1), $\alpha \in (-m, 2-m) \cup (3-m, 0)$ and $\gamma = \frac{2-m-\alpha}{2}$. Then the following estimates*

$$\int_{Q_R} |D'^2 u|^2 r^\alpha \zeta dx dt \leq \tag{5.3.8}$$

$$\leq \frac{1}{E}\Bigg\{ A^2_{\alpha,m}(1+M^2_\gamma) \int_{Q_R} |\varepsilon \dot{u} - \Delta u|^2 r^\alpha \zeta dx dt + C_0(\eta)|S| \times$$

$$\times \Bigg[\frac{(m-2+2\gamma)(m-1)+mM^2_\gamma}{m} R^{\alpha-2} \int_{Q_R} |\nabla u|^2 dx dt +$$

$$+(m-1)[(m+1+2\gamma)M^2_\gamma + (m-2+2\gamma)/2]R^{\alpha-4} \times$$

$$\times \int_{Q_R} |u|^2 dx dt \Bigg]\Bigg\} + CR^\alpha \int_{Q_R} |\varepsilon \dot{u} - \Delta u|^2 \, dx dt,$$

$$\int_{Q_R} |D'^2 u|^2 r^\alpha \zeta dx dt \leq \tag{5.3.9}$$

$$\leq \frac{1}{E}\Bigg\{ A^2_{\alpha,m}(1+M^2_\gamma) \int_{Q_R} |\varepsilon \dot{u} - \Delta u|^2 r^\alpha \zeta dx dt + \frac{C_0(\eta)|S|}{\lambda} R^\alpha \times$$

$$\times \Bigg[\frac{(m-2+2\gamma)(m-1)+mM^2_\gamma}{m} +$$

$$+ \frac{(m-1)\left((m-1+2\gamma)M^2_\gamma + (m-2+2\gamma)/2\right)}{\lambda} \Bigg] \times$$

$$\times \int_{Q_R} |\Delta u|^2 dx dt \Bigg\} + CR^\alpha \int_{Q_R} |\varepsilon \dot{u} - \Delta u|^2 dx dt$$

and

$$\int_{Q_R} |D'^2 u|^2 r^\alpha \zeta dx dt \leq \frac{1}{E}\Bigg\{ A^2_{\alpha,m}(1+M^2_\gamma) + \frac{C_0(\eta)|S|}{\lambda} \times \tag{5.3.10}$$

$$\times \Bigg[\frac{(m-2+2\gamma)(m-1)+mM^2_\gamma}{m} +$$

$$+ \frac{(m-1)\left((m+1+2\gamma)M^2_\gamma + (m-2+2\gamma)/2\right)}{\lambda} \Bigg]\Bigg\} \times$$

$$\times \int_{Q_R} |\varepsilon \dot{u} - \Delta u|^2 r^\alpha \zeta dx dt + CR^\alpha \int_{Q_R} |\varepsilon \dot{u} - \Delta u|^2 dx dt$$

are true where ε and T are arbitrary positive values, and all other constants are defined at the end of the formulation of theorem 4.2.1 and by (5.3.7) (C is independent of ε and R and ζ is a cut-off function).

Proof. We can assume initially u is as smooth as we wish. Let $w(t,x)$ be a solution in Q of the following boundary value problem

$$\varepsilon \dot{w} + \Delta w = -\Delta u r^{\alpha} \zeta, \tag{5.3.11}$$

$$w|_{t=T} = w|_{\partial B_R} = 0.$$

Multiply equation (5.3.11) by $\Delta u \zeta$ and integrate once by parts with respect to t and twice with respect to x to give

$$\int_Q (\varepsilon \dot{u} - \Delta u)\Delta w \zeta dx dt = \int_Q |\Delta u|^2 r^{\alpha} \zeta dx dt + \ldots,$$

where the nonwritten terms are those containing the derivatives of ζ. Applying the Hölder inequality, we get

$$\int_Q |\Delta u|^2 r^{\alpha} \zeta dx dt \le \left(\int_Q |\varepsilon \dot{u} - \Delta u|^2 r^{\alpha} \zeta dx dt \right)^{\frac{1}{2}} \times \tag{5.3.12}$$

$$\times \left(\int_Q |\Delta w|^2 r^{-\alpha} dx dt \right)^{\frac{1}{2}} + C \int_Q |\varepsilon \dot{u} - \Delta u|^2 dx dt.$$

It is trivial that w also satisfies the inequalities (5.3.2) and (5.3.3) (one only has to exchange t with $T-t$). Therefore

$$\begin{aligned} \int_Q |\Delta w|^2 r^{-\alpha} dx dt &\le A^2_{\alpha,m} \int_Q |\varepsilon \dot{w} + \Delta w|^2 r^{-\alpha} dx dt = \\ &= A^2_{\alpha,m} \int_Q |\Delta u|^2 r^{\alpha} \zeta dx dt. \end{aligned}$$

Now from (5.3.12) and theorem 4.2.1 after rescaling ($x \leftarrow Rx, t \leftarrow R^2 t$) we get the results of the theorem. We should also take into account that $\zeta \le 1$ and therefore

$$\int_Q |\Delta u|^2 r^{\alpha} dx dt \ge \int_Q |\Delta u|^2 r^{\alpha} \zeta dx dt.$$

□

Theorem 5.3.2. *Suppose that u and α satisfy the conditions of theorem 5.3.1, but only the second of the conditions (5.3.1): $u = 0$, when $t = 0$. Then the estimate*

$$\int_{Q_R} \left|D'^2 u\right|^2 r^{\alpha} \zeta dx dt \le \left(1 + M_{\gamma}^2 + \eta\right) A^2_{\alpha,m} \int_{Q_R} |\varepsilon \dot{u} - \Delta u|^2 r^{\alpha} \zeta dx dt + \tag{5.3.13}$$

$$+ C \left\{ \left(\int_{Q_R} \left|D'^2 u\right|^2 r^{\alpha} \zeta dx dt \right)^{\frac{m}{m+2\gamma}} \left[\int_{Q_R} |Du|^2 dx dt \right]^{\frac{2\gamma}{m+2\gamma}} + \int_{Q_R} |Du|^2 dx dt \right\}$$

is true, where C is independent of ε.

Proof. Take a function w which satisfies the equation (5.3.11) and the same boundary conditions. Multiply both sides of the differential equation by $\Delta u\zeta$. After integration over Q_R we derive

$$\int_{Q_R} (\varepsilon\dot{w} + \Delta w)\Delta u\zeta\,dxdt = -\int_{Q_R} |\Delta u|^2 r^\alpha \zeta^2\,dxdt.$$

After two integrations by parts on the left–hand side with respect to x, we get

$$\int_{Q_R} [\varepsilon\Delta\dot{w}\zeta u + \Delta w\Delta u\zeta]\,dxdt = -\int_{Q_R} |\Delta u|^2 r^\alpha \zeta\,dxdt -$$
$$-2\varepsilon\int_{Q_R} \nabla\dot{w}\nabla\zeta u\,dxdt - \varepsilon\int_{Q_R} \dot{w}\Delta\zeta u\,dxdt.$$

Integrating on the left-hand side by parts once with respect to t and on the right-hand side in the second integral once with respect to x, we get

$$\int_{Q_R} (\varepsilon\dot{u} - \Delta u)\Delta w\zeta\,dxdt = \int_{Q_R} |\Delta u|^2 r^\alpha \zeta^2\,dxdt -$$
$$-\varepsilon\int_{Q_R} \dot{w}u\cdot\Delta\zeta\,dxdt - 2\varepsilon\int_{Q_R} \dot{w}\nabla\zeta\nabla u\,dxdt.$$

Therefore

$$\int_{Q_R} |\Delta u|^2 r^\alpha \zeta^2\,dxdt = \int_{Q_R} (\varepsilon\dot{u} - \Delta u)\Delta w\zeta\,dxdt +$$
$$+\int_{Q_R} (\varepsilon\dot{w} + \Delta w)u\Delta\zeta\,dxdt + 2\int_{Q_R} (\varepsilon\dot{w} + \Delta w)\nabla\zeta\nabla u\,dxdt -$$
$$-\int_{Q_R} u\Delta w\Delta\zeta\,dxdt - 2\int_{Q_R} \Delta w\nabla u\cdot\nabla\zeta\,dxdt.$$

Let us now estimate the integrals on the right-hand side. After applying (2.2.15) we come to the following relations ($\eta_1 > 0$ is arbitrary):

1) $$\left|\int_{Q_R} (\varepsilon\dot{u} - \Delta u)\Delta w\zeta\,dxdt\right| \le \frac{1}{4\eta_1}\int_{Q_R} |\varepsilon\dot{u} - \Delta u|^2 r^\alpha \zeta^2\,dxdt +$$
$$+\eta_1\int_{Q_R} |\Delta w|^2 r^{-\alpha}\,dxdt;$$

2) $$\left|\int_{Q_R} (\varepsilon\dot{w} + \Delta w)u\Delta\zeta\,dxdt\right| \le \eta\int_{Q_R} |\varepsilon\dot{w} + \Delta w|^2 r^{-\alpha}\,dxdt +$$
$$+\frac{1}{4\eta}\int_{Q_R} |u|^2|\Delta\zeta|^2 r^\alpha\,dxdt \le \eta\int_{Q_R} |\Delta u|^2 r^\alpha\zeta^2\,dxdt + C\int_{Q_R} |u|^2\,dxdt;$$

3) $$\left|2\int_{Q_R} (\varepsilon\dot{w} + \Delta w)\nabla\zeta\nabla u\,dxdt\right| \le \eta\int_{Q_R} |\Delta u|^2 r^\alpha\zeta^2\,dxdt + C\int_{Q_R} |\nabla u|^2\,dxdt.$$

Then by (5.3.2) and (5.3.3) we get

$$\begin{aligned}\int_{Q_R} |\Delta u|^2 r^\alpha \zeta^2 dxdt &\le \eta_1 A^2_{\alpha,m} \int_{Q_R} |\Delta u|^2 r^\alpha \zeta^2 dxdt + \\ +\frac{1}{4\eta_1}\int_{Q_R} |\varepsilon\dot{u} - \Delta u|^2 r^\alpha \zeta^2 dxdt &+ \eta \int_{Q_R} |\Delta u|^2 r^\alpha \zeta^2 dxdt + \\ &+ C\int_{Q_R} |Du|^2 dxdt.\end{aligned}$$

Taking $\eta_1 = 2^{-1} A^{-2}_{\alpha,m}$ we obtain the inequality

$$\begin{aligned}\int_{Q_R} |\Delta u|^2 r^\alpha \zeta^2 dxdt &\le \left(A^2_{\alpha,m} + \eta\right) \int_{Q_R} |\varepsilon\dot{u} - \Delta u|^2 r^\alpha \zeta^2 dxdt + \\ &+ C\int_{Q_R} |Du|^2 dxdt.\end{aligned}$$

After using theorem 4.2.4. the proof of this theorem is concluded.

□

5.4 Coupled parabolic systems; necessary conditions for "blow up".

This section is devoted to the problem of nonregularity of solutions for quasilinear parabolic systems. For one second order parabolic equation the "blow up" of solutions was considered in many papers and a lot very important results were obtained. It is not necessary to give a survey of these results in this book, but we would like to note only that a majority of these results are based on some variant of the maximum princliple for one second order parabolic equation. For general parabolic systems this method can not be applied.
Therefore we have choosen the sufficient conditions of sections 2.3 and 5.2 for the regularity of weak solutions to get necessary conditons for the loss of regularity. As seen in sections 2.5 and 5.2, these regularity conditions are sharp or unavoidable. We further show numerically that the loss of regularity really happens for solution of so - called chemotaxis system. It is clear the necessary conditions can not be sufficient for all systems. The last situation takes place, for example, for the solution of the two - dimensional semiconductor system in which under natural conditions, the weak solutions are regular for all times (Gajewski and Gröger [1]). We would like to state that sometimes the "blow up" effect in strongly coupled systems can be shown by simplifying the system. Here it is necessary to mention an interesting result of Jäger and Luckhaus [1] concerning the two - dimensional chemotaxis system. For the coupled system

$$\begin{aligned}u_t - \Delta u &= v^p\\ v_t - \Delta v &= u^p\end{aligned}$$

the conditions of "blow up" were found by Escobedo and Herrero [1].
If we reformulate theorem 5.2.3 we get

Theorem 5.4.1. *Let conditions* 1) – 4) *of section* 5.1 *be satisfied. The conditions*

$$K^2\left(1+\frac{m-2}{m+1}\right)[1+(m-2(m-1)]\frac{m}{2}\geq 1 \quad \text{for} \quad m\geq 3, \tag{5.4.1}$$

$$2K^2\geq 1 \quad \text{for} \quad m=2 \tag{5.4.2}$$

are necessary conditions for loss of Hölder continuity of weak solutions for the problem (5.1.1), (5.1.6).

These conditions (5.4.1), (5.4.2) can be very restrictive. In fact since the coefficients of (5.1.1) depend on the first derivatives of the solution, the constant K can include the max of the derivatives of the solution. Hence, in some cases the growth of the max for the solutions will be determined through the growth of the maximum of the absolute value for the derivatives.
Therefore, we now present a simpler condition, which we will apply later for strongly coupled systems

Theorem 5.4.2. *If the assumptions* 1) – 3) *of* 5.1 *are satisfied,* $q_0 = 2, m \geq 3$, $2-m<\alpha\leq 0$ *and the inequality*

$$K\frac{-\alpha+m-2}{\alpha+m-2}\sqrt{2-\frac{\alpha(m-2)}{m-1}}<1 \tag{5.4.3}$$

is true, then the weak solution u of (5.1.1), (5.1.6) *satisfies the estimate*

$$\sup_{x_0\in\Omega'}\int_{\Omega'}|u|^2\,|x-x_0|^\alpha dx\Big|_{t=T}+\sup_{x_0\in\Omega'}\int_{\Omega'}|Du|^2\,|x-x_0|^\alpha dxdt<C\ . \tag{5.4.4}$$

Proof. As usual we can start with smooth functions. Subtract two succesive equations (5.1.7) and consider the difference $w_n=u_n-u_{n-1}$. Multiply each equation by $v_{n+1}=w_{n+1}r^\alpha\zeta$. Integrating by parts once we have

$$\frac{\varepsilon}{2}\int_{B_\delta}|w_{n+1}|^2\,r^\alpha\zeta dx\Big|_{t=T}+\int_{Q_\delta}Dw_{n+1}D(w_{n+1}r^\alpha\zeta)dxdt= \tag{5.4.5}$$

$$=\int_{Q_\delta}D_iv_{n+1}\left[D_iw_n-\varepsilon\frac{\partial a_i}{\partial p_j}D_jw_n\right]dxdt\ .$$

Consider at first the right-hand side term. Multiplying and dividing under the sign of the integral by $r^{\alpha/2}$ we get

$$\left|\int_{Q_\delta}D_iv_{n+1}\left[D_iw_n-\varepsilon\frac{\partial a_i}{\partial p_j}D_jw_n\right]dxdt\right|\leq$$

$$\leq\left[\int_{Q_\delta}|Dv_{n+1}|^2\,r^{-\alpha}dxdt\right]^{1/2}\left[\int_{Q_\delta}\left\|(I-\varepsilon\overline{A})Dw_n\right\|^2r^\alpha dxdt\right]^{1/2}.$$

Taking ε from the relation (1.2.12) and (1.2.13) we get the estimate

$$\left| \int_{Q_\delta} D_i v_{n+1} \left[D_i w_n - \varepsilon \frac{\partial a_i}{\partial p_j} D_j w_n \right] dx dt \right| \leq$$

$$\leq K \left[\int_{Q_\delta} |Dw_n|^2 r^\alpha \zeta dx dt + C \int_{Q_\delta} |Dw_n|^2 dx dt \right]^{1/2} \left[\int_{Q_\delta} |Dv_{n+1}|^2 r^{-\alpha} dx dt \right]^{1/2} .$$

Applying the inequalities (2.2.15) and (2.1.17′) give

$$\left| \int_{Q_\delta} D_i v_{n+1} \left[D_i w_n - \varepsilon \frac{\partial a_i}{\partial p_j} D_j w_n \right] dx dt \right| \leq$$

$$\leq \eta \int_{Q_\delta} |Dv_{n+1}|^2 r^{-\alpha} dx dt + \frac{K^2}{4\eta} \int_{Q_\delta} |Dw_n|^2 r^\alpha \zeta dx dt + C \int_{Q_\delta} |Dw_n|^2 dx dt.$$

From (5.4.5) it follows

$$\int_{Q_\delta} D' w_{n+1} D'(w_{n+1} r^\alpha \zeta) dx dt \leq \tag{5.4.6}$$

$$\leq \eta \int_{Q_\delta} |D' v_{n+1}|^2 r^{-\alpha} dx dt + \frac{K^2}{4\eta} \int_{Q_\delta} |D' w_n|^2 r^\alpha \zeta dx dt \ldots \quad ,.$$

where the unwritten terms can be estimated by $\int_{Q_\delta} |Dw_n|^2 dx dt$ and $\int_{Q_\delta} |Dw_{n+1}|^2 dx dt$. This can be done as in the proof of lemma 5.2.1. We now must estimate the integral

$$J_{n+1} = \int_{Q_\delta} |D' v_{n+1}|^2 r^{-\alpha} dx dt \ .$$

Using the expansion of the type (2.3.2) we come to the identity

$$J = \sum_{s,k} \int_0^T \int_0^\delta \left[\left| v'_{s,k} \right|^2 + s(s+m-2) \left| v_{s,k} \right|^2 r^{-2} \right] r^{-\alpha+m-1} dr dt$$

(we have omitted for now the index $n+1$).
After simple calculations we obtain

$$J = \sum_{s,k} \int_0^T \int_0^\delta \left\{ \left| w'_{s,k} \right|^2 + \left[s(s+m-2) - \alpha(m-2) \right] \left| w_{s,k} \right|^2 r^{-2} \right\} r^{\alpha+m-1} \zeta dr dt + \ldots \ .$$

Then analagous to lemma 2.3.1 we'll have

$$J \leq \int_0^T \int_0^\delta \left[\left| w'_{0,1} \right|^2 - \alpha(m-2) \left| w_{0,1} \right|^2 r^{-2} \right] r^{\alpha+m-1} \zeta dr dt + \tag{5.4.7}$$

$$+ \left[1 - \frac{(m-2)\alpha}{m-1} \right] \sum_{s \geq 1, k} \int_0^T \int_0^\delta \left[\left| w'_{s,k} \right|^2 + s(s+m-2) \left| w_{s,k} \right|^2 r^{-2} \right] r^{\alpha+m-1} \zeta dr dt +$$

$$+ C \int_{Q_\delta} |Dw|^2 dx dt \ .$$

In the same way, using formula (2.2.11), we get

$$\int\limits_{Q_\delta} D'wD'(wr^\alpha\zeta)dxdt =$$

$$= \int\limits_0^T\int\limits_0^\delta \left[\left|w'_{0,1}\right|^2 - \frac{\alpha(\alpha+m-2)}{2}\left|w_{0,1}\right|^2 r^{-2}\right] r^{\alpha+m-1}\zeta drdt +$$

$$+ \sum_{s\geq 1,k}\int\limits_0^T\int\limits_0^\delta \left\{\left|w'_{s,k}\right|^2 + \left[s(s+m-2) - \frac{\alpha(\alpha+m-2)}{2}\right]\left|w_{s,k}\right|^2 r^{-2}\right\} r^{\alpha+m-1}\zeta drdt + \dots .$$

Since $\alpha(\alpha+m-2)<0$, we arrive at

$$(5.4.8)\qquad \int\limits_{Q_\delta} D'wD'(wr^\alpha\zeta)dxdt \geq$$

$$\geq \int\limits_0^T\int\limits_0^\delta \left[\left|w'_{0,1}\right|^2 - \frac{\alpha(\alpha+m-2)}{2}\left|w_{0,1}\right|^2 r^{-2}\right] r^{\alpha+m-1}\zeta drdt +$$

$$+ \sum_{s\geq 1,k}\int\limits_0^T\int\limits_0^\delta \left\{\left|w'_{s,k}\right|^2 + s(s+m-2)\left|w_{s,k}\right|^2 r^{-2}\right\} r^{\alpha+m-1}\zeta drdt + \dots .$$

Now compare the first integrals on the right-hand side of (5.4.7) or (5.4.8). Suppose $0<p<1$. Then

$$\int\limits_0^\delta \left[\left|w'_{0,1}\right|^2 - \frac{\alpha(\alpha+m-2)}{2}\left|w_{0,1}\right|^2 r^{-2}\right] r^{\alpha+m-1}\zeta drdt =$$

$$= p\int\limits_0^\delta \left|w'_{0,1}\right|^2 r^{\alpha+m-1}\zeta drdt +$$

$$+ \int\limits_0^\delta \left[(1-p)\left|w'_{0,1}\right|^2 - \frac{\alpha(\alpha+m-2)}{2}\left|w_{0,1}\right|^2 r^{-2}\right] r^{\alpha+m-1}\zeta drdt.$$

According to the Hardy inequality (2.1.8) we get

$$\int\limits_0^\delta \left[\left|w'_{0,1}\right|^2 - \frac{\alpha(\alpha+m-2)}{2}\left|w_{0,1}\right|^2 r^{-2}\right] r^{\alpha+m-1}\zeta drdt \geq$$

$$\geq p\int\limits_0^\delta \left|w'_{0,1}\right|^2 r^{\alpha+m-1}drdt + \frac{\alpha+m-2}{2}\left[(1-p)\frac{\alpha+m-2}{2} - \alpha\right]\int\limits_0^\delta \left|w_{0,1}\right|^2 r^{\alpha+m-3}drdt + \dots$$

If we take

$$p = [-\alpha(m-2)]^{-1}\left[(1-p)\frac{\alpha+m-2}{2} - \alpha\right]\frac{\alpha+m-2}{2} = \frac{\alpha+m-2}{-\alpha+m-2},$$

we have

$$\int\limits_0^\delta \left[\left|w'_{0,1}\right|^2 - \frac{\alpha(\alpha+m-2)}{2}\left|w_{0,1}\right|^2 r^{-2}\right] r^{\alpha+m-1}\zeta drdt \geq$$

$$\geq \frac{\alpha+m-2}{-\alpha+m-2}\int_0^\delta \left[\left|w_{0,1}'\right|^2-\alpha(m-2)\left|w_{0,1}\right|^2 r^2\right] r^{\alpha+m-1}\zeta drdt+\dots \ .$$

Therefore we can find from (5.4.8) the following relation

$$\int_Q DwD(wr^\alpha\zeta)drdt \geq$$

$$\geq \frac{\alpha+m-2}{-\alpha+m-2}\int_0^T\int_0^\delta \left[\left|w_{0,1}'\right|^2-\alpha(m-2)\left|w_{0,1}\right|^2 r^{-2}\right] r^{\alpha+m-1}\zeta drdt+$$

$$+\sum_{s\geq 1,k}\int_0^T\int_0^\delta \left\{\left|w_{s,k}'\right|^2+s(s+m-2)\left|w_{s,k}\right|^2 r^{-2}\right\} r^{\alpha+m-1}\zeta drdt+\int_Q |w|^2 r^\alpha \zeta dxdt+\dots \ .$$

If we estimate the last term on the right-hand side by (2.1.17′), apply (5.4.7) and (5.4.6) we get the following inequality

$$\left(\frac{\alpha+m-2}{-\alpha+m-2}-\eta\right)\int_0^T\int_0^\delta \left[\left|w_{n+1,0,1}'\right|^2-\alpha(m-2)\left|w_{n+1,0,1}\right|^2 r^{-2}\right] r^{\alpha+m-1}\zeta drdt+$$

$$+\sum_{s\geq 1,k}\int_0^T\int_0^\delta \left[\left|w_{n+1,s,k}'\right|^2+s(s+m-2)\left|w_{n+1,s,k}\right|^2 r^{-2}\right] r^{\alpha+m-1}\zeta drdt\leq$$

$$\leq \eta\left[1-\frac{(m-2)\alpha}{m-1}\right]\int_{Q_\delta}\left|Dw_{n+1}\right|^2 r^\alpha dxdt+\frac{K^2}{4\eta}\int_{Q_\delta}\left|Dw_n\right|^2 r^\alpha\zeta dxdt+\dots \ .$$

Since

$$\frac{\alpha+m-2}{-\alpha+m-2}\leq 1 \ ,$$

we get the inequality

$$\left(\frac{\alpha+m-2}{-\alpha+m-2}-\eta\right)\int_{Q_\delta}\left|Dw_{n+1}\right|^2 r^\alpha\zeta dxdt\leq$$

$$\leq \eta\left[1-\frac{(m-2)\alpha}{m-1}\right]\int_{Q_\delta}\left|Dw_{n+1}\right|^2 r^\alpha\zeta dxdt+\frac{K^2}{4\eta}\int_{Q_\delta}\left|Dw_n\right|^2 r^\alpha\zeta dxdt+\dots \ .$$

Taking

$$\eta=\frac{1}{2}\left(\frac{\alpha+m-2}{-\alpha+m-2}\right)\frac{1}{2-\alpha(m-2)/(m-1)}$$

we have

$$\int_{Q_\delta}\left|Dw_{n+1}\right|^2 r^\alpha\zeta dxdt\leq q_1\int_{Q_\delta}\left|Dw_n\right|^2 r^\alpha\zeta dxdt+Cq^n \ ,$$

where

$$0\leq q_1=\left[\frac{-\alpha+m-2}{\alpha+m-2}\right]^2\left[2-\frac{(m-2)\alpha}{m-1}\right]K^2<1 \ .$$

Therefore

$$\int_{Q_\delta} |Dw_n|^2 r^\alpha \zeta dx dt \le C q_0^2 ,$$

where $0 < q_0 < 1$ and w_n converges in the space H_α with the norm

$$\|u\|^2 = \sup_{x_0 \in Q', \delta \le \delta_0} \int_{Q_\delta} |Du|^2 r^\alpha dx dt .$$

Then from (5.4.5) we can prove the convergency of

$$\int_{B_\delta} |w_n|^2 r^\alpha \zeta dx \Big|_{t=T}$$

and reach the end of the proof.

□

Corollary 5.4.1 *A necessary condition for the unboundedness of the expression*

$$E[u] = \sup_{x_0 \in \Omega} \int_\Omega |u|^2 |x - x_0|^\alpha dx \Big|_{t=T} + \sup_{x \in \Omega} \int_\Omega |Du|^2 |x - x_0|^\alpha dx dt \tag{5.4.9}$$

under the assumptions 1)-3) of 5.1 is the following

$$\begin{cases} K^{\frac{-\alpha+m-2}{\alpha+m-2}} \sqrt{2 - \frac{\alpha(m-2)}{m-1}} \ge 1, & m \ge 3, \\ 2K^2 \ge 1, & m = 2. \end{cases} \tag{5.4.10}$$

Remark 5.4.1. *From* (5.4.3) *it follows that* $K < 1$.

We shall now consider systems of the type

$$\partial_t u^{(k)} - \left[\sum_{i=0}^m D_i a_i^{(k)}(x, t, Du) + \sum_{i,j=0}^m \sum_{h,l=1}^N b_{i,j}^{h,k,l} D_i \left[u^{(h)} D_j u^{(l)} \right] \right] = 0, \tag{5.4.11}$$

where $b_{i,j}^{h,k,l}$ are constants.
Suppose there exists a weak solution $U(t, x)$ of the system (5.4.11) for which the expression $E[U]$ ((5.4.9)) is finite in the cylinder $Q_0(\eta) = (0, T_0 - \eta) \times \Omega$. If

$$\lim_{\eta \to 0} E[U] = \infty , \tag{5.4.12}$$

then T_0 will be the "blow up" time for the fixed solution $U(t, x)$. If we substitute $U(t, x)$ in the part of the last term on the left side of (5.4.11) we get

$$\partial_t u^{(k)} - \left[\sum_{i=0}^m D_i a_i^{(k)}(x, t, Du) + \sum_{i,j=0}^m \sum_{h,l=1}^N b_{i,j}^{h,k,l} D_i \left[U^{(h)} D_j u^{(l)} \right] \right] = 0 . \tag{5.4.13}$$

It is clear, if

$$\sup_Q |U(t, x)| < \varepsilon , \tag{5.4.14}$$

where ε is a sufficient small number, then all conditions of the criterium (5.4.10) for the system (5.4.13) are satisfied. So, to find a T_0 for which (5.4.12) is true we must construct the matrix A ((1.1.4)) for the system (5.4.13) and to apply (5.4.10). The time T_0 will be found from the growth rate of the left side of (5.4.14).

The same consideration can be applied to the criterium for (5.4.1), (5.4.2). But in this case the modul of the Hölder continuity for $U(t,x)$ should tend to infinity. Here it should be taken into account that the coefficients for the system (5.4.13) should satisfy the condition (5.1.2). This means if $U(t,x)$ is losing differentiability at some point (t_0, x_0), then the value t_0 also gives the first value of t at which the loss of regularity happens.

As a special case of (5.4.13) we consider the system

(5.4.15)
$$\begin{aligned} \partial_t u &= \Delta u - \nabla(v\nabla u) \ , \\ \delta\partial_t v &= \alpha\Delta v - \beta v + \gamma u \ \text{ in } \Omega \ , \end{aligned}$$

(5.4.16)
$$\begin{aligned} u(0) = u_0 > 0, \quad v(0) = v_0 \geq 0 \ , \\ \nu\nabla u = \nu\nabla v = 0 \ \text{ on } \ \partial\Omega. \end{aligned}$$

Here α,β and δ are positive constants and ν is the outward normal to Ω.

With $\gamma > 0$, the system (5.4.15) is used to model the dynamics of a population (concentration u) moving in Ω and driven by the gradient of a chemotactic agents (concentration v) produced by the population itself. If $\gamma < 0$, the system can be interpreted as a special case of a semiconductor model for the drift of electrons (concentration u) under the electric field produced by the antigradient of the electrostatic potential. The behaviour of solutions of (5.4.15) depends strongly on the sign of the parameter γ. In the semiconductor case ($\gamma < 0$) the system has an unique global solution for arbitrary smooth $u_0 > 0$. Moreover, this solution decays exponentially in time to the unique equilibrium state

(5.4.17)
$$u^* = \overline{u}_0, \quad v^* = \frac{\gamma}{\beta}\overline{u}_0 \ ,$$

where

(5.4.18)
$$\overline{w} = \frac{1}{|\Omega|}\int w d\Omega \ .$$

The situation turns out to be more compilicated if $\gamma > 0$. It is clear that for small $\gamma\overline{u}_0$ there exists a smooth solution. However, for sufficiently large $\gamma\overline{u}_0$ it can be expected the solution may explode in finite time. A hint for this situation was given by Luckhaus, Jäger [1].

Here we illustrate the "blow up" situation by some numerical evidence. Our main aim is to check the criterium (5.4.10) empirically.

In order to calculate K^2 we must specify the matrix (1.1.4). Since we are interested in necessary conditions for loss of regularity, we can consider a simplified matrix, yielding upper (resp. lower) bounds for λ (resp. Λ) and σ. Thus, omitting $i,j = 0$ in (1.1.4), we find

$$A = \begin{pmatrix} 1 & 0 & 0 & 0 \\ 0 & 1 & 0 & 0 \\ -U & 0 & \alpha & 0 \\ 0 & -U & 0 & \alpha \end{pmatrix}, \quad U = U(t) = \|U(t,\cdot)\|_\infty$$

and

$$\lambda = \frac{1}{2}(\alpha + 1 - \rho), \quad \Lambda = \frac{1}{2}(\alpha + 1 + \rho), \quad \rho = \sqrt{\alpha - 1 + U^2}, \quad \sigma = \frac{U}{2}\left(\frac{U}{2} + \rho\right) .$$

To solve, numerically, the system in the chemotaxis case ($\gamma > 0$) we used a modified version of the well-tried code ToSCA (see Gajewski [1]), which was originally designed to solve van Roosbroeck semiconductor equations ($\gamma < 0$) in two-dimensional domains Ω with the help of the finite elements method.
During our numerical calculations we fixed the data as follows:

(5.4.19)
$$m = 2, \quad \Omega = (0, a) \times (0, a), \quad a = \pi \ ;$$
$$\alpha = 1, \quad \beta = 0.1, \quad \delta = 1, \quad \gamma = 1 \ ;$$
$$v_0 = 0 \ .$$

Take

(5.4.20)
$$u_0 = \mu^* + \varepsilon\left(e^{-(x^2+y^2)} - 1\right), \quad \varepsilon = 0.1, \quad \mu^* = \alpha\left(\frac{\pi}{a}\right)^2 + \beta = 1.1$$

or

(5.4.21)
$$u_0 = \mu^* + \varepsilon e^{-(x^2+y^2)} \ .$$

From the homogeneous Neumann boundary conditions follows that the solutions of (5.4.15), realizing the initial values (5.4.20) (resp. (5.4.21)), satisfy

$$\overline{u}(t) = \overline{u}_0 = u^*, \quad \mu = \gamma\overline{u}_0 = 1.0557$$

(resp.)

$$\overline{u}(t) = \overline{u}_0, \quad \mu = 1.1557 \ .$$

Our numerical results can be summarized as follows :

(i) the solution of the problem (5.1.15), with the parameters (5.4.19), (5.4.20) exists globally and satisfies the following relation

$$\lim_{t \to \infty} \|u(t) - u^*\|_2 \to 0 \ ,$$

where $\|\cdot\|_2$ denotes the norm in $L_2(\Omega)$;

(ii) the solution of the problem (5.4.15), with the parameters (5.4.19), (5.4.21) blows up in finite time T_0 so that

$$\lim_{t \to T_0} \|u(t) - u^*\|_2 \to \infty \ .$$

The results of this section were mainly obtained in a joint preprint with Gajewski and Jäger (Gajewski, Jäger, Koshelev [1]).

5.5 The Liouville theorem for parabolic systems.

We considered theorems of the Liouville type for elliptic systems in section 3.1. In this section we prove two analogous theorems for a quasilinear nondegenerate parabolic system with bounded nonlinearities. For one linear parabolic equation the Liouville theorem was proved first by Appel[1]. Later Hirshman [1] improved the proof and derived new conditions for the Liouville theorem. In the monograph of Landis [1] a variant of the Liouville theorem for one second order parabolic equation is proved. The only Liouville theorem concerning parabolic systems found by the author in the literature was a paper by Stara, John and Daneček [1]. In this paper the regularity of solutions for semilinear parabolic system is connected with the validity of the Liouville theorem.
Take a smooth function $\tilde{w}(t,x)$ equal to zero on $[0,T)\times\partial B_1$ and for $t=T$. Then for $\alpha = 2-m-2\gamma$ $(0<\gamma<1/2)$ from theorem 4.3.1 (inequality (4.3.4)) follows the relation

$$\int_{B_1} |\tilde{w}|^2 r^{-\alpha} dx + \int_{B_1} |D'\tilde{w}|^2 r^{-\alpha} dx + \int_{B_1} \left|D'^2\tilde{w}\right|^2 r^{-\alpha} dx \leq$$
$$\leq C \int_{B_1} |\Delta\tilde{w}|^2 r^{-\alpha} dx \ .$$

After rescaling with respect to x we get

$$\delta^{-4}\int_{B_\delta} |w|^2 r^{-\alpha} dx + \delta^{-2}\int_{B_\delta} |D'w|^2 r^{-\alpha} dx + \int_{B_\delta} |D'^2 w|^2 r^{-\alpha} dx \leq C\int_{B_\delta} |\Delta w|^2 r^{-\alpha} dx,$$

where $w(t,x) = \tilde{w}(t,x/\delta)$ Applying lemmas 5.2.1 and 5.3.1 (inequalities (5.2.3) and (5.3.3)), then integrating over t we get

$$\delta^{-4}\int_{Q_\delta} |w|^2 r^{-\alpha} dxdt + \delta^{-2}\int_{Q_\delta} |D'w|^2 r^{-\alpha} dxdt + \int_{Q_\delta} \left|D'^2 w\right|^2 r^{-\alpha} dxdt \leq \tag{5.5.1}$$
$$\leq C\int_{Q_\delta} |\varepsilon\dot{w} + \Delta w|^2 r^{-\alpha} dxdt \ ,$$

where C is independent of δ and ε. It is trivial that you can extend the inequality for all functions which has a finite integral on the right-hand side .
Suppose that a smooth function $u(t,x)$ is defined on $R^1_+ \times R^m$ and satisfies the following condition:
there exists a nonnegative nondecreasing function $T(\delta)$, which is defined on $[0,+\infty)$ with $\lim\limits_{\delta\to+\infty} T(\delta) = +\infty$ and for which the equality

$$\lim_{\delta\to+\infty} \frac{1}{\delta^2} \int_0^{T(\delta)} \int_{B_\delta} \left|D'u - D'u\Big|_{x=0}\right|^2 r^{\alpha} dxdt = 0 \tag{5.5.2}$$

holds. For example, the function u with bounded derivatives on R^+ satisfies the condition (5.5.2) with $T(\delta) = \delta^{\varrho}$ and $\varrho < 2\gamma$ $(\alpha = 2-m-2\gamma)$. It is easy to see that the equality

$$(5.5.3) \qquad \lim_{\delta\to+\infty} \frac{1}{\delta^4} \int_0^{T(\delta)} \int_{B_\delta} |u - u|_{x=0} - u_i|_{x=0}\, x_i|^2 r^\alpha dx dt = 0$$

is true for any function $T(\delta)$ satisfying (5.5.2).
In fact, from Hardy inequality (2.1.8) it follows that

$$\int_{B_\delta} |u - u|_{x=0} - u_i|_{x=0}\, x_i|^2 r^{\alpha-2} dx \le \frac{1}{\gamma^2} \int_{B_\delta} |D'u - D'u|_{x=0}|^2 r^\alpha dx .$$

Since for $x \in B_\delta$, we have $r < \delta$, and gives

$$\frac{1}{\delta^4} \int_0^{T(\delta)} \int_{B_\delta} |u - u|_{x=0} - u_i|_{x=0}\, x_i|^2 r^\alpha dx dt \le \frac{1}{\delta^2} \int_0^{T(\delta)} \int_{B_\delta} |u - u|_{x=0} - u_i|_{x=0}\, x_i|^2 r^{\alpha-2} dx dt .$$

From the previous inequality and from (5.5.2), the relation (5.5.3) follows.

Theorem 5.5.1. *Suppose the system (5.1.1) has the form*

$$(5.5.4) \qquad \partial_t u - \sum_i^m D_i a_i(D'u) = 0,$$

with

$$(5.5.5) \qquad a_i(0) = 0 \quad (i = 1, ..., m)$$

and the homogeneous initial condition $u|_{t=0} = 0$ is true. Let conditions 1)-3) of 5.1 be satisfied for all $\xi_k^{(0)} = 0$ and the weak solution of (5.5.4) $u \in \overset{\circ}{W}{}^{(0,1)}_{2,loc}$. Also let the conditions (5.5.2) be satisfied, then the inequality

$$(5.5.6) \qquad K' E_m \left(1 + \frac{m-2}{m+1}\right)^{1/2} \left(\frac{m}{2}\right)^{1/2} < 1$$

is true, K is defined for the matrix A′ (1.2.18) with $l = 1$ and

$$(5.5.7) \quad E_m^2 = \begin{cases} 1 + (m-1)(m-2)\left[1 + \frac{16(m-2)^4}{(2m-5)\left(m-1+\frac{m-2}{2}+\frac{1}{m-2}\right)}\right], & m \ge 3, \\ 1, & m = 2 \end{cases}$$

then $u \equiv 0$.

Proof. From (5.5.6) it follows that the inequality

$$K'[1 + (m-2)(m-1)]^{1/2} \left(\frac{m}{2}\right)^{1/2} \left(1 + \frac{m-2}{m+1}\right)^{1/2} < 1$$

is satisfied. Then, according to theorem 5.2.3 it follows that $u(t, x)$ belongs to $W^{1,2}_{2,\alpha,loc}$ with sufficiently small γ. Therefore we can write the system (5.5.4) in the form

$$(5.5.8) \qquad \varepsilon \dot{u} - \Delta u = -L_\varepsilon(u),$$

where ε is a positive constant and

$$L_\varepsilon(u) = \Delta u - \varepsilon D_i a_i(D'u) . \tag{5.5.9}$$

Multiply both sides of (5.5.8) by $\Delta w\zeta$, where $w(t,x)$ is the solution of the boundary value problem (5.2.1), (5.2.2) with

$$f = -\Delta u r^\alpha \zeta(r) \ . \tag{5.5.10}$$

The cut-off function ζ in addition to (2.2.1) should be smooth and satisfy the inequality $|D^s\zeta| < C\delta^{-s}$ $(s = 1, 2, 3 \ldots)$.
Then

$$\int\limits_{Q_\delta} (\varepsilon\dot{u} - \Delta u)\Delta w\zeta dx dt = -\int\limits_{Q_\delta} L_\varepsilon(u)\Delta w\zeta dx dt . \tag{5.5.11}$$

From (4.3.4) the inequality

$$\int\limits_{Q_\delta} |D'^2 w|^2 r^{-\alpha} dx dt \le \left(E_m^2 + \eta\right) \int\limits_{Q_\delta} |\Delta w|^2 r^\alpha dx dt \tag{5.5.12}$$

follows.
Consider now the left-hand side of (5.5.11). Integrating by parts and using the boundary conditions in (5.2.2) we get the following equality

$$\int\limits_Q (\varepsilon\dot{u} - \Delta u)\Delta w\zeta dx dt =$$

$$= -\int\limits_Q (\varepsilon\dot{w} + \Delta w)\Delta u\zeta dx dt - 2\varepsilon\int\limits_Q \nabla u\nabla\zeta\dot{w} dx dt - \varepsilon\int\limits_Q \dot{w}u\Delta\zeta dx dt$$

(we omit for now the index δ in Q_δ).
Using the relations (5.2.2), (5.5.10) and integrating by parts in the second and third integrals on the right–hand side, we get

$$\int\limits_Q (\varepsilon\dot{u} - \Delta u)\Delta w\zeta dx dt = \int\limits_Q |\Delta u|^2 r^\alpha \zeta^2 dx dt - 2\varepsilon\int\limits_Q \dot{u}\nabla(w\nabla\zeta) dx dt + \varepsilon\int\limits_Q \dot{u}w\Delta\zeta dx dt.$$

From the system (5.5.4) we can find $\dot{u}$ and substitute it in the right–hand side. After integrating by parts with respect to x in the second and the third integral we get the equality

$$\int\limits_Q (\varepsilon\dot{u} - \Delta u)\Delta w\zeta dx dt = \int\limits_Q |\Delta u|^2 r^\alpha \zeta^2 dx dt - 2\varepsilon\int\limits_Q D_i \left[a_i(D'u) - \right. \tag{5.5.13}$$

$$\left. - a_i\left(D'u|_{x=0}\right)\right] \nabla(w\nabla\zeta) dx dt + \varepsilon\int\limits_Q D_i\left[a_i(D'u) - a_i\left(D'u|_{x=0}\right)\right] w\Delta\zeta dx dt =$$

$$= \int\limits_Q |\Delta u|^2 r^\alpha \zeta^2 dx dt + 2\varepsilon\int\limits_Q \left[a_i(D'u) - a_i\left(D'u|_{x=0}\right)\right] D_i\left[\nabla(w\nabla\zeta)\right] dx dt -$$

$$-\varepsilon\int\limits_Q \left[a_i(D'u) - a_i\left(D'u|_{x=0}\right)\right] D_i(w\Delta\zeta) dx dt .$$

The second and the third integrals on the right-hand side can be estimated in the same manner. Let us estimate for example the third integral. From the basic conditions 1) - 3) of section 5.1 it follows that

$$|a_i(D'u) - a_i\,(D'u|_{x=0})\,| \le C|D'u - D'u|_{x=0}|\,.$$

Then multiplying and dividing under the sign of the third integral by $r^{\alpha/2}$ and using the Hölder inequality we get

$$\int_Q [a_i(D'u) - a_i\,(D'u|_{x=0})]D_i(w\Delta\zeta)dxdt \le$$
$$\le C(\eta)\frac{1}{\delta^2}\int_Q |D'u - D'u|_{x=0}|^2 r^\alpha dxdt + \eta\delta^2\int_Q |\nabla(w\Delta\zeta)|^2 r^{-\alpha}dxdt.$$

It follows from (5.5.1) and (5.5.10) that

$$\int_{Q_\delta} |\nabla(w\Delta\zeta)|^2 r^{-\alpha}dxdt \le C\delta^{-2}\int_{Q_\delta} |\Delta u|^2 r^\alpha\zeta^2 dxdt\,. \tag{5.5.14}$$

In the same manner we'll have

$$\int_{Q_\delta} |\nabla\,[\nabla(w\nabla\zeta)]\,|^2 r^\alpha dxdt \le C\delta^{-2}\int_{Q_\delta} |\Delta u|^2 r^\alpha\zeta^2 dxdt. \tag{5.5.15}$$

Taking into account that η is positive and arbitrarily small with the help of (5.5.13) we get the inequality

$$\int_{Q_\delta}(\varepsilon\dot u - \Delta u)\Delta w\zeta dx \ge \tag{5.5.16}$$
$$\ge (1-\eta)\int_{Q_\delta} |\Delta u|^2 r^\alpha\zeta^2 dxdt - C(\eta)\frac{1}{\delta^2}\int_{Q_\delta} |D'u - D'u|_{x=0}|^2 r^\alpha dxdt\,.$$

Consider now the right-hand side of (5.5.11). Integrating two times by parts with respect to x we'll get

$$\int_Q(-\Delta u + \varepsilon D_i a_i)\Delta w\zeta dxdt = \tag{5.5.17}$$
$$= -\int_Q D_i\,[u_i - u_i|_{x=0} - \varepsilon(a_i - a_i|_{x=0})]\,w_{kk}\zeta dxdt =$$
$$= \int_Q w_{ik}(-u_{ik} + \varepsilon D_k a_i)\zeta dxdt + \int_Q \Delta w\,[u_i - u_i|_{x=0} - \varepsilon(a_i - a_i|_{x=0})]\,\zeta_i dxdt -$$
$$- \int_Q w_{ik}\,[u_i - u_i|_{x=0} - \varepsilon(a_i - a_i|_{x=0})]\,\zeta_k dxdt.$$

The second and the third integrals on right-hand side can be estimated in the same manner as the second and third integrals on the right-hand side of (5.5.13) to give, for example,

$$\Big|\int_Q \Delta w\, [u_i - u_i|_{x=0} - \varepsilon(a_i - a_i|_{x=0})]\, \zeta_i dxdt\Big| \leq$$

$$\leq \eta \int_Q |\Delta u|^2 r^\alpha \zeta^2 dxdt + C(\eta)\frac{1}{\delta^2}\int_Q |D'u - D'u|_{x=0}|^2 r^\alpha dxdt.$$

Now consider the first integral on the right–hand side of (5.5.17)

$$I = \int_Q w_{ik}(-u_{ik} + \varepsilon D_k a_i)\zeta dxdt\,. \tag{5.5.18}$$

Take, according to (1.2.12) and (1.2.13)

$$\varepsilon = \begin{cases} (\sigma' + \lambda'^2)^{-1}\lambda & \text{for } \sigma' \geq \frac{\lambda'(\Lambda'-\lambda')}{2} \\ 2(\Lambda' + \lambda')^{-1} & \text{for } \sigma \leq \frac{\lambda'(\Lambda'-\lambda')}{2} \end{cases}, \tag{5.5.19}$$

$$I \leq K' \left(\int_Q |D'^2 w|^2 r^{-\alpha} dxdt\right)^{1/2} \left(\int_Q |D'^2 u|^2 r^\alpha \zeta^2 dxdt\right)^{1/2}.$$

Using (5.5.12) we have for small $\gamma > 0$ $(\alpha = 2 - m - 2\gamma)$

$$I \leq (K'E_m + \eta)\left[\int_Q |\Delta w|^2 r^\alpha dxdt\right]^{1/2} \left(\int_Q |D'^2 u|^2 r^\alpha \zeta^2 dxdt\right)^{1/2}.$$

Then from (5.2.10) (lemma 5.2.2) and (5.5.10), we obtain the following estimate

$$I \leq (K'E_m + \eta)\left(\frac{m}{2}\right)^{1/2} \left(\int_Q |\Delta u|^2 r^\alpha \zeta^2 dxdt\right)^{1/2} \left(\int_Q |D'^2 u|^2 r^\alpha \zeta^2 dxdt\right)^{1/2}. \tag{5.5.20}$$

Now

$$\int_Q |D'^2 u|^2 r^\alpha \zeta^2 dxdt = \int_Q |D'^2 z|^2 r^\alpha \zeta^2 dxdt\,,$$

where

$$z = u - u|_{x=0} - u_i|_{x=0}\, x_i\,. \tag{5.5.21}$$

Performing elementary calculations we have

$$\int_Q |D'^2 u|^2 r^\alpha \zeta^2 dxdt = \sum_{j,l=1}^m \int_Q |(z\zeta)_{jl} - \zeta_{jl} z - 2\zeta_j z_l|^2\, r^\alpha dxdt,$$

$$\int_Q |D'^2 u|^2 r^\alpha \zeta^2 dxdt \leq (1+\eta)\int_Q |D'^2(z\zeta)|^2 r^\alpha dxdt +$$

$$+C(\eta)\left[\int_Q |z|^2 r^\alpha |\zeta''|^2 dxdt + \int_Q |D'u - D'u|_{x=0}|^2 r^\alpha |\zeta'|^2 dxdt\right].$$

Then from the properties of ζ we get

$$\int_Q |D'^2 u|^2 r^\alpha \zeta^2 dxdt \leq (1+\eta)\int_Q |D'^2(z\zeta)|^2 r^\alpha dxdt +$$

$$+C(\eta)\delta^{-2}\left(\delta^{-2}\int_Q |z|^2 r^\alpha dxdt + \int_Q |D'u - D'u|_{x=0}|^2 r^\alpha dxdt\right).$$

Now, it can easily be seen from the definition of ζ the function $z(t,x)$ defined by (5.5.21) satisfies the equalities

$$z\zeta|_{x=0} = D'(z\zeta)|_{x=0} = z\zeta|_{\partial B_\delta} = D'(z\zeta)|_{\partial B_\delta} = 0.$$

Therefore for almost all t the inequality (4.1.4)

$$\int_{B_\delta} |D'^2(z\zeta)|r^\alpha dx \leq \left[1 + \frac{m-2}{m+1} + 0(\gamma)\right]\int_{B_\delta} |\Delta(z\zeta)|^2 r^\alpha dx$$

is valid and we have for sufficiently small $\gamma > 0$ the estimate

(5.5.22) $$\int_{Q_\delta} |D'^2 u|^2 r^\alpha \zeta^2 dxdt \leq (1+\eta)\left(1+\frac{m-2}{m+1}\right)\int_{Q_\delta} |\Delta u|^2 r^\alpha \zeta^2 dxdt +$$

$$+C(\eta)\delta^{-2}\left(\delta^{-2}\int_{Q_\delta} |u - u|_{x=0} - u_i|_{x=0}\, x_i|^2 r^\alpha dxdt + \int_{Q_\delta} |D'u - D'u|_{x=0}|^2 r^\alpha dxdt\right).$$

Then from (5.5.18) and (5.5.20) it follows, that

(5.5.23) $$\int_Q w_{ik}(-u_{ik} + \varepsilon D_k a_i)\zeta dxdt \leq$$

$$\leq \left\{K' E_m (m/2)^{1/2}\left[1 + \frac{m-2}{m+1}\right]^{1/2} + \eta\right\}\int_Q |\Delta u|^2 r^\alpha \zeta^2 dxdt +$$

$$+\frac{C(\eta)}{\delta}\left(\int_Q |\Delta u|^2 r^\alpha \zeta^2 dxdt\right)^{1/2} \times$$

$$\times\left[\delta^{-2}\int_Q |u - u|_{x=0} - u_i|_{x=0}\, x_i|^2 r^\alpha dxdt + \int_Q |Du - Du|_{x=0}|^2 r^\alpha dxdt\right]^{1/2}$$

Applying (5.5.11) , (5.5.16) we have

$$(1-\eta)\int\limits_Q |\Delta u|^2 r^\alpha \zeta^2 dxdt \le$$

$$\le \left\{K'E_m(m/2)^{1/2}\left[1+\frac{m-2}{m+1}\right]^{1/2}+\eta\right\}\int\limits_Q |\Delta u|^2 r^\alpha \zeta^2 dxdt +$$

$$+\frac{C(\eta)}{\delta^2}\left[\frac{1}{\delta^2}\int\limits_Q |u-u|_{x=0}-u_i|_{x=0}\, x_i|^2 r^\alpha dxdt + \int\limits_Q |D'u - D'u|_{x=0}|^2 r^\alpha dxdt\right].$$

Using the inequality (5.5.6) we obtain

$$\int\limits_{Q_\delta} |\Delta u|^2 r^\alpha \zeta^2 dxdt \le C\left[\frac{1}{\delta^4}\int\limits_{Q_\delta} |u-u|_{x=0}-u_i|_{x=0}\, x_i|^2 r^\alpha dxdt + \right.$$

$$\left. + \frac{1}{\delta^2}\int\limits_{Q_\delta} |D'u - D'u|_{x=0}|^2 r^\alpha dxdt\right].$$

Let δ tend to $+\infty$. Then from (5.5.2) it follows that the right–hand side tends to zero and $\int\limits_{R^1_+\times R^m} |\Delta u|^2 r^\alpha \zeta^2 dxdt = 0$.

From the system (5.1.1) $\dot{u}=0$ and $\Delta u = 0$ follows, then the inequality (5.5.22) gives $|D^2 u| = 0$ and, therefore, according to the initial condition (5.5.5) it follows that $u \equiv 0$. □

Chapter 6

The Navier-Stokes system; strong solutions

6.1 Some inequalities for solutions of a stationary Stokes system.

Consider initially the stationary Stokes system

$$\left\{ \begin{array}{lcl} \Delta u + \nabla p & = & f, \\ \operatorname{div} u & = & 0 \end{array} \right. \tag{6.1.1}$$

with the boundary condition

$$u|_{\partial\Omega} = 0. \tag{6.1.2}$$

We assume the mean value of p is equal to zero. Using the inequalities (4.2.39) we derive estimates with explicit sometimes sharp constants for the solution of the problem (6.1.1), (6.1.2). The Stokes system is very extensively discussed in many books and papers and a survey of results can be found in the books of Ladyzhenskaya [1], von Wahl [1]. The main theorems, which we are using here, are the theorems of Solonnikov [2]. In this theorems the principal estimates for the solution in $W_p^{1,2}(\Omega)$ and in $W_p^{1,2}(Q)$ are given and we apply the constructions used by Ladyzhenskaya and Solonnikov (see also Koshelev [2]).

Suppose $\Omega \subset R^m$ is a bounded domain and $\partial\Omega$ is sufficiently smooth and let (u, p) be a weak (in Hopf sense) solutiuon of (6.1.1).

Theorem 6.1.1. *Let $f \in L_{2,\alpha}(\Omega)$ and $\alpha = 2 - m - 2\gamma$ $(0 < \gamma < 1)$ satisfy* (2.3.29). *Then for the weak solution of system* (6.1.1) *the inequalities*

$$\int\limits_{B_R} |\nabla p|^2 r^\alpha \zeta dx \leq \tag{6.1.3}$$

$$\leq \left[1 - \frac{\alpha(m-2)}{m-1} + \eta\right] \left[1 - \frac{\alpha(\alpha+m-2)}{2(m-1)}\right]^{-2} \int\limits_{B_R} |f|^2 r^\alpha \zeta dx +$$

$$+C\left(\int\limits_{B_R} |p|^2 dx + \int\limits_{B_R} |Du|^2 dx\right),$$

$$(6.1.4)\quad \int_{B_R} |D'^2u|^2 r^\alpha \zeta dx \le$$

$$\le \left\{1+\left[1-\frac{\alpha(m-2)}{m-1}\right]^{\frac{1}{2}}\left[1-\frac{\alpha(\alpha+m-2)}{2(m-1)}\right]^{-1}+\eta\right\}^2 \times$$

$$\times(1+M_\gamma^2)\int_{B_R}|f|^2 r^\alpha\zeta dx + C\left[\left(\int_{B_R}|D'^2u|^2 r^\alpha\zeta dx\right)^{\frac{m}{m+2\gamma}}\left(\int_{B_R}|Du|^2dx\right)^{\frac{2\gamma}{m+2\gamma}}+\right.$$

$$\left.+\int_{B_R}|p|^2dx+\int_{B_R}|Du|^2dx\right]$$

are true. Here x_0 is an arbitrary point inside Ω, $R < dist(x_0, \partial\Omega), \eta = const > 0$ is arbitrary and M_γ is defined by (4.2.20).

The proof of this theorem is analogous to the proof which was given in section 2.3 for the solutions of the elliptic systems. We will sketch this proof. According to the above mentioned result of Solonnikov[2] we can initially assume both f and the solution u, p are as smooth as we wish. Take a point $x_0 \in \Omega$ and consider a ball $B_R(x_0)$ with $R <$ dist $(x_0, \partial\Omega)$.
Let $Y_{s,k}(\Theta)$ be a complete orthonormal set of spherical functions and

$$(6.1.5)\qquad p(x)=\sum_{s=0}^{+\infty}\sum_{k=1}^{k_s}p_{s,k}(r)Y_{s,k}(\Theta).$$

Construct the function

$$(6.1.6)\qquad v(x)=\sum_{s=0}^{+\infty}\sum_{k=1}^{k_s}v_{s,k}(r)Y_{s,k}(\Theta),$$

where

$$v_{0,1}(r) = -\int_r^R p'_{0,1}(\varrho)\varrho^\alpha d\varrho,$$
$$v_{s,k}(r) = p_{s,k}(r)r^\alpha \quad (s\ge 1).$$

Take the function

$$(6.1.7)\qquad w(x)=v(x)\zeta(r),$$

where the cut-off function $\zeta(r)$ is determined by (2.2.1), (2.2.2), (2.2.20) and is monotone.
Multiplying the Stokes system (6.1.1) by ∇w and taking into account that

$$\int_{B_R}\Delta u\nabla w dx = -\int_{B_R}\Delta(div u)w dx = 0,$$

we come to the equality

$$\int_{B_R}\nabla p\nabla w dx = \int_{B_R} f\nabla w dx.$$

Integrating by parts and substituting the expansions (6.1.5), (6.1.6) for p and w we have

$$\int_{B_R} \nabla p \nabla w dx = \int_0^R p_{0,1}'^2 r^{\alpha+m-1} dr + \sum_{s\geq 1,k} \int_0^R \{|p_{s,k}'|^2 +$$
$$+\left[s(s+m-2) - \frac{\alpha(\alpha+m-2)}{2}\right] |p_{s,k}|^2 r^{-2}\} r^{\alpha+m-1} \zeta dr + \dots,$$

where the unwritten terms contain only integrals without singularity.
From this immediately follows

$$\int_{B_R} \nabla p \nabla w dx \geq \int_0^R |p_{0,1}'|^2 r^{\alpha+m-1} dr + \sum_{s\geq 1,k} \int_0^R \left[|p_{s,k}'|^2 +\right.$$
$$\left. + s(s+m-2) \min_{s\geq 1} \frac{s(s+m-2) - \alpha(\alpha+m-2)/2}{s(s+m-2)} |p_{s,k}|^2 r^{-2} \right] r^{\alpha+m-1} \zeta dr + \dots$$

Finally

$$\int_{B_R} \nabla p \nabla w dx \geq \left[1 - \frac{\alpha(\alpha+m-2)/2}{m-1}\right] \int_{B_R} |\nabla p|^2 r^\alpha \zeta dx - C \int_{B_R} |p|^2 dx.$$

On the other hand

$$\int_{B_R} \nabla p \nabla w dx = \int_{B_R} f \nabla w dx \leq \left(\int_{B_R} |f|^2 r^\alpha \zeta dx \right)^{1/2} \left(\int_{B_R} |\nabla w|^2 r^{-\alpha} \zeta^{-1} dx \right)^{1/2}.$$

Comparing the last two relations we come to the inequality

$$\left[1 - \frac{\alpha(\alpha+m-2)}{2(m-1)}\right] \int_{B_R} |\nabla p|^2 r^\alpha \zeta dx \leq$$
$$\leq \left(\int_{B_R} |f|^2 r^\alpha \zeta dx \right)^{1/2} \left(\int_{B_{3R/4}} |\nabla w|^2 r^{-\alpha} \zeta^{-1} dx \right)^{1/2} + C \int_{B_R} |p|^2 dx.$$

By this same method it was shown that

$$\int_{B_{3R/4}} |\nabla w|^2 r^{-\alpha} \zeta^{-1} dx \leq \left[1 - \frac{\alpha(m-2)}{m-1}\right] \int_{B_R} |\nabla p|^2 r^\alpha \zeta dx + C \int_{B_R} |p|^2 dx.$$

Therefore one of the statements of the theorem is proved.
Take now in the Stokes system (6.1.1) ∇p to the right-hand side and apply the inequality (4.2.39). After minor calculations we come to the inequality (6.1.4). From theorem 6.1.1 obviously follows:

Theorem 6.1.2. *Let the conditions of theorem* 6.1.1 *be satisfied, and assume* $\gamma > 0$ *is small. Then the following estimates for the solutions of system* (6.1.1)

$$\int_{B_R} |\nabla p|^2 r^\alpha \zeta dx \leq \left[1 + \frac{(m-2)^2}{m-1} + 0(\gamma)\right] \int_{B_R} |f|^2 r^\alpha \zeta dx + C \int_{B_R} |p|^2 dx,$$

$$(6.1.8)\qquad \int_{B_R} |D'^2u|^2 r^\alpha \zeta dx \leq$$

$$\leq \left\{1+\left[1+\frac{(m-2)^2}{m-1}+0(\gamma)\right]^{1/2}\right\}^2\left(1+\frac{m-2}{m+1}\right)\int_{B_R}|f|^2 r^\alpha \zeta dx +$$

$$+C\left[\left(\int_{B_R}|D'^2u|^2 r^\alpha \zeta dx\right)^{\frac{m}{m+2\gamma}}\left(\int_{B_R}|Du|^2 dx\right)^{\frac{2\gamma}{m+2\gamma}}+\right.$$

$$\left.+\int_{B_R}|Du|^2dx+\int_{B_R}|p|^2dx\right]$$

are true.

We shall see later (Koshelev [20]) (6.1.7) is sharp. It is now necessary, for the problem (6.1.1), (6.1.2), to get the estimates in the neighbourhood of the boundary $\partial\Omega$. For this purpose assume that a piece of the boundary is flat and has the equation $x_m = 0$. Thus, in this neighbourhood the domain Ω lies in the half-space $x_m < 0$. Take a point $M_0(x_1^{(0)},\dots,x_m^{(0)})$ in Ω and consider the ball $B_{R_0}(M_0)$ such that

$$(6.1.9)\qquad R_0 > |x_m^{(0)}|.$$

Consider also a parallelepiped Π_-

$$(6.1.10)\qquad \left.\begin{array}{l} x_k^{(0)}-R\leq x_k \leq x_k^{(0)}+R, \qquad (k=1,\dots,m-1),\\ x_m^{(0)}-R\leq x_m\leq 0,\ (R>R_0)\end{array}\right\}.$$

Suppose for a moment that $x_k^{(0)}=0$ $(k=1,\dots,m-1)$ and $R=\pi$. The principal part of the estimates is independent of these assumptions. We can assume initially all the functions f, p and u are smooth. Expand $f(x)$ in Π in the following Fourier series

$$(6.1.11)\qquad \left\{\begin{array}{ll} f^{(k)}(x) & =\sum_k f_n^{(k)}(x_m)\cos n_1x_1\dots\cos n_{k-1}x_{k-1}\times\\ & \times\sin n_kx_k\cos n_{k+1}x_{k+1}\dots\cos n_{m-1}x_{m-1}\ (1\leq k\leq m-1),\\ f^{(m)}(x) & =\sum_n f_n^{(m)}(x_m)\cos n_1x_1\dots\cos n_{m-1}x_{m-1},\end{array}\right.$$

where $n=(n_1,\dots,n_{m-1},)$ and all n_k are nonnegative integers. Take

$$(6.1.12)\qquad p(x) = \sum_n p_n(x_m)\cos n_1x_1\dots\cos n_{m-1}x_{m-1},$$

$$(6.1.13)\qquad \left\{\begin{array}{ll} u^{(k)}(x) & =\sum_n u_n^{(k)}(x_m)\cos n_1x_1\dots\cos n_{k-1}x_{k-1}\times\\ & \times\sin n_kx_k\cos n_{k+1}x_{k+1}\dots\cos n_{m-1}x_{m-1}\ ,\\ u^{(m)}(x) & =\sum_n u_n^{(m)}(x_m)\cos n_1x_1\dots\cos n_{m-1}x_{m-1}.\end{array}\right.$$

Let

$$u|_{x_m=0}=0.$$

Then

(6.1.14) $$u_n^{(k)}(0) = 0, \quad (k = 1, \dots, m).$$

We see that the functions $u^{(k)}(x)$ $(k < m)$ also satisfy the following boundary conditions

(6.1.15) $$u^{(k)}|_{x_k=0,\pi} = 0, \quad u_{x_j}^{(k)}|_{x_j=0,\pi} = 0 \;\; (j \neq k, j \neq m).$$

The functions $u(x), p(x)$ satisfy the Stokes system (6.1.1) in Π^- if the following equalities

(6.1.16) $$\begin{cases} \ddot{u}_n^{(k)} - |n|^2 u_n^{(k)} + n_k p_n &= f_n^{(k)}, (k = 1, \dots, m-1), \\ \ddot{u}_n^{(m)} - |n|^2 u_n^{(m)} + \dot{p}_n &= f_n^{(m)}, \\ \dot{u}_n^{(m)} - \sum\limits_{s=1}^{m-1} n_s u_n^{(s)} &= 0, \end{cases}$$

are satisfied. Here

$$|n|^2 = \sum_{s=1}^{m-1} n_s^2$$

and dots over $u_n^{(k)}$ denote derivatives with respect to x_m. Multiply the first $m-1$ equations of (6.1.16) respectively by n_k. After summation and using the last equation (6.1.16) we have

(6.1.17) $$\dddot{u}_n^{(m)} - |n|^2 \dot{u}_n^{(m)} + |n|^2 p_n = -\sum_{k=1}^{m-1} n_k f_n^{(k)}.$$

If we differentiate the second equation of (6.1.16) with respect to x_m and subtract the relation (6.1.17) from the result we have

(6.1.18) $$\ddot{p}_n - |n|^2 p_n = \dot{f}_n^{(m)} + \sum_{k=1}^{m-1} n_k f_n^{(k)} \equiv F_n^-(x_m).$$

The bounded solution of this equation for $x_m < 0$ is given by

(6.1.19) $$p_n = p_n^-(x_m) = -\frac{1}{2|n|} \int\limits_{-\infty}^{x_m} F_n^-(\xi_m) e^{|n|(\xi_m - x_m)} d\xi_m +$$

$$+\frac{1}{2|n|} \int\limits_{0}^{x_m} F_n^-(\xi_m) e^{|n|(x_m - \xi_m)} d\xi_m + C_- e^{|n| x_m}.$$

Here $F_n^-(x_m)$ is a function which coincides with $F_n(x_m)$ on $x_m > -\pi$ and is continuously expanded on $x_m < -\pi$. We assume all the functions are absolutely summable on $(-\infty, 0]$.

We also consider the equation (6.1.18) in $x_m > 0$ with such suitable right-hand side $F_n^+(x_m)$ that $p_n(x_m)$ is continuous and absolutely summable on the whole strip $-\infty < x_m < +\infty$. The solution for $x_m > 0$ is

(6.1.20) $$p_n = p_n^+(x_m) = -\frac{1}{2|n|} \int\limits_{0}^{x_m} F_n^+(\xi_m) e^{|n|(\xi_m - x_m)} d\xi_m +$$

$$+\frac{1}{2|n|} \int\limits_{+\infty}^{x_m} F_n^+(\xi_m) e^{|n|(x_m - \xi_m)} d\xi_m + C_+ e^{-|n| x_m}.$$

Substitute $x_m = 0$ in (6.1.19) and (6.1.20). Assume that $C_- = C_+$ and

$$\int_{-\infty}^{0} F_n^-(\xi_m)e^{|n|\xi_m}d\xi_m = \int_{0}^{+\infty} F_n^+(\xi_m)e^{-|n|\xi_m}d\xi_m.$$

Then

$$p_n^-(0) = p_n^+(0).$$

We see that F_n^- should be extended symmetrically on $x_m > 0$. According to (6.1.18) the functions $f_n^{(m)}(x_m)$ and $f_n^{(k)}(x_m)(k = 1,\dots,m-1)$ should be extended respectively in asymmetric and symmetric ways. Integrating once by parts gives

$$(6.1.21) \qquad p_n = p_n^-(x_m) = \frac{1}{2}\int_{-\infty}^{x_m} f_n^{(m)}(\xi_m)e^{|n|(\xi_m - x_m)}d\xi_m +$$

$$+\frac{1}{2}\int_{0}^{x_m} f_n^{(m)}(\xi_m)e^{|n|(x_m-\xi_m)}d\xi_m +$$

$$+\frac{1}{2|n|}\sum_{k=1}^{m-1} n_k\left[\int_{0}^{x_m} f_n^{(k)}(\xi_m)e^{|n|(x_m-\xi_m)}d\xi_m -\right.$$

$$\left.-\int_{-\infty}^{x_m} f_n^{(k)}(\xi_m)e^{|n|(\xi_m-x_m)}d\xi_m\right] + Ce^{-|n||x_m|} \quad (x_m < 0).$$

The analogous formula following from (6.1.20) holds true for $x_m > 0$.
Denote by $\tilde{f}_n^{(k)}(\lambda)$ the Fourier transform of the functions $f_n^{(k)}(x_m)$ $(k = 1,\dots,m)$ on $-\infty < x_m < +\infty$.
We have

$$f_n^{(k)}(x_m) = \frac{2}{\sqrt{\pi}}\int_{0}^{+\infty} \tilde{f}_n^{(k)}(\lambda)\cos\lambda x_m d\lambda, \quad k = 1,\dots,m-1,$$

$$f_n^{(m)}(x_m) = \frac{2}{\sqrt{\pi}}\int_{0}^{+\infty} \tilde{f}_n^{(m)}(\lambda)\sin\lambda x_m d\lambda, \quad k = 1,\dots,m-1.$$

For example

$$Z_m = -\frac{2}{\sqrt{\pi}}\sum_n \cos n_1x_1\dots\cos n_{m-1}x_{m-1}\int_{0}^{+\infty} \tilde{f}_n^{(m)}(\lambda)\frac{\sin\lambda x_m}{\lambda^2+|n|^2}d\lambda$$

belongs to $W_2^{(2)}(\Pi)$ and therefore its boundary values are defined for $\partial\Pi$. Moreover they can be estimated by the norm of $f^{(m)}$ in $\mathcal{L}_2(\Pi)$.
Then according to (4.2.39) (theorem 4.2.3)

$$\int_{B_{R_0}}\left|\frac{\partial p}{\partial x_m}\right|^2 r^\alpha\zeta dx \le (1+M_\gamma^2+\eta)m\int_{B_{R_0}}|f|^2r^\alpha\zeta dx +$$

$$+C\left[\left(\int_{B_{R_0}}\left|\frac{\partial p}{\partial x_m}\right|^2 r^\alpha\zeta dx\right)^{\frac{m}{m+2\gamma}}\left(\int_{B_{R_0}}|f|^2dx\right)^{\frac{2\gamma}{m+2\gamma}} + \int_{B_{R_0}}|f|^2dx\right].$$

Differentiating $p(x)$ with respect to $x_k (k = 1, \ldots, m-1)$ we get the same estimates. Then

$$(6.1.22) \qquad \int_{B_{R_0}} |\nabla p|^2 r^\alpha \zeta dx \le m^2(1 + M_\gamma^2 + \eta) \int_{B_{R_0}} |f|^2 r^\alpha \zeta dx +$$

$$+C\left[\left(\int_{B_{R_0}} |\nabla p|^2 r^\alpha \zeta dx\right)^{\frac{m}{m+2\gamma}} \left(\int_{B_{R_0}} |f|^2 dx\right)^{\frac{2\gamma}{m+2\gamma}} + \int_{B_{R_0}} |f|^2 dx\right].$$

Denote

$$B_{R_0}^+ = B_{R_0} \cap (x_m > 0), \quad B_{R_0}^- = B_{R_0} \cap (x_m < 0).$$

As far as $x_0 \in B_{R_0}^-$ then for $x \in B_{R_0}^+$

$$|\bar{x} - x_0| \le |x - x_0|,$$

where $\bar{x}$ is symmetric to x with respect to $x_m = 0$. For $\alpha = 2 - m - 2\gamma < 0$ we have

$$|\bar{x} - x_0|^\alpha \ge |x - x_0|^\alpha$$

and from the monotonicity of ζ follows

$$\zeta(|\bar{x} - x_0|) \ge \zeta(|x - x_0|).$$

Since f and p are expanded on B_{R_0} in symmetric and antisymmetric ways we have

$$(6.1.23) \qquad \int_{B_{R_0}} |f|^2 r^\alpha \zeta dx = \int_{B_{R_0}^+} |f|^2 r^\alpha \zeta dx + \int_{B_{R_0}^-} |f|^2 r^\alpha \zeta dx \le 2 \int_{B_{R_0}^-} |f|^2 r^\alpha \zeta dx.$$

The same estimate is true for the integral $\int_{B_{R_0}} |\nabla p|^2 r^\alpha \zeta dx$. Taking into account that

$$\int_{B_{R_0}} |\nabla p|^2 r^\alpha \zeta dx \ge \int_{B_{R_0}^-} |\nabla p|^2 r^\alpha \zeta dx$$

we come, with the help of (6.1.22), to

$$(6.1.24) \qquad \int_{B_{R_0}^-} |\nabla p|^2 r^\alpha \zeta dx \le 2m^2(1 + M_\gamma^2 + \eta) \int_{B_{R_0}^-} |f|^2 r^\alpha \zeta dx +$$

$$+C\left[\left(\int_{B_{R_0}^-} |\nabla p|^2 r^\alpha \zeta dx\right)^{\frac{m}{m+2\gamma}} \left(\int_{B_{R_0}^-} |f|^2 dx\right)^{\frac{2\gamma}{m+2\gamma}} + \int_{B_{R_0}^-} |f|^2 dx\right].$$

Theorem 6.1.3. *If $f \in L_{2,\alpha}(\Omega)$ with $\alpha = 2 - m - 2\gamma$ $(0 < \gamma < 1)$ and satisfying* (2.3.29) *then the solution of the boundary value problem* (6.1.1), (6.1.2) *satisfies the inequalities*

$$(6.1.25) \qquad \int_{\Omega_R} |\nabla p|^2 r^\alpha \zeta dx \le \left(N_\gamma^2 + \eta\right) \int_{\Omega_R} |f|^2 r^\alpha \zeta dx +$$

$$+C\left[\left(\int_{\Omega_R} |\nabla p|^2 r^\alpha \zeta dx\right)^{\frac{m}{m+2\gamma}} \left(\int_{\Omega_R} |f|^2 dx\right)^{\frac{2\gamma}{m+2\gamma}} + \int_{\Omega_R} |f|^2 dx\right]$$

and

$$(6.1.26)\qquad \int\limits_{\Omega_R} |D'^2u|^2 r^\alpha \zeta dx \le$$

$$\le \frac{A^2_{\alpha,m}}{\nu^2}\left(1+M^2_\gamma+\eta\right)(1+N_\gamma)^2 \int\limits_{\Omega_R} |f|^2 r^\alpha \zeta dx +$$

$$+C\left[\left(\int\limits_{\Omega_R} |D'^2u|^2 r^\alpha \zeta dx\right)^{\frac{m}{m+2\gamma}} \left(\int\limits_{\Omega_R} |Du|^2 dx\right)^{\frac{2\gamma}{m+2\gamma}} + \int\limits_{\Omega_R} |Du|^2 dx + \right.$$

$$\left. + \left(\int\limits_{\Omega_R} |\nabla p|^2 r^\alpha \zeta dx\right)^{\frac{m}{m+2\gamma}} \left(\int\limits_{\Omega_R} |f|^2 dx\right)^{\frac{2\gamma}{m+2\gamma}} + \int\limits_{\Omega_R} |f|^2 dx\right],$$

where

$$N^2_\gamma = \max\left\{\left[1-\frac{\alpha(m-2)}{m-1}\right]\left[1-\frac{\alpha(\alpha+m-2)}{2(m-1)}\right]^{-2}, 2m^2(1+M^2_\gamma)\right\},$$

$\Omega_R = \Omega \cap B_R(x_0)$, *$R$ sufficiently small,* $r = |x-x_0|, x_0 \in \bar{\Omega}, C$ *is independent of* x_0 *and* η *is an arbitrary small positive number.*

Proof. The inequality (6.1.25) follows by comparing the estimates (6.1.3) and (6.1.24). In fact, it is enough to compare the coefficients in front of $\int |f|^2 r^\alpha \zeta dx$ and take the largest one. To get (6.1.26), the system (6.1.1) should be written in the form

$$\Delta u = f - \nabla p$$

and the boundary conditions (6.1.2) are to be used.
In the interior the inequality follows from the estimate (6.1.4). In the boundary strip the solution should be continued in an asymmetric way and estimated with the help of (6.1.23). The proof of the theorem can now be completed by some minor calculations. □

For $\gamma \approx 0$ we have

Theorem 6.1.4. *If the conditions of theorem* 6.1.3 *are satisfied and the positive* γ *is small, then for the solution of problem* (6.1.1), (6.1.2) *the following estimates*

$$(6.1.27)\qquad \int\limits_{\Omega_R} |\nabla p|^2 r^\alpha \zeta dx \le 2m^2\left[1+\frac{m-2}{m+1}+0(\gamma)\right]\int\limits_{\Omega_R} |f|^2 r^\alpha \zeta dx +$$

$$+C\left[\left(\int\limits_{\Omega_R} |\nabla p|^2 r^\alpha \zeta dx\right)^{\frac{m}{m+2\gamma}} \left(\int\limits_{\Omega_R} |f|^2 dx\right)^{\frac{2\gamma}{m+2\gamma}} + \int\limits_{\Omega_R} |f|^2 dx\right],$$

$$\int\limits_{\Omega_R} |D'^2u|^2 r^\alpha \zeta dx \le 2\left[1+\frac{m-2}{m+1}+0(\gamma)\right]\left[1+\sqrt{2}m\left(1+\frac{m-2}{m+1}\right)^{1/2}\right]^2 \times$$

$$\times \int_{\Omega_R} |f|^2 r^\alpha \zeta dx + C\left[\left(\int_{\Omega_R} |D'^2 u|^2 r^\alpha \zeta dx\right)^{\frac{m}{m+2\gamma}} \left(\int_{\Omega_R} |Du|^2 dx\right)^{\frac{2\gamma}{m+2\gamma}} + \int_{\Omega_R} |Du|^2 dx + \right.$$

$$\left. + \left(\int_{\Omega_R} |p|^2 r^\alpha \zeta dx\right)^{\frac{m}{m+2\gamma}} \left(\int_{\Omega_R} |f|^2 dx\right)^{\frac{2\gamma}{m+2\gamma}} + \int_{\Omega_R} |f|^2 dx\right]$$

are true.

Now assume $2\gamma = \varepsilon > 0$ and turn to the system (6.1.1) and inequalities (6.1.13), (6.1.14). Consider initially the vector-function $u_\varepsilon(x) = \left\{u_\varepsilon^{(1)}(x), ..., u_\varepsilon^{(m)}(x)\right\}$ where

(6.1.28)
$$u_\varepsilon^{(i)}(x) = \frac{1}{(\varepsilon - 1)(\varepsilon + m + 1)} x_i x_m r^{-1+\varepsilon} - \\ - \frac{m + \varepsilon}{(\varepsilon^2 - 1)(\varepsilon + m + 1)} \delta_{im} r^{1+\varepsilon}, \quad (i = 1, ..., m).$$

After simple calculations we get

(6.1.29)
$$div u_\varepsilon = 0.$$

We also have

(6.1.30)
$$\Delta u_\varepsilon^{(i)} = x_i x_m r^{-3+\varepsilon} + [m - 2 + O(\varepsilon)] \delta_{im} r^{-1+\varepsilon}, (i = 1, ..., m).$$

Consider the scalar function

(6.1.31)
$$p_\varepsilon(x) = -x_m r^{-1+\varepsilon}.$$

After differentiation we have

(6.1.32)
$$(\nabla p_\varepsilon)^{(i)} = (1 - \varepsilon) x_i x_m r^{-3+\varepsilon} - \delta_{im} r^{-1+\varepsilon}, \quad (i = 1, ..., m).$$

The formulas (6.1.30) and (6.1.32) can be written in the form

(6.1.33)
$$\Delta u_\varepsilon^{(i)} = x_i x_m r^{-3+\varepsilon} + (m - 2) \delta_{im} r^{-1+\varepsilon} + O(\varepsilon) r^{-1+\varepsilon}$$

and

(6.1.34)
$$(\nabla p_\varepsilon)^{(i)} = x_i x_m r^{-3+\varepsilon} - \delta_{im} r^{-1+\varepsilon} + O(\varepsilon) r^{-1+\varepsilon}.$$

Denote the vector-functions $G_\varepsilon(x) = \{G_\varepsilon^{(1)}(x), ..., G_\varepsilon^{(m)}(x)\}$ and $H_\varepsilon(x) = \{H_\varepsilon^{(1)}(x), ..., H_\varepsilon^{(m)}(x)\}$ with the help of equalities

(6.1.35)
$$G_\varepsilon^{(i)} = x_i x_m r^{-3+\varepsilon} - m^{-1} \delta_{im} r^{-1+\varepsilon} + O(\varepsilon) r^{-1+\varepsilon}$$

and

(6.1.36)
$$H_\varepsilon^{(i)} = \delta_{im} r^{-1+\varepsilon} + O(\varepsilon) r^{-1+\varepsilon} (i = 1, ..., m).$$

Applying the equality (2.5.17) and making transformations gives

(6.1.37)
$$\|G_\varepsilon\|^2 \equiv \int_B |G_\varepsilon|^2 r^{2-m-\varepsilon} dx = \left[\frac{m-1}{\varepsilon m^2} + O(1)\right] |S|$$

and

(6.1.38)
$$\|H_\varepsilon\|^2 \equiv \int_B |H_\varepsilon|^2 r^{2-m-\varepsilon} dx = \left[\frac{1}{\varepsilon} + O(1)\right] |S|.$$

Theorem 6.1.5. *The first inequality of* (6.1.8) *is sharp for all* $m \geq 2$. *The functions* u, p *and* f , *for which the equality*

$$\int_B |\nabla p|^2 r^{2-m-\varepsilon} dx = \left[1 + \frac{(m-2)^2}{m-1} + O(\varepsilon)\right] \int_B |f|^2 r^{2-m-\varepsilon} dx \tag{6.1.39}$$

is realized are the following

$$u = \left[\frac{m-2}{m^2-3m+3} + O(\varepsilon)\right] u_\varepsilon(x), \quad p = [1 + O(\varepsilon)] p_\varepsilon(x) \tag{6.1.40}$$

and

$$f = \left[\frac{m-2}{m^2-3m+3} + O(\varepsilon)\right] \Delta u_\varepsilon + \nabla p_\varepsilon \tag{6.1.41}$$

where $u_\varepsilon(x)$ *and* $p_\varepsilon(x)$ *are determined respectively by* (6.1.28) *and* (6.1.31). *With the same* u, p *and* f, *the second inequality of* (6.1.8) *turns to an equality with an addititional term* $O(\varepsilon)$ *before the norm of* f *for big dimensions* m.

Proof. From (6.1.30), (6.1.31), (6.1.35) and (6.1.36) follows

$$\Delta u_\varepsilon = G_\varepsilon + \left[\frac{(m-1)^2}{m} + O(\varepsilon)\right] H_\varepsilon, \tag{6.1.42}$$

$$\nabla p_\varepsilon = G_\varepsilon - \left[\frac{(m-1)^2}{m} + O(\varepsilon)\right] H_\varepsilon \tag{6.1.43}$$

and (6.1.41) gives

$$f = \left[\frac{m-2}{m^2-3m+3} + 1 + O(\varepsilon)\right] G_\varepsilon + \\ + \left[\frac{(m-2)(m-1)}{m^2-3m+3} - 1 + O(\varepsilon)\right] \frac{m-1}{m} H_\varepsilon. \tag{6.1.44}$$

Simple calculations with the help of (6.1.37) and (6.1.38) lead to

$$\int_B |f|^2 r^{2-m-\varepsilon} dx = \frac{(m-1)^2 |S|}{\varepsilon m(m^2-3m+3)} + O(1) \tag{6.1.45}$$

and

$$\int_B |\nabla p|^2 r^{2-m-\varepsilon} dx = \frac{(m-1)|S|}{\varepsilon m} + O(1). \tag{6.1.46}$$

This proves (6.1.39) and therefore the sharpness of the first inequality of (6.1.8) is established. The same considerations give

$$\int_B |\Delta u|^2 r^{2-m-\varepsilon} dx \left(\int_B |f|^2 r^{2-m-\varepsilon} dx\right)^{-1} = \frac{(m-2)^2}{m-1} + O(\varepsilon).$$

Hence the second inequality of (6.1.8) is also sharp for large dimensions m. □

Remark 6.1.1. The same situation as in previous theorem occurs here. The pressure $p_\varepsilon = -x_m r^{-1+\varepsilon}$ passes from continuous values for $\varepsilon > 0$ to bounded values for $\varepsilon = 0$ and is unbounded for $\varepsilon < 0$. □

Remark 6.1.2. The result of this theorem was obtained after considering the expression

$$\sup_{\alpha,\beta} \frac{\beta^2\|\nabla p_\varepsilon\|^2}{\|\alpha\Delta u_\varepsilon + \beta\nabla p_\varepsilon\|^2} \tag{6.1.47}$$

for different real α and β. The equality (6.1.39) follows for

$$\alpha = \left[\frac{m-2}{m^2-3m+3} + O(\varepsilon)\right]\beta. \quad \square \tag{6.1.48}$$

Consider the expression

$$\sup_{\alpha,\beta} \frac{\alpha^2\|\Delta u_\varepsilon\|^2}{\|\alpha\Delta u_\varepsilon + \beta\nabla p_\varepsilon\|^2}. \tag{6.1.49}$$

The sup will be attained for

$$\beta = [m-2+O(\varepsilon)]\alpha. \tag{6.1.50}$$

Take the function $f(x)$ (6.1.1) equal to

$$f = \alpha\Delta u_\varepsilon + \beta\nabla p_\varepsilon \tag{6.1.51}$$

with α, β satisfying (6.1.50). For the solution $(u,p) = (\alpha u_\varepsilon, \beta p_\varepsilon)$ we get

$$\|\Delta u\|^2 = \frac{m-1}{m}\left[1 + \frac{(m-2)^2}{m-1} + O(\varepsilon)\right]\|f\|^2. \tag{6.1.52}$$

The corresponding relation for ∇p satisfies the equality

$$\|\nabla p\|^2 = \left[\frac{(m-2)^2}{m-1} + O(\varepsilon)\right]\|f\|^2. \tag{6.1.53}$$

Notice the coefficient before $\|f\|^2$ on the right-hand side of (6.1.53) is less than the one in (6.1.39) but for sufficiently large dimensions they are equal. By the way if we consider the sum $\|\Delta u\|^2 + \|\nabla p\|^2$ then for both cases (6.1.40), (6.1.41), (6.1.48) and (6.1.50),(6.1.51) we have:

$$\|\Delta u\|^2 + \|\nabla p\|^2 = \left[1 + 2\frac{(m-2)^2}{m-1} + O(\varepsilon)\right]\|f\|^2$$

in the first case and

$$\|\Delta u\|^2 + \|\nabla p\|^2 = \left\{\frac{m-1}{m}\left[1 + \frac{(m-2)^2}{m-1}\right] + \frac{(m-2)^2}{m-1} + O(\varepsilon)\right\}\|f\|^2$$

in the second one. For large dimensions m, both coefficients on the right-hand side before $\|f\|^2$ are the same as in (6.1.8).
It is reasonable to consider the ratio

$$\frac{\alpha^2\|\Delta u_\varepsilon\|^2+\beta^2\|\nabla p_\varepsilon\|^2}{\|\alpha\Delta u_\varepsilon+\beta\nabla p_\varepsilon\|^2}. \tag{6.1.54}$$

The inf and sup of this ratio attend respectively for

$$\beta=-\left[\sqrt{m^2-3m+3}+O(\varepsilon)\right]\alpha \tag{6.1.55}$$

and

$$\beta=\left[\sqrt{m^2-3m+3}+O(\varepsilon)\right]\alpha. \tag{6.1.56}$$

Simple calculations show that for the case (6.1.56) we have

$$\alpha^2\|\Delta u_\varepsilon\|^2+\beta^2\|\nabla p_\varepsilon\|^2=$$
$$=\left[1+\frac{(m-2)^2}{m-1}+\frac{m-2}{(m-1)^{1/2}}\sqrt{1+\frac{(m-2)^2}{m-1}}+O(\varepsilon)\right]\|\alpha\Delta u_\varepsilon+\beta\nabla p_\varepsilon\|^2$$

and

$$\alpha^2\|\Delta u_\varepsilon\|^2+\beta^2\|\nabla p_\varepsilon\|^2=$$
$$\left[1+\frac{(m-2)^2}{m-1}-\frac{m-2}{(m-1)^{1/2}}\sqrt{1+\frac{(m-2)^2}{m-1}}+O(\varepsilon)\right]\|\alpha\Delta u_\varepsilon+\beta\nabla p_\varepsilon\|^2$$

for the case (6.1.55).

6.2 Coercivity estimates for the nonstationary Stokes system

Consider now the nonstationary Stokes system

$$\begin{cases}\dot u-\nu\Delta u+\nabla p=f \quad (\nu=\text{ const}>0),\\ \text{div }u=0\end{cases} \tag{6.2.1}$$

with boundary conditions

$$u|_{\partial\Omega}=u|_{t=0}=0. \tag{6.2.2}$$

Initially we consider the inner estimates.
Suppose $f\in L_2\{(0,T);L_{2,\alpha}(\Omega)\}$ and $Q_R=(0,T)\times B_R$, where $R<$ dist $(x_0,\partial\Omega)$. It is trivial for estimate (6.1.3) to hold if we change B_R for Q_R. Then, dividing the first equation of (6.2.1) by ν and applying (5.3.13), we come to

Theorem 6.2.1. *The solution of the problem* (6.2.1), (6.2.2) *satisfies the inequalities*

(6.2.3) $$\int_{Q_R} |\nabla p|^2 r^\alpha \zeta dxdt \le \left[1-\frac{\alpha(m-2)}{m-1}+\eta\right]\left[1-\frac{\alpha(\alpha+m-2)}{2(m-1)}\right]^{-2} \times$$
$$\times \int_{Q_R} |f|^2 r^\alpha \zeta dxdt + C\int_{Q_R} |f|^2 dxdt,$$

(6.2.4) $$\int_{Q_R} |D'^2 u|^2 r^\alpha \zeta dxdt \le$$
$$\le \frac{1}{\nu^2}\left\{1+\left[1-\frac{\alpha(m-2)}{m-1}\right]^{\frac{1}{2}}\left[1-\frac{\alpha(\alpha+m-2)}{2(m-1)}\right]^{-1}+\eta\right\}^2 \times$$
$$\times(1+M_\gamma^2)A^2_{\alpha,m}\int_{Q_R} |f|^2 r^\alpha \zeta dxdt + C\left[\int_{Q_R} |f|^2 dxdt + \right.$$
$$\left. + \left(\int_{Q_R} |D'^2 u|^2 r^\alpha \zeta dxdt\right)^{\frac{m}{m+2\gamma}}\left(\int_{Q_R} |Du|^2 dxdt\right)^{\frac{2\gamma}{m+2\gamma}} + \int_{Q_R} |Du|^2 dxdt\right].$$

Here C is independent on ν, M_γ and $A_{\alpha,m}$ are respectiely defined by (4.2.20) and (5.3.7).
For small $\gamma > 0$ we have

Theorem 6.2.2. *Let the conditions of theorem* 6.2.1 *be satisfied, and let* $\gamma > 0$ *be small. Then the following estimates for the solutions of the system* (6.2.1)

(6.2.5) $$\int_{B_R} |\nabla p|^2 r^\alpha \zeta dxdt \le \left[1+\frac{(m-2)^2}{m-1}+0(\gamma)\right]\int_{Q_R} |f|^2 r^\alpha \zeta dxdt +$$
$$+C\int_{Q_R} |p|^2 dxdt,$$

(6.2.6) $$\int_{Q_R} |D'^2 u|^2 r^\alpha \zeta dxdt \le \frac{1}{\nu^2\gamma}[2+0(\gamma)]\int_{Q_R} |f|^2 r^\alpha \zeta dxdt +$$
$$+C\left[\int_{Q_R} |f|^2 dxdt + \left(\int_{Q_R} |D'^2 u|^2 r^\alpha \zeta dxdt\right)^{\frac{1}{1+\gamma}} \times\right.$$
$$\left. \times\left(\int_{Q_R} |Du|^2 dxdt\right)^{\frac{\gamma}{1+\gamma}} + \int_{Q_R} |Du|^2 dxdt\right], (m=2)$$

(6.2.7) $$\int_{Q_R} |D'^2 u|^2 r^\alpha \zeta dxdt \le \frac{m}{2\nu^2}\left\{1+\left[1+\frac{(m-2)^2}{m-1}+0(\gamma)\right]^{1/2}\right\} \times$$

$$\times\left(1+\frac{m-2}{m+1}\right)\int\limits_{Q_R}|f|^2r^\alpha\zeta\,dxdt+$$

$$+C\left[\int\limits_{Q_R}|f|^2dxdt+\left(\int\limits_{Q_R}|D'^2u|^2r^\alpha\zeta\,dxdt\right)^{\frac{m}{m+2\gamma}}\times\right.$$

$$\left.\times\left(\int\limits_{Q_R}|Du|^2dxdt\right)^{\frac{2\gamma}{m+2\gamma}}+\int\limits_{Q_R}|Du|^2dxdt\right]\quad(m\geq 3)$$

are true, where C is independent of ν.

Remark 6.2.1. *From the proof of theorems* 6.1.1 *and* 6.1.2 *and from theorem* 6.1.5 *follow that* (6.2.5) *is sharp.*

Theorem 6.2.3. *If $f\in L_2\{(0,T);L_{2,\alpha}(\Omega)\}$ with $\alpha=2-m-2\gamma(0<\gamma<1)$ satisfying (2.3.29), then the solution of system* (6.2.1) *with the boundary condition* (6.2.2) *satisfies the estimates*

$$\int\limits_{Q_R}|\nabla p|^2r^\alpha\zeta\,dxdt\leq 2m^2\left(N_\gamma^2+\eta\right)\int\limits_{Q_R}|f|^2r^\alpha\zeta\,dxdt+\tag{6.2.8}$$

$$+C\left[\left(\int\limits_{Q_R}|\nabla p|^2r^\alpha\zeta\,dxdt\right)^{\frac{m}{m+2\gamma}}\left(\int\limits_{Q_R}|f|^2dxdt\right)^{\frac{2\gamma}{m+2\gamma}}+\int\limits_{Q_R}|f|^2dxdt\right]$$

and

$$\int\limits_{Q_R}|D'^2u|^2r^\alpha\zeta\,dxdt\leq\tag{6.2.9}$$

$$\leq\frac{2A^2_{\alpha,m}}{\nu^2}(1+M_\gamma^2+\eta)\left(1+N_\gamma^2\right)^2\int\limits_{Q_R}|f|^2r^\alpha\zeta\,dxdt+$$

$$+C\left[\left(\int\limits_{Q_R}|D'^2u|^2r^\alpha\zeta\,dxdt\right)^{\frac{m}{m+2\gamma}}\left(\int\limits_{Q_R}|Du|^2dxdt\right)^{\frac{2\gamma}{m+2\gamma}}+\left(\int\limits_{Q_R}|\nabla p|^2r^\alpha\zeta\,dxdt\right)^{\frac{m}{m+2\gamma}}\times\right.$$

$$\left.\times\left(\int\limits_{Q_R}|f|^2r^\alpha\zeta\,dxdt\right)^{\frac{2\gamma}{m+2\gamma}}+\int\limits_{Q_R}|f|^2dxdt+\int\limits_{Q_R}|Du|^2dxdt\right].$$

Here $r=|x-x_0|$ and $Q_R=(0,T)\times B_R(x_0)\cap\Omega$ with a sufficiently small R. The constant C is independent of x_0 and ν.

The proof is completely analogous to that of theorem 6.1.3 with only difference being one must refer to estimate (5.3.13) . □

For small $\gamma>0$ the last theorem can be formulated in a more explicit way.

Theorem 6.2.4. *If the conditions of theorem 6.2.3 are satisfied, then the solution of the problem* (6.2.1), (6.2.2) *satisfies the following inequalities:*

(6.2.10)
$$\int_{Q_R} |\nabla p|^2 r^\alpha \zeta dxdt \leq 2m^2 \left[1 + \frac{m-2}{m+1} + 0(\gamma)\right] \int_{Q_R} |f|^2 r^\alpha \zeta dxdt +$$
$$+C\left[\left(\int_{Q_R} |\nabla p|^2 r^\alpha \zeta dxdt\right)^{\frac{m}{m+2\gamma}} \left(\int_{Q_R} |f|^2 dx\right)^{\frac{2\gamma}{m+2\gamma}} + \int_{Q_R} |f|^2 dxdt\right],$$

(6.2.11)
$$\int_{Q_R} |D'^2 u|^2 r^\alpha \zeta dxdt \leq$$
$$\leq \frac{m}{\nu^2}\left[1 + \frac{m-2}{m+1} + 0(\gamma)\right]\left[1 + \sqrt{2}m\left(1 + \frac{m-2}{m+1}\right)^{1/2}\right]^2 \int_{Q_R} |f|^2 r^\alpha \zeta dxdt +$$
$$+C\left[\left(\int_{Q_R} |D'^2 u|^2 r^\alpha \zeta dxdt\right)^{\frac{m}{m+2\gamma}} \left(\int_{Q_R} |Du|^2 dxdt\right)^{\frac{2\gamma}{m+2\gamma}} + \left(\int_{Q_R} |\nabla p|^2 r^\alpha \zeta dxdt\right)^{\frac{2m}{m+2\gamma}} \times\right.$$
$$\left.\times \left(\int_{Q_R} |f|^2 dxdt\right)^{\frac{2\gamma}{m+2\gamma}} + \int_{Q_R} |Du|^2 dxdt + \int_{Q_R} |f|^2 dxdt\right], \quad (m \geq 3),$$

(6.2.12)
$$\int_{Q_R} |D'^2 u|^2 r^\alpha \zeta dxdt \leq$$
$$\leq \frac{2(1+2\sqrt{2})^2}{\nu^2 \gamma}[1 + 0(\gamma)] \int_{Q_R} |f|^2 r^\alpha \zeta dxdt +$$
$$+C\left[\left(\int_{Q_R} |D'^2 u|^2 r^\alpha \zeta dxdt\right)^{\frac{1}{1+\gamma}} \left(\int_{Q_R} |Du|^2 dxdt\right)^{\frac{\gamma}{1+\gamma}} + \int_{Q_R} |Du|^2 dxdt + \right.$$
$$\left. + \left(\int_{Q_R} |\nabla p|^2 r^\alpha \zeta dxdt\right)^{\frac{1}{1+\gamma}} \left(\int_{Q_R} |f|^2 dxdt\right)^{\frac{\gamma}{1+\gamma}} + \int_{Q_R} |f|^2 dxdt\right] \quad (m = 2).$$

6.3 The universal iterative process for the Navier-Stokes system and the pointwise boundedness of its iterations

The problem of the existence of regular (continuous, Hölder continuous) solutions for the nonstationary Navier-Stokes system is one of the most important problems in modern mathematical physics. This problem is closely connected with two main issues: the uniqueness theorem and the ability to apply approximate methods to

numerical analysis and practical computations. The estimates for the pointwise values of the velocity fields are essential for different numerical applications (see, for example, Temam [1]). We consider here mainly the multidimensional nonstationarity problem for finite (not necessary small) time. In principal theorems of Ladyzhenskaya [1] and Solonnikov [2],[3] the constants, which estimate the pointwise values of velocities are implicit even for small t or for small Reynolds numbers. In this and the next section we show, with the help of inequalities (6.2.8) and (6.2.9) (Koshelev [18]) and for small Reynolds numbers, the explicit bounds for pointwise velocity estimates can be obtained and the existence of unique strong solution can be proved.
Let Ω be a bounded domain in R^m with $\partial\Omega \in C^{1,\kappa}$. Consider in $Q = (0,T) \times \Omega$ the Navier-Stokes system

$$\dot{u} - \nu \Delta u + u^{(k)} D_k u + \nabla p + f(t,x) = 0, \tag{6.3.1}$$

$$\operatorname{div} u = 0 \tag{6.3.2}$$

with initial and boundary conditions

$$u|_{t=0} = 0, \quad u|_{\partial\Omega} = 0. \tag{6.3.3}$$

Here

$$u(t,x) = \{u^{(1)}(t,x), ..., u^{(m)}(t,x)\}$$

is the unknown velocity vector-function, $\nu = \text{const} > 0$ (viscosity), $f(t,x)$ denotes the vector of external forces, p is the pressure, which is normed by the equality $\int\limits_\Omega p dx = 0$, and the summation runs as usual over repeated indicies k from 1 to m.
The Hopf or weak solution of (6.3.1) - (6.3.3), which belongs to $L_2\left\{(0,T); W_2^{(1)}(\Omega)\right\}$ and satisfying the equality

$$\int\limits_Q \left(-u\,\dot{v} + \nu D'u D'v - u^{(k)} D_k v u\right) dx dt + \int\limits_\Omega uv dx|_{t=T} + \int\limits_Q f v dx dt = 0, \tag{6.3.4}$$

We consider in this and next sections the existence of Hölder continuous in both x and t solutions of (6.3.1)-(6.3.3). For this purpose we apply the universal iterative process

$$\varepsilon\, \dot{u}_{n+1} - \nu \Delta u_{n+1} + \varepsilon \nabla p_{n+1} = -\nu \Delta u_n - \varepsilon[-\nu \Delta u_n + u_n^{(k)} D_k u_n + f] \tag{6.3.5}$$

where $\varepsilon = \text{const} \in (0,1]$(to be determined later) and all u_n satisfy the conditions (6.3.2) and (6.3.3).The initial iteration should be smooth enough and the equations (6.3.5) can be also written in the weak form. Taking into account some numerical applications we prove that under natural assumptions this method converges in some weighted spaces. At the same time we obtain an estimate in the Hölder norm for the solution. The constants, which are present in the inequality, have an explicit form for small Reynold's numbers (to be obtained later). Notice that for weak continuous solution the uniquiness theorem is true.
Denote the distance between two points (t,x) and (t_0,x_0) in Q by r_t. Since $\alpha < 0$ for $0 < \gamma < 1$, we have

$$r_t^\alpha = (|x - x_0|^2 + |t - t_0|^2)^{\frac{\alpha}{2}} \le |x - x_0|^\alpha.$$

The space of functions defined in Q with the norm

$$(6.3.6) \qquad \|u\|'_{\alpha,1,l} = \left(\sup_{(t_0,x_0)\in Q} \int_Q \left(|\dot{u}|^2 + |D^l u|^2 \right) r_t^{\alpha} dx dt \right)^{\frac{1}{2}}.$$

we shall denote by $H'_{\alpha,1,l}$ It is obvious for $\alpha \le 0$

$$(6.3.7) \qquad \|u\|'_{\alpha,1,l} \le \|u\|^{1,l}_{2,\alpha},$$

where the norm on the right-hand side is defined by (5.1.5).
Consider Q as a set in R^{m+1}. It is clear if Ω satisfies (2.1.11), then Q also satisfies the same inequality with the factor δ^{m+1} (instead of δ^m) on the right-hand side. If we take $l = 1$, $\alpha = 1 - m - 2\gamma$ $(0 < \gamma < 1/2)$ and $p = 2$, then from (2.1.12) we come to the inequality

$$\frac{|u(t_1,x_1) - u(t_2,x_2)|}{(|x_1 - x_2|^2 + |t_1 - t_2|^2)^{\frac{\gamma}{2}}} \le \frac{2^{\frac{m+3}{2}}}{\gamma A_Q^{\frac{1}{2}}} \|u\|'_{\alpha,1,1}.$$

Applying(6.3.7) we get an estimate

$$(6.3.8) \qquad \frac{|u(t_1,x_1) - u(t_2,x_2)|}{(|x_1 - x_2|^2 + |t_1 - t_2|^2)^{\frac{\gamma}{2}}} \le \frac{2^{\frac{m+3}{2}}}{\gamma A_Q^{\frac{1}{2}}} \|u\|^{1,1}_{2,\alpha}$$

where (t_1, x_1) and (t_2, x_2) are arbitrary points from $\overline{Q}$, the norm on the right-hand side is also taken over Q and $\alpha = 1 - m - 2\gamma$ $(0 < \gamma < 1/2)$. The inequalities (2.1.21) and (6.3.7) can be applied to $Q \subset R^{m+1}$ with $\alpha = 1 - m - 2\gamma$ $(0 < \gamma < \frac{1}{2})$. Then we get the relation

$$(6.3.9) \qquad |u(t_0,x_0)| \le \eta \|u\|^{1,1}_{2,\alpha} + C\eta^{-\frac{m+1}{1-m-\alpha}} \left(\int_Q |u|^2 dx dt \right)^{\frac{1}{2}}.$$

Integrating from zero to T both sides of the inequality (2.1.17′)

$$\int_{\Omega_R(x_0)} |u|^2 |x - x_0|^{\alpha} dx \le \eta \int_{\Omega_R(x_0)} |D' u|^2 |x - x_0|^{\alpha} dx + C(\eta) \int_{\Omega} |u|^2 dx,$$

we get the relation

$$(6.3.10) \qquad \int_{Q_R} |u|^2 |x - x_0|^{\alpha} dx dt \le \eta \int_{Q_R} |D' u|^2 |x - x_0|^{\alpha} dx dt + C(\eta) \int_Q |u|^2 dx dt$$

where $Q_R = Q_R(x_0) = (0,T) \times \Omega_R(x_0)$ and η can be taken a sufficiently small for a sufficiently small R. Applying the estimate $\|u\|^{1,1}_{2,\alpha} \le \|u\|^{1,2}_{2,\alpha}$ which follows from (5.1.4) and using (6.3.9) we get

$$(6.3.11) \qquad \sup_Q |u| \le \eta \|u\|^{1,2}_{2,\alpha} + C\eta^{-\frac{m+1}{1-m-\alpha}} \left(\int_Q |u|^2 dx dt \right)^{\frac{1}{2}},$$

where $\alpha = 1 - m - 2\gamma$ $(0 < \gamma < \frac{1}{2})$.
In the space $H^{1,2}_{2,\alpha}(Q)$ we can take

$$\|u\|^{1,2}_{2,\alpha} = \left[\sup_{x_0\in\Omega}\int_{Q_R}\left(|\dot u|^2 + |D'^2 u|^2\right) r^\alpha \zeta\, dxdt + \int_Q |u|^2 dxdt\right]^{\frac{1}{2}}$$

as the equivalent norm. It is easy to see in this case from (6.3.11),(6.3.3) follows the relation

$$\sup_Q |u|^2 \le \eta \sup_{x_0\in\Omega}\int_{Q_R}\left(|\dot u|^2 + |D'^2 u|^2\right) r^\alpha \zeta\, dxdt + \tag{6.3.12}$$
$$+ C\eta^{-\frac{m+1}{1-m-\alpha}} \int_Q |D'u|^2 dxdt.$$

If we apply the estimates (6.2.8) and (6.2.9) for the linearized system

$$\dot u - \nu\Delta u + \nabla p + f(t,x) = 0 \tag{6.3.13}$$

with the same conditions (6.3.2) and (6.3.3) for $\alpha = 1 - m - 2\gamma$, $0 < \gamma < \frac{1}{2}$, we arrive at the following relations

$$\int_{Q_R} |\nabla p|^2 r^\alpha \zeta\, dxdt \le 2m^2(1 + \overline{M}^2_\gamma + \eta)\int_{Q_R} |f|^2 r^\alpha \zeta\, dxdt + \tag{6.3.14}$$
$$+ C\left[\left(\int_{Q_R} |\nabla p|^2 r^\alpha \zeta\, dxdt\right)^{\frac{m}{m+1+2\gamma}} \left(\int_{Q_R} |p|^2 dxdt\right)^{\frac{1+2\gamma}{m+1+2\gamma}} + \right.$$
$$\left. + \int_{Q_R} |p|^2 dxdt + \int_Q |D'u|^2 dxdt\right]$$

and

$$\int_{Q_R} |D'^2 u|^2 r^\alpha \zeta\, dxdt \le \frac{B^2_{\alpha,m}}{\nu^2}\int_{Q_R} |f|^2 r^\alpha \zeta\, dxdt + \tag{6.3.15}$$
$$+ C\left[\left(\int_{Q_R} |D'^2 u|^2 r^\alpha \zeta\, dxdt\right)^{\frac{m}{m+1+2\gamma}} \left(\int_{Q_R} |D'u|^2 dxdt\right)^{\frac{1+2\gamma}{m+1+2\gamma}} + \right.$$
$$\left. + \frac{1}{\nu^2}\left(\int_{Q_R} |\nabla p|^2 r^\alpha \zeta\, dxdt\right)^{\frac{m}{m+1+2\gamma}} \left(\int_{Q_R} |p|^2 dxdt\right)^{\frac{1+2\gamma}{m+1+2\gamma}} + \frac{1}{\nu^2}\int_{Q_R} |p|^2 dxdt\right],$$

where

$$B^2_{\alpha,m} = 2A^2_{\alpha,m}\left(1 + \overline{M}^2_\gamma + \eta\right)\left(1 + \sqrt{2}m\sqrt{1 + \overline{M}^2_\gamma}\right)^2. \tag{6.3.16}$$

In this formula γ is changed to $\gamma+1/2$. Therefore $\overline{M}_\gamma = M_{\gamma+1/2}$. But we omit the bar and now denote M_γ by (6.3.17). Here $r = |x - x_0|$ and x_0 is an arbitrary point in Ω. The cut-off function $\zeta(r), 0 \le \zeta(r) \le 1$, and the constants M_γ and $B_{\alpha,m}$ according to previous notation are determined for $\alpha = 1 - m - 2\gamma (0 < \gamma < 1/2)$, by (2.2.1),(2.2.2), (4.2.20) and (5.3.7). Therefore, we have

$$(6.3.17) \qquad M_\gamma^2 = \frac{(m-1+2\gamma)\left\{(\frac{3}{2}+\gamma)^2 + [2-(\frac{1}{2}-\gamma)^2]m\right\}}{(m+\frac{3}{2}+\gamma)^2(\frac{1}{2}-\gamma)^2},$$

$$(6.3.18) \qquad A^2_{\alpha,m} = \begin{cases} 1 - \frac{\alpha}{2+\alpha} - \frac{2+\alpha}{(1+\frac{\alpha}{4})^2\alpha}, & m = 2, \\ \frac{m}{m+\alpha}, & m > 2. \end{cases}$$

Using the Young inequality (2.2.15) in the form

$$|ab| \le \eta |a|^{\frac{m+1+2\gamma}{m}} + C(\eta)|b|^{\frac{m+1+2\gamma}{1+2\gamma}},$$

from (6.3.14) and (6.3.15) we get

$$(6.3.19) \qquad \int_{Q_R} |\nabla p|^2 r^\alpha \zeta dxdt \le 2m^2(1 + M_\gamma^2 + \eta) \int_{Q_R} |f|^2 r^\alpha \zeta dxdt + C \int_Q |p|^2 dxdt,$$

$$(6.3.20) \qquad \int_{Q_R} |D'^2 u|^2 r^\alpha \zeta dxdt \le \frac{B^2_{\alpha,m}}{\nu^2} \int_{Q_R} |f|^2 r^\alpha \zeta dxdt +$$
$$+C\left[\int_Q |D'u|^2 dxdt + \frac{1}{\nu^2}\int_Q |p|^2 dxdt\right]$$

and the relations (6.3.13),(6.3.19),(6.3.20) brings us to the inequality

$$(6.3.21) \quad \int_{Q_R} |\dot{u}|^2 r^\alpha \zeta dxdt \le 3\left[m B^2_{\alpha,m} + 2m^2(1 + M_\gamma^2 + \eta) + 1\right] \int_{Q_R} |f|^2 r^\alpha \zeta dxdt +$$
$$+C\left[\nu^2 \int_Q |D'u|^2 dxdt + \int_Q |p|^2 dxdt\right],$$

where u is the solution of (6.3.13),(6.3.2) and (6.3.3). Here we used the Cauchy inequality and the relation

$$(6.3.22) \qquad |\Delta u|^2 \le m |D'^2 u|^2.$$

Lemma 6.3.1. *Let $u(t,x)$ and $p(t,x)$ determine a weak solution of problem (6.3.13), (6.3.2), (6.3.3) and for $\alpha = 1 - m - 2\gamma, 0 < \gamma < 1/2$, the condition*

$$\sup_{x_0 \in \Omega} \int_{Q_R} |f|^2 r^\alpha \zeta dxdt < +\infty$$

be satisfied. Then $u \in H^{1,2}_\alpha(Q)$ *and the inequalities*

$$(6.3.23)\quad \int\limits_{Q_R} |\nabla p|^2 r^\alpha \zeta dx dt \le 2m^2(1+M_\gamma^2+\eta)\int\limits_{Q_R} |f|^2 r^\alpha\zeta dxdt + C\int\limits_{Q}|f|^2 dxdt,$$

$$(6.3.24)\quad \int\limits_{Q_R} |D'^2 u|^2 r^\alpha\zeta dxdt \le \frac{B^2_{\alpha,m}}{\nu^2}\int\limits_{Q_R}|f|^2 r^\alpha\zeta dxdt + \frac{C}{\nu^2}\int\limits_{Q}|f|^2 dxdt.$$

$$(6.3.25)\quad \int\limits_{Q_R} |\,\dot u\,|^2 r^\alpha\zeta dxdt \le 3\left[mB^2_{\alpha,m}+2m^2(1+M^2_\gamma+\eta)+1\right]\int\limits_{Q_R}|f|^2 r^\alpha\zeta dxdt + +C\int\limits_{Q}|f|^2 dxdt$$

are true.

Proof. In the monography of Ladyzhenskaya [1] for the solutions of (6.3.13),(6.3.2),(6.3.3) the following estimates

$$(6.3.26)\quad \int\limits_{Q}|p|^2 dxdt \le C\int\limits_{Q}|f|^2 dxdt,$$

$$(6.3.27)\quad \int\limits_{Q}|D'u|^2 dxdt \le \frac{C}{\nu^2}\int\limits_{Q}|f|^2 dxdt,$$

$$(6.3.28)\quad \int\limits_{Q}|D'^2 u|^2 dxdt \le \frac{C}{\nu^2}\int\limits_{Q}|f|^2 dxdt,$$

whith C independent of ν were proved. Estimating the corresponding terms on the right-hand sides (6.3.19)-(6.3.21) with the help of (6.3.26)-(6.3.28) we get (6.3.23)-(6.3.25). □

Remark 6.3.1. Note that from (6.3.10),(6.3.24) and (6.3.27) follows the inequality

$$(6.3.29)\quad \int\limits_{Q_R}|D'^2 u|^2 r^\alpha\zeta dxdt \le \frac{\eta}{\nu^2}\int\limits_{Q_R}|f|^2 r^\alpha\zeta dxdt + \frac{C}{\nu^2}\int\limits_{Q}|f|^2 dxdt.$$

Lemma 6.3.2. *If*

$$(6.3.30)\quad \sup_{x_0\in\Omega}\int\limits_{Q}|f|^2|x-x_0|^\alpha dxdt < +\infty$$

and initial iteration $u_0(t,x)$ *belongs to* $H^{1,2}_{2,\alpha}(Q)$, *then the iterations of the process (6.3.5) satisfy the inequality*

$$(6.3.31)\quad \int\limits_{Q}|D'^2 u_{n+1}|^2 dxdt \le \eta\frac{\varepsilon^2}{\nu^2}\left(\sup_{x_0\in\Omega}\int\limits_{Q_R}\left(|\,\dot u_n\,|^2+|D'^2u_n|^2\right) r^\alpha\zeta dxdt\right)^2 + +C(1-\varepsilon)^2\int\limits_{Q}|D'u_n|^2 dxdt + C\frac{\varepsilon^2}{\nu^2}\left[\left(\int\limits_{Q}|D'u_n|^2 dxdt\right)^2+\int\limits_{Q}|f|^2 dxdt\right].$$

Proof. Since $u_0 \in H^{1,2}_{2,\alpha}(Q)$, then from the inequalities (6.3.11),(6.3.24) and lemma 6.3.1 follows that all $u_n \in H^{1,2}_{2,\alpha}(Q)$ satisfy both the equalities (6.3.5) and conditions (6.3.2) and (6.3.3). Let us multiply equations (6.3.5) by u_{n+1} and integrate over Q. Integrating by parts and taking into account (6.3.2) and (6.3.3), we get the equality

$$\frac{\varepsilon}{2}\int\limits_{\Omega} |u_{n+1}|^2 dx|_{t=T} + \nu \int\limits_{Q} |D'u_{n+1}|^2 dxdt =$$
$$= \int\limits_{Q} \left[\nu(1-\varepsilon)D_k u_n + \varepsilon u_n^{(k)} u_n\right] D_k u_{n+1} dxdt - \varepsilon \int\limits_{Q} f u_{n+1} dxdt.$$

If we omit the positive term on the left-hand side and apply the Hölder inequality to terms on the right-hand side, we get

$$\nu \int\limits_{Q} |D'u_{n+1}|^2 dxdt \le$$
$$\le \left\{ \int\limits_{Q} \left[\nu^2(1-\varepsilon)^2 |D'u_n|^2 + 2\varepsilon(1-\varepsilon)\nu u_n u_n^{(k)} D_k u_n + \varepsilon^2 \left(\sum_{k=1}^{m} |u_n^{(k)}|^2 \right)^2 \right] dxdt \right\}^{\frac{1}{2}} \times$$
$$\times \left(\int\limits_{Q} |D'u_{n+1}|^2 dxdt \right)^{\frac{1}{2}} + \varepsilon \left(\int\limits_{Q} |f|^2 dxdt \right)^{\frac{1}{2}} \left(\int\limits_{Q} |u_{n+1}|^2 dxdt \right)^{\frac{1}{2}}.$$

It is easy to see with the help of (6.3.2) and (6.3.3) we get the relation

$$\int\limits_{Q} u_n u_n^{(k)} D_k u_n dxdt = 0 \tag{6.3.32}$$

and the Poincaré inequality

$$\int\limits_{Q} |u_{n+1}|^2 dxdt \le C^2 \int\limits_{Q} |D'u_{n+1}|^2 dxdt. \tag{6.3.33}$$

Applying the inequality (2.2.15) after minor calculations we get

$$\nu \left(\int\limits_{Q} |D'u_{n+1}|^2 dxdt \right)^{\frac{1}{2}} \le$$
$$\le \sqrt{2} \left(\int\limits_{Q} \left[\nu^2(1-\varepsilon)^2 |D'u_n|^2 + \varepsilon^2 |u_n|^4 \right] dxdt \right)^{\frac{1}{2}} + C\varepsilon \left(\int\limits_{Q} |f|^2 dxdt \right)^{\frac{1}{2}}.$$

Enlarging the right-hand side, we obtain

$$\nu \left(\int\limits_{Q} |D'u_{n+1}|^2 dxdt \right)^{\frac{1}{2}} \le$$
$$\le \sqrt{2} \left[\nu^2(1-\varepsilon)^2 \int\limits_{Q} |D'u_n|^2 dxdt + \varepsilon^2 \sup_{Q} |u_n|^2 \int\limits_{Q} |u_n|^2 dxdt \right]^{1/2} + C\varepsilon \left(\int\limits_{Q} |f|^2 dxdt \right)^{\frac{1}{2}}.$$

With the help of (6.3.12) and (6.3.33) we get (6.3.31). □

We now obtain some necessary estimates for further investigations. Dividing both sides of (6.3.5) by ε we get

$$\dot{u}_{n+1} - \frac{\nu}{\varepsilon} \triangle u_{n+1} + \nabla p_{n+1} = -(1-\varepsilon)\frac{\nu}{\varepsilon} \triangle u_n - (u_n^{(k)} D_k u_n + f),$$

Applying (6.3.24),(6.3.25) and (6.3.28) and (6.3.21) we get

$$\text{(6.3.34)} \qquad \int_{Q_R} |D'^2 u_{n+1}|^2 r^\alpha \zeta dx dt \le B^2_{\alpha,m} \int_{Q_R} |f_n|^2 r^\alpha \zeta dx dt + C \int_Q |f_n|^2 dx dt,$$

$$\text{(6.3.35)} \qquad \frac{\varepsilon^2}{\nu^2} \int_{Q_R} |\dot{u}_{n+1}|^2 r^\alpha \zeta dx dt \le$$

$$\le 3\left[m B^2_{\alpha,m} + 2m^2(1 + M^2_\gamma + \eta) + 1\right] \int_{Q_R} |f_n|^2 r^\alpha \zeta dx dt + C \int_Q |f_n|^2 dx dt,$$

$$\text{(6.3.36)} \qquad \int_{Q_R} |D'^2 u_{n+1}|^2 dx dt \le C \int_Q |f_n|^2 dx dt,$$

$$\text{(6.3.37)} \qquad \int_{Q_R} |D' u_{n+1}|^2 r^\alpha \zeta dx dt \le \eta \int_{Q_R} |f_n|^2 r^\alpha \zeta dx dt + C \int_Q |f_n|^2 dx dt$$

where $f_n = (1-\varepsilon) \triangle u_n + \frac{\varepsilon}{\nu}\left(u_n^{(k)} D_k u_n + f\right)$. If we square under the integrals sign, we get

$$\int_{Q_R} |f_n|^2 r^\alpha \zeta dx dt =$$

$$= \int_{Q_R} \Big[(1-\varepsilon)^2 |\triangle u_n|^2 + \frac{2(1-\varepsilon)\varepsilon}{\nu} \triangle u_n u_n^{(k)} D_k u_n + \frac{2(1-\varepsilon)\varepsilon}{\nu} \triangle u_n f +$$

$$+ \frac{\varepsilon^2}{\nu^2} |u_n^{(k)} D_k u_n|^2 + 2\frac{\varepsilon^2}{\nu^2} u_n^{(k)} D_k u_n f + \frac{\varepsilon^2}{\nu^2} |f|^2\Big] r^\alpha \zeta dx dt.$$

From (6.3.12) with the help of the Hölder and Young inequalities we obtain

$$|2(1-\varepsilon)\frac{\varepsilon}{\nu} \int_{Q_R} \triangle u_n u_n^{(k)} D_k u_n r^\alpha \zeta dx dt| \le$$

$$\le C\sqrt{\frac{\varepsilon}{\nu}} \sup_Q |u_n| \left((1-\varepsilon)^2 \int_{Q_R} |D'^2 u_n|^2 r^\alpha \zeta dx dt\right)^{\frac{1}{2}} \left(\frac{\varepsilon}{\nu} \int_{Q_R} |D' u_n|^2 r^\alpha \zeta dx dt\right)^{\frac{1}{2}} \le$$

$$\le \eta \frac{\varepsilon^2}{\nu^2} \sup_{x_0 \in \Omega} \left(\int_{Q_R} \left(|\dot{u}_n|^2 + |D'^2 u_n|^2\right) r^\alpha \zeta dx dt\right)^2 +$$

$$+ C\frac{\varepsilon^2}{\nu^2} \left(\int_Q |D' u_n|^2 dx dt\right)^2 + C(1-\varepsilon)^2 \int_{Q_R} |D'^2 u_n|^2 r^\alpha \zeta dx dt,$$

$$\int_{Q_R} |u_n^{(k)} D_k u_n|^2 r^\alpha \zeta dx dt \le \sup_Q |u_n|^2 \int_{Q_R} |D' u_n|^2 r^\alpha \zeta dx dt \le$$

$$\le \eta \sup_{x_0\in\Omega} \left(\int_{Q_R} \left(|\dot{u}_n|^2 + |D'^2 u_n|^2\right) r^\alpha \zeta dx dt \right)^2 + C \left(\int_Q |D' u_n|^2 dx dt \right)^2$$

and

$$|\int_{Q_R} u_n^{(k)} D_k u_n f r^\alpha \zeta dx dt| \le C \sup_Q |u_n| \left(\int_{Q_R} |D' u_n|^2 r^\alpha \zeta dx dt \right)^{\frac{1}{2}} \left(\int_{Q_R} |f|^2 r^\alpha \zeta dx dt \right)^{\frac{1}{2}} \le$$

$$\le \eta \sup_{x_0\in\Omega} \left(\int_{Q_R} \left(|\dot{u}_n|^2 + |D'^2 u_n|^2\right) r^\alpha \zeta dx dt \right)^2 + \eta \sup_{x_0\in\Omega} \int_{Q_R} |f|^2 r^\alpha \zeta dx dt +$$

$$+ C \left(\int_Q |D' u_n|^2 dx dt \right)^2 .$$

Combining the last inequalities we get for $\varepsilon \in (0,1]$ the relation

$$\int_{Q_R} |f_n|^2 r^\alpha \zeta dx dt \le \int_{Q_R} \left[(1-\varepsilon)|\triangle u_n| + \frac{\varepsilon}{\nu}|f|\right]^2 r^\alpha \zeta dx dt + \tag{6.3.38}$$

$$+\eta \frac{\varepsilon^2}{\nu^2} \sup_{x_0\in\Omega} \left(\int_{Q_R} \left(|\dot{u}_n|^2 + |D'^2 u_n|^2\right) r^\alpha \zeta dx dt \right)^2 + \eta \frac{\varepsilon^2}{\nu^2} \sup_{x_0\in\Omega} \int_{Q_R} |f|^2 r^\alpha \zeta dx dt +$$

$$+C \left[\frac{\varepsilon^2}{\nu^2} \left(\int_Q |D' u_n|^2 dx dt \right)^2 + (1-\varepsilon)^2 \int_{Q_R} |D'^2 u_n|^2 r^\alpha \zeta dx dt \right].$$

In the same way we get

$$\int_Q |f_n|^2 dx dt \le \int_Q \left[(1-\varepsilon)|\triangle u_n| + \frac{\varepsilon}{\nu}|f|\right]^2 dx dt + \tag{6.3.39}$$

$$+\eta \frac{\varepsilon^2}{\nu^2} \sup_{x_0\in\Omega} \left(\int_{Q_R} \left(|\dot{u}_n|^2 + |D'^2 u_n|^2\right) r^\alpha \zeta dx dt \right)^2 + \eta \frac{\varepsilon^2}{\nu^2} \int_Q |f|^2 dx dt +$$

$$+C \left[\frac{\varepsilon^2}{\nu^2} \left(\int_Q |D' u_n|^2 dx dt \right)^2 + (1-\varepsilon)^2 \int_Q |D'^2 u_n|^2 dx dt \right].$$

It is easy to estimate the integrals $\int_Q [\nu(1-\varepsilon)|\triangle u_n| + \varepsilon|f|]^2 \, dx dt$ and $\int_{Q_R} [\nu(1-\varepsilon)|\triangle u_n| + \varepsilon|f|]^2 \, r^\alpha \zeta dx dt$, which are present in (6.3.38) and (6.3.39). For example,

$$\int_Q [\nu(1-\varepsilon)|\,\Delta\, u_n| + \varepsilon|f|]^2\, dxdt \leq \tag{6.3.40}$$
$$\leq 2\varepsilon^2 \int_Q |f|^2 dxdt + 2\nu^2(1-\varepsilon)^2 m \int_Q |D'^2 u_n|^2 dxdt.$$

Here we applied the inequality (6.3.22). In the same way we arrive at

$$\int_{Q_R} [\nu(1-\varepsilon)|\,\Delta\, u_n| + \varepsilon|f|]^2\, r^\alpha \zeta dxdt \leq \tag{6.3.41}$$
$$\leq (1+\eta)\varepsilon^2 \int_{Q_R} |f|^2 r^\alpha \zeta dxdt + C\nu^2(1-\varepsilon)^2 \int_{Q_R} |D'^2 u_n|^2 r^\alpha \zeta dxdt.$$

Applying (6.3.38) - (6.3.41) to (6.3.34) - (6.3.35) we get

$$\int_{Q_R} |D'^2 u_{n+1}|^2 r^\alpha \zeta dxdt \leq \tag{6.3.42}$$
$$\leq C(1-\varepsilon)^2 \sup_{x_0\in\Omega} \int_{Q_R} |D'^2 u_n|^2 r^\alpha \zeta dxdt + \frac{\varepsilon^2}{\nu^2} B^2_{\alpha,m} \sup_{x_0\in\Omega} \int_{Q_R} |f|^2 r^\alpha \zeta dxdt +$$
$$+ C(1-\varepsilon)^2 \int_Q |D'^2 u_n|^2 dxdt + C\frac{\varepsilon^2}{\nu^2} \int_Q |f|^2 dxdt + C\frac{\varepsilon^2}{\nu^2} \left(\int_Q |D' u_n|^2 dxdt \right)^2 -$$
$$+ \eta \frac{\varepsilon^2}{\nu^2} \left(\sup_{x_0\in\Omega} \int_{Q_R} \left(|\, \dot u_n\,|^2 + |D'^2 u_n|^2\right) r^\alpha \zeta dxdt \right)^2,$$

$$\tfrac{\varepsilon^2}{\nu^2} \int_{Q_R} |\, \dot u_{n+1}\,|^2 r^\alpha \zeta dxdt \leq C(1-\varepsilon)^2 \sup_{x_0\in\Omega} \int_{Q_R} |D'^2 u_n|^2 r^\alpha \zeta dxdt + \tag{6.3.43}$$
$$+ C^2_{\alpha,m} \tfrac{\varepsilon^2}{\nu^2} \sup_{x_0\in\Omega} \int_{Q_R} |f|^2 r^\alpha \zeta dxdt +$$
$$+ \eta \tfrac{\varepsilon^2}{\nu^2} \left(\sup_{x_0\in\Omega} \int_{Q_R} \left(|\, \dot u_n\,|^2 + |D'^2 u_n|^2\right) r^\alpha \zeta dxdt \right)^2 +$$
$$+ C(1-\varepsilon)^2 \int_Q |D'^2 u_n|^2 dxdt + C\tfrac{\varepsilon^2}{\nu^2} \left(\int_Q |D' u_n|^2 dxdt \right)^2 + C\tfrac{\varepsilon^2}{\nu^2} \int_Q |f|^2 dxdt$$

where

$$C^2_{\alpha,m} = 3[m B^2_{\alpha,m} + 2m^2(1 + M^2_\gamma + \eta) + 1]. \tag{6.3.44}$$

Analogous considerations brings us to

$$\int_{Q_R} |D' u_{n+1}|^2 r^\alpha \zeta dxdt \leq \tag{6.3.45}$$
$$\leq C(1-\varepsilon)^2 \sup_{x_0\in\Omega} \int_Q |D'^2 u_n|^2 r^\alpha \zeta dxdt + \eta \frac{\varepsilon^2}{\nu^2} \sup_{x_0\in\Omega} \int_Q |f|^2 r^\alpha \zeta dxdt +$$

$$+C(1-\varepsilon)^2 \int_Q |D'^2 u_n|^2 dx dt + C\frac{\varepsilon^2}{\nu^2} \int_Q |f|^2 dx dt + C\frac{\varepsilon^2}{\nu^2} \left(\int_Q |D' u_n|^2 dx dt \right)^2 +$$

$$+\eta \frac{\varepsilon^2}{\nu^2} \left(\sup_{x_0 \in \Omega} \int_{Q_R} \left(|\dot{u}_n|^2 + |D'^2 u_n|^2 \right) r^\alpha \zeta dx dt \right)^2$$

and

$$(6.3.46) \quad \int_Q |D'^2 u_{n+1}|^2 dx dt \leq C(1-\varepsilon)^2 \int_Q |D'^2 u_n|^2 dx dt + \frac{C\varepsilon^2}{\nu^2} \int_Q |f|^2 dx dt +$$

$$+C\frac{\varepsilon^2}{\nu^2} \left(\int_Q |D' u_n|^2 dx dt \right)^2 + \eta \frac{\varepsilon^2}{\nu^2} \left(\sup_{x_0 \in \Omega} \int_{Q_R} \left(|\dot{u}_n|^2 + |D'^2 u_n|^2 \right) r^\alpha \zeta dx dt \right)^2.$$

Summing up (6.3.31), (6.3.43),(6.3.45),(6.3.46) we get

$$(6.3.47) \quad \int_Q |D' u_{n+1}|^2 dx dt + \int_Q |D'^2 u_{n+1}|^2 dx dt +$$

$$+ \int_{Q_R} \left(\frac{\varepsilon^2}{\nu^2} |\dot{u}_{n+1}|^2 + |D' u_{n+1}|^2 + |D'^2 u_{n+1}|^2 \right) r^\alpha \zeta dx dt \leq$$

$$\leq C(1-\varepsilon)^2 \int_Q |D'^2 u_n|^2 dx dt + C\frac{\varepsilon^2}{\nu^2} \int_Q |f|^2 dx dt + C\frac{\varepsilon^2}{\nu^2} \left(\int_Q |D' u_n|^2 dx dt \right)^2 +$$

$$+\eta \frac{\varepsilon^2}{\nu^2} \left(\sup_{x_0 \in \Omega} \int_{Q_R} \left(|\dot{u}_n|^2 + |D'^2 u_n|^2 \right) r^\alpha \zeta dx dt \right)^2 + C(1-\varepsilon)^2 \int_Q |D' u_n|^2 dx dt +$$

$$+C(1-\varepsilon)^2 \sup_{x_0 \in \Omega} \int_{Q_R} |D'^2 u_n|^2 r^\alpha \zeta dx dt + \frac{\varepsilon^2}{\nu^2} E_{\alpha,m}^2 \sup_{x_0 \in \Omega} \int_{Q_R} |f|^2 r^\alpha \zeta dx dt,$$

where

$$(6.3.48) \qquad\qquad E_{\alpha,m}^2 = B_{\alpha,m}^2 + C_{\alpha,m}^2.$$

Denote

$$(6.3.49) \qquad X_n = \int_{Q'} |D' u_n|^2 dx dt + \int_Q |D'^2 u_n|^2 dx dt +$$

$$+ \sup_{x_0 \in \Omega} \int_{Q_R} \left(\frac{\varepsilon^2}{\nu^2} |\dot{u}_n|^2 + |D' u_n|^2 + |D'^2 u_n|^2 \right) r^\alpha \zeta dx dt.$$

It is obvious for $a \geq 0, b \geq 0, æ > 0$ the relation

$$(a+b)^2 \leq \max\{1, \frac{1}{æ^2}\}(æa+b)^2$$

is true. Therefore we get

$$(6.3.50)\qquad \left(\sup_{x_0\in\Omega}\int_{Q_R}\left(|\dot u_n|^2+|D'^2u_n|^2\right)r^\alpha\zeta\,dxdt\right)^2\le$$

$$\le\max\{1,\frac{\nu^4}{\varepsilon^4}\}\left(\sup_{x_0\in\Omega}\int_{Q_R}\left(\frac{\varepsilon^2}{\nu^2}|\dot u_n|^2+|D'^2u_n|^2\right)r^\alpha\zeta\,dxdt\right)^2.$$

Take the supremum of the left-hand side with respect to $x_0\in\Omega$. Then from (6.3.50) follows the inequality

$$(6.3.51)\qquad X_{n+1}\le C(1-\varepsilon)^2X_n+\frac{\varepsilon^2}{\nu^2}\max\{C_1,\eta\max\{1,\frac{\nu^4}{\varepsilon^4}\}\}X_n^2+$$
$$+\frac{\varepsilon^2}{\nu^2}E^2_{\alpha,m}\sup_{x_0\in\Omega}\int_{Q_R}|f|^2r^\alpha\zeta\,dxdt+C\frac{\varepsilon^2}{\nu^2}\int_{Q_R}|f|^2dxdt.$$

In (6.3.51) $C=C(\eta), C_1=C_1(\eta)$ are independent of ν and ε. It is possible to show $C_1=\frac{C}{\eta^\beta}$, where $\beta>0$ and $\beta\to+\infty$ for $\gamma\to 0$. Assume $\nu\ge 1$, then $\max\{1,\frac{\nu^4}{\varepsilon^4}\}=\frac{\nu^4}{\varepsilon^4}$. Minimizing the function $\varphi(\eta)=\max\{\frac{C}{\eta^\beta},\eta\frac{\nu^4}{\varepsilon^4}\}$ we arrive at $\min\varphi(\eta)=\varphi\left(\left(C\frac{\varepsilon^4}{\nu^4}\right)^{\frac{1}{\beta+1}}\right)=C\left(\frac{\nu^4}{\varepsilon^4}\right)^{\frac{\beta}{\beta+1}}$. Take $\eta=\left(C\frac{\varepsilon^4}{\nu^4}\right)^{\frac{1}{\beta+1}}$. After applying this estimate with the help of (6.3.51) we get

$$(6.3.52)\qquad X_{n+1}\le C_2(1-\varepsilon)^2X_n+C_3\left(\frac{\nu^2}{\varepsilon^2}\right)^{\frac{\beta-1}{\beta+1}}X_n^2+$$
$$+\frac{\varepsilon^2}{\nu^2}E^2_{\alpha,m}\sup_{x_0\in\Omega}\int_{Q_R}|f|^2r^\alpha\zeta\,dxdt+C_4\frac{\varepsilon^2}{\nu^2}\int_{Q_R}|f|^2dxdt.$$

This inequality can be written in the form

$$(6.3.53)\qquad X_{n+1}\le qX_n+aX_n^2+b$$

where

$$q=C_2(1-\varepsilon)^2,\quad a=C_3\left(\frac{\nu^2}{\varepsilon^2}\right)^{\frac{\beta-1}{\beta+1}},$$

$$b\equiv\frac{\varepsilon^2}{\nu^2}|||f|||^2=\frac{C_4\varepsilon^2}{\nu^2}\int_Q|f|^2dxdt+\frac{\varepsilon^2}{\nu^2}E^2_{\alpha,m}\sup_{x_0\in\Omega}\int_{Q_R}|f|^2r^\alpha\zeta\,dxdt$$

and the constants C_2,C_3,C_4 are independent of ε and ν.
Take $\varepsilon=1$, then $q=0$. In this case from (6.3.53) follows the inequality

$$(6.3.54)\qquad X_{n+1}\le aX_n^2+b.$$

The quantity $\Re$, satisfying the equality

$$(6.3.55)\qquad \Re^2=\nu^{-\frac{4}{\beta+1}}|||f|||^2$$

which we shall call the strong Reynolds number. The usual Reynolds number is denoted by

(6.3.56) $$\Re_0 = ||f||\nu^{-1}.$$

Assume that the external force is so small that

(6.3.57) $$4C_3\Re < 1.$$

Under this condition the equation

(6.3.58) $$az^2 - z + b = 0$$

has two different positive roots. Let z^* be one of them. Take the initial approximation $u_0(t,x)$ in such a way that $X_0 < z^*$. Then from (6.3.54) we get

$$X_1 \leq aX_0^2 + b < az^{*^2} + b = z^*.$$

Since $X_1 < z^*$ then $X_n < z^*$ for any n and we come to the following

Theorem 6.3.1. *Let $\nu \geq 1$, the external forces f satisfy the inequality (6.3.57) and $\int_Q |f|^2 r^\alpha dxdt < \infty$ for $\alpha = 1 - m - 2\gamma$, $\gamma \in (0, 1/2)$ and all $x_0 \in \overline{\Omega}$. If the initial iteration $u_0(t,x)$ satisfies the relation $X_0 \leq z^*$, where z^* is one of the roots of (6.3.58), then the norms in $W_2^{0,2}(Q)$ and $H_{2,\alpha}^{1,2}(Q)$ for all iterations (6.3.5) are bounded. At the same time the estimate*

(6.3.59) $$X_n \leq z^*$$

is true.

□

Remark 6.3.2. Let $\xi = 4C_3\nu^{-\frac{4}{\beta+1}}|||f|||^2 \ll 1$ and z^* is the smallest root of (6.3.58). Then

$$z^* = \frac{1-\sqrt{1-\xi}}{2a} = \frac{1}{4a}(\xi + o(\xi)) = \frac{|||f|||^2}{\nu^2} + o\left(\frac{1}{\nu^2}|||f|||^2\right).$$

Hence, the estimate

$$X_n \leq \frac{1}{\nu^2}|||f|||^2 + o\left(\frac{1}{\nu^2}|||f|||^2\right)$$

is true.

Remark 6.3.3. In the proof of theorem 6.3.1 we had $\varepsilon = 1$. If we take $\varepsilon \neq 1$, then instead of (6.3.57) we must suppose the following conditions

$$1 - C_2(1-\varepsilon)^2 > 0, \quad 4C_3\Re < \psi(\varepsilon)$$

where

$$\psi(\varepsilon) = \varepsilon^{-\frac{4}{\beta+1}}\left[1 - C_2(1-\varepsilon)^2\right]^2$$

are satisfied.

The $\max\psi(\varepsilon)$ for $\varepsilon \in \left(1 - \frac{1}{\sqrt{C_2}}, 1\right]$ attends for

$$\varepsilon = \varepsilon_0 = 1 - \frac{\beta+1}{\beta}\left[1 - \sqrt{1 - \frac{4\beta}{C_2(\beta+1)^2}}\right].$$

Notice for $\gamma \to 0$ we have $\beta \to +\infty$. Then for small $\gamma > 0$ we have

$$\varepsilon_0 = 1 - \frac{1}{C_2(\beta+1)} + o\left(\frac{1}{\beta}\right)$$

and

$$\max\psi(\varepsilon) = \psi(\varepsilon_0) = 1 + \frac{2}{C_2\beta^2} + o\left(\frac{1}{\beta^2}\right).$$

Hence, for small $\gamma > 0$ the optimal choice of the iterative parameter ε gives

$$4C_3\Re < 1 + \frac{2}{C_2\beta^2} + o\left(\frac{1}{\beta^2}\right).$$

This condition differs very little from (6.3.57).

6.4 Convergence of the iterative process in $H^{1,2}_{2,\alpha}(Q)$ and existence of Hölder continuous solution

Theorem 6.3.1 allows to prove of the convergence of the process (6.3.5) in the norm of $H^{1,2}_{2,\alpha}(Q)$ for $\alpha = 1 - m - 2\gamma, \gamma \in (0, 1/2)$.
Denote $w_n = u_n - u_{n-1}, s_n = p_n - p_{n-1}$. Subtracting two succesive iterative equations (6.3.5) we obtain w_n and s_n satisfy the following relations

(6.4.1) $$\varepsilon\, \dot{w}_{n+1} - \nu \bigtriangleup w_{n+1} + \varepsilon\nabla s_{n+1} = \\ = -\nu(1-\varepsilon) \bigtriangleup w_n - \varepsilon\left(w_n^{(k)} D_k u_{n-1} + u_n^{(k)} D_k w_n\right),$$

(6.4.2) $$\operatorname{div} w_{n+1} = 0,$$

(6.4.3) $$w_{n+1}|_{t=0} = 0; \quad w_{n+1}|_{\partial\Omega} = 0.$$

Applying (6.3.24),(6.3.25),(6.3.27)-(6.3.29) for the solution w_{n+1} of the problem (6.4.1)-(6.4.3) we get estimates

(6.4.4) $$\int_Q |D' w_{n+1}|^2 dx dt \le C \int_Q |g_n|^2 dx dt,$$

(6.4.5) $$\int_Q |D'^2 w_{n+1}|^2 dx dt \le C \int_Q |g_n|^2 dx dt,$$

$$(6.4.6)\qquad \int_{Q_R} |D'^2 w_{n+1}|^2 r^\alpha \zeta dxdt \le B^2_{\alpha,m} \int_{Q_R} |g_n|^2 r^\alpha \zeta dxdt + C\int_Q |g_n|^2 dxdt,$$

$$(6.4.7)\qquad \frac{\varepsilon^2}{\nu^2}\int_{Q_R} |\, \dot w_{n+1}\,|^2 r^\alpha \zeta dxdt \le C^2_{\alpha,m} \int_{Q_R} |g_n|^2 r^\alpha \zeta dxdt + C\int_Q |g_n|^2 dxdt,$$

$$(6.4.8)\qquad \int_{Q_R} |D' w_{n+1}|^2 r^\alpha \zeta dxdt \le \eta \int_{Q_R} |g_n|^2 r^\alpha \zeta dxdt + C\int_Q |g_n|^2 dxdt,$$

where

$$g_n = (1-\varepsilon)\,\Delta\, w_n + \frac{\varepsilon}{\nu}(w_n^{(k)} D_k u_{n-1} + u_n^{(k)} D_k w_n).$$

Summing up (6.4.4)-(6.4.8) we get

$$(6.4.9)\qquad \int_{Q_R} \left(\frac{\varepsilon^2}{\nu^2}|\, \dot w_{n+1}\,|^2 + |D' w_{n+1}|^2 + |D'^2 w_{n+1}|^2\right) r^\alpha \zeta dxdt +$$

$$+\int_Q \left(|D' w_{n+1}|^2 + |D'^2 w_{n+1}|^2\right) dxdt \le$$

$$\le E^2_{\alpha m}\int_{Q_R} |g_n|^2 r^\alpha \zeta dxdt + C_1 \int_Q |g_n|^2 dxdt.$$

Estimate the integrals on the right-hand side of this inequality, we assume the conditions of theorem 6.3.1 are satisfied. Then we obtain

$$(6.4.10)\qquad \sup_{x_0\in\Omega} \int_{Q_R} \left(|\, \dot u_n\,|^2 + |D'^2 u_n|^2\right) r^\alpha \zeta dxdt \le$$

$$\le \frac{\nu^2}{\varepsilon^2}\int_{Q_R} \left(\frac{\varepsilon^2}{\nu^2}|\, \dot u_n\,|^2 + |D'^2 u_n|^2\right) r^\alpha \zeta dxdt \le \frac{\nu^2}{\varepsilon^2} z^*,$$

$$(6.4.11)\qquad \sup_{x_0\in\Omega} \int_{Q_R} |D' u_n|^2 r^\alpha \zeta dxdt \le z^*,$$

$$(6.4.12)\qquad \int_{Q_R} |D' u_n|^2 dxdt \le z^*$$

where z^* is one of the roots of (6.3.58). Then from (6.3.12),(6.4.10),(6.4.12) we get

$$(6.4.13)\qquad \sup_Q |u_n|^2 \le \max\left\{\frac{\nu^2}{\varepsilon^2}\eta, C_* \eta^{\frac{m+1}{1-m-\alpha}}\right\} z^*,$$

where C_* is a constant from the inequality (6.3.12). Similarly we get the relations

$$(6.4.14)\qquad \sup_Q |w_n|^2 \le \eta \sup_{x_0\in\Omega} \int_{Q_R} \left(|\, \dot w_n\,|^2 + |D'^2 w_n|^2\right) r^\alpha \zeta dxdt +$$

$$+C_* \int_Q |D' w_n|^2 dxdt \le$$

$$\le \max\left\{\frac{\nu^2}{\varepsilon^2}\eta, C_* \eta^{-\frac{m+1}{1-m-\alpha}}\right\} Y_n,$$

where

$$Y_n = \sup_{x_0\in\Omega} \int\limits_{Q_R} \left(\frac{\varepsilon^2}{\nu^2} | \dot{w}_n |^2 + |D' w_n|^2 + |D'^2 w_n|^2 \right) r^\alpha \zeta dx dt +$$

$$+ \int\limits_{Q} \left(|D' w_n|^2 + |D'^2 w_n|^2 \right) dx dt.$$

It is obvious

$$\int\limits_{Q} |D' w_n|^2 dx dt \le Y_n, \tag{6.4.15}$$

$$\int\limits_{Q} |D'^2 w_n|^2 dx dt \le Y_n, \tag{6.4.16}$$

$$\int\limits_{Q_R} |D' w_n|^2 r^\alpha \zeta dx dt \le Y_n \tag{6.4.17}$$

and

$$\int\limits_{Q_R} |D'^2 w_n|^2 r^\alpha \zeta dx dt \le Y_n. \tag{6.4.18}$$

The inequalities (6.4.10)-(6.4.18) allow us to get the necessary estimates for the integrals of the right-hand side of (6.4.9). In fact

$$\int\limits_{Q} |g_n|^2 dx dt \le 3(1-\varepsilon)^2 \int\limits_{Q} | \triangle w_n|^2 dx dt + \tag{6.4.19}$$

$$+3\frac{\varepsilon^2}{\nu^2} \left[\int\limits_{Q} |w_n^{(k)} D_k u_{n-1}|^2 dx dt + \int\limits_{Q} |u_n^{(k)} D_k w_n|^2 dx dt \right] \le$$

$$\le 3m(1-\varepsilon)^2 Y_n + 3\frac{\varepsilon^2}{\nu^2} \left[\sup_Q |w_n|^2 \int\limits_{Q} |D' u_{n-1}|^2 dx dt + \sup_Q |u_n|^2 \int\limits_{Q} |D' w_n|^2 dx dt \right] \le$$

$$\le 3m(1-\varepsilon)^2 Y_n + 6\frac{\varepsilon^2}{\nu^2} \max\left\{ \frac{\nu^2}{\varepsilon^2}\eta, C_* \eta^{-\frac{m+1}{1-m-\alpha}} \right\} z^* Y_n.$$

Here we apply the relations (6.3.22),(6.4.12)-(6.4.16). Taking into account the inequalities (6.3.22),(6.4.11),(6.4.13),(6.4.14),(6.4.17) and (6.4.18) we get the following estimate

$$\int\limits_{Q_R} |g_n|^2 r^\alpha \zeta dx dt \le 3(1-\varepsilon)^2 \int\limits_{Q_R} | \triangle w_n|^2 r^\alpha \zeta dx dt + \tag{6.4.20}$$

$$+3\frac{\varepsilon^2}{\nu^2} \left[\sup_Q |w_n|^2 \int\limits_{Q_R} |D' u_{n-1}|^2 r^\alpha \zeta dx dt + \sup_Q |u_n|^2 \int\limits_{Q_R} |D' w_n|^2 r^\alpha \zeta dx dt \right] \le$$

$$\le 3 \left[m(1-\varepsilon)^2 + 2\frac{\varepsilon^2}{\nu^2} \max\left\{ \frac{\nu^2}{\varepsilon^2}\eta, C_* \eta^{-\frac{m+1}{1-m-\alpha}} \right\} z^* \right] Y_n.$$

Now from the inequalities (6.4.9),(6.4.19) and (6.4.20) follows the estimate

$$\int_{Q_R} \left(\frac{\varepsilon^2}{\nu^2} | \dot{w}_{n+1} |^2 + |D' w_{n+1}|^2 + |D'^2 w_{n+1}|^2 \right) r^\alpha \zeta dx dt +$$
$$+ \int_Q \left(|D' w_{n+1}|^2 + |D'^2 w_{n+1}|^2 \right) dx dt \le \overline{q} Y_n,$$

where

$$\overline{q} = 3 \max\{E^2_{\alpha,m}, C_1\} \left[m(1-\varepsilon)^2 + 2 \max\left\{ \frac{\nu^2}{\varepsilon^2} \eta, C_* \eta^{-\frac{m+1}{1-m-\alpha}} \right\} z^* \right]$$

and C_1 is a constant from (6.4.9). If we take the sup of the right-hand side with respect to $x_0 \in \Omega$, we get

(6.4.21) $$Y_{n+1} \le \overline{q} Y_n.$$

Minimizing the function $\phi(\eta) = \max\{\eta, C \frac{\varepsilon^2}{\nu^2} \eta^{-\frac{m+1}{1-m-\alpha}}\}$ for $\alpha = 1 - m - 2\gamma$ we get

$$\min \phi(\eta) = C_2 \left(\frac{\varepsilon}{\nu} \right)^{\frac{4\gamma}{m+1+2\gamma}}.$$

Take $\varepsilon = 1$ and denote

(6.4.22) $$K = 6 C_2 \max\{E^2_{\alpha,m}, C_1\}.$$

Since $Y_n \ge 0$ from the condition $\overline{q} < 1$ and (3.21) it follows that

$$Y_n \le \overline{q}^n Y_0$$

which means the sequence converges to zero in $H^{1,2}_{2,\alpha}(Q)$ as a geometric progression. From the completeness of $H^{1,2}_{2,\alpha}(Q)$ we obtain the convergence in $H^{1,2}_{2,\alpha}(Q)$ of $u_n(t,x)$. From the embedding of $H^{1,2}_{2,\alpha}(Q)$ in $C^{0,\delta}(Q)$ we get the convergence in $C(Q)$ and $C^{0,\gamma}(Q)$. It is clear the limit $u^*(t,x)$ belongs both to $W^{1,2}_2(Q)$ and $C(Q)$. Hence we can pass to the limit in the integral identity

$$\int_Q (-u_{n+1} \dot{v} + \nu D' u_{n+1} D' v) dx dt = \int_Q u_n^{(k)} D_k v u_n dx dt - \int_Q f v dx dt.$$

So u^* is the weak solution of the problem (6.3.1)-(6.3.3). Since $u^*(t,x) \in C^{0,\delta}(Q)$ then $u^*(t,x)$ is an unique continuous solution of the considered problem and we get

Theorem 6.4.1. *Let $\nu \ge 1$, $\int_Q |f|^2 r^\alpha dx dt < \infty$ for $\alpha = 1 - m - 2\gamma, \gamma \in (0, 1/2)$ and all $x_0 \in \overline{\Omega}$, the inequality (6.3.57) and the relation*

(6.4.23) $$K \nu^{-\frac{4\gamma}{m+1+2\gamma}} z^* < 1,$$

where K as defined by (6.4.22) is satisfied. If for the initial iteration the inequality $X_0 < z^*$ holds, then the sequence $u_n(t,x)$ converges in $H^{1,2}_{2,\alpha}(Q), W^{1,2}_2(Q), C^{0,\delta}(Q), C(Q)$ to an unique Hölder continuous solution $u^*(t,x)$ of the problem (6.3.1)-(6.3.3). For the solution $u^*(t,x)$ the inequality

$$X^* < z^*$$

is true, where

$$X^* = \sup_{x_0 \in \Omega} \int\limits_{Q_R} \left(\frac{1}{\nu^2} | \dot{u}^* |^2 + |D' u^*|^2 + |D'^2 u^*|^2 \right) r^\alpha \zeta dx dt \, -$$

$$+ \int\limits_{Q} \left(|D' u^*|^2 + |D'^2 u^*|^2 \right) dx dt.$$

References

Agmon S., Douglis A., Nirenberg L.
[1] Estimates near the boundary for solutions of elliptic partial differential equations satisfying general boundary value conditions I. - Comm.Pure and Appl. Math., 1959, v.12, p.623 -727.

Almgren F.J.
[1] Existence and regularity almost everywhere of solutions for elliptic variational problems among surfaces of varying topological type and singularity structure. - Ann. of Math., 1968, v87, 321-391.

Appel P.
[1] Sur l'equation $z_{xx} - z_y = 0$ et la théorie de la chaleur. Journal de Mathématiques Pures et Appliquées,1892, v8, 187-216.

Bernstein S.
[1] Selected works, v.3, Academy of Sciences USSR, Moscow, 1960.

Besov O.
[1] Investigations of a class of function spaces in connection with imbedding and extension theorems. (Russian) Trudy Mat. Inst. Steklov. 60, 1961, 42-81.

Besov O., I'lin V., Nikolskii S.
[1] Integral representations of functions and imbedding theorems. V.H. Winston and Sons, New-York, Toronto, London, Sydney, 1979.

Browder P.
[1] Problemes non-lineares. Les Presse de l'Université Montreal, 1966.

Caccioppoli R.
[1] Limitazioni integrali per le soluzioni di un'equazione lineare ellittica a derivate parziali. Gior. Mat. Battaglini,186-212, 1950-1951.

Caffarelli L., Kohn R. and Nirenberg L.
[1] Partial regularity of suitable weak solutions of the Navier-Stokes equations. Commun. on Pure and Appl. Math., v.35, 1982 771-831.

Campanato S.
[1] Hölder continuity of the solutions of some non-linear elliptic systems. - Adv. in Math., 1983, v.48, 16-43.

[2] On the nonlinear parabolic systems in divergence form. Hölder continuity and Partial Hölder continuity of the solutions. Annali di Matematica pura ed applicata (IV), vol CXXXVII,1984, pp. 83-122, Bologna.

Chelkak S. and Koshelev A.
[1] About regularity of solutions for the nonstationary Navier-Stokes system. Stuttgart University, Math. Institut A, preprint №94 – 4, 1994.

Chistyakov V.
[1] Convergence of an iteration process for quasilinear parabolic equations. English Translation: Soviet Math.(Iz VUZ) 28, 1984, N1 48-53.

Cordes H.O.
[1] Über die erste Randwertaufgabe bei quasilinearen Differentialgleichungen zweiter Ordnung in mehr als zwei Variablen. - Math. Ann., 1956, Bd.131, 278-312.

De Giorgi E.
[1] Sulla differenziabilitá e l'analicitá delle estremali degli integrali multipli regolari. Memorie ACC: SCI. Torino. ser. 3, 1957, 3 , 25-43.

[2] Un esempio di estremali discontinue per problemi variazionali di tipo ellittico. Boll. UMI, 1968, 4, 135-138.

Escobedo M. and Herrero M.
[1] Boundness and Blow Up for Semilinear Reaction-Diffusion System. Journal of Differential Equations, 89, 1991.

Frehse J.
[1] On the boundness of weak solutions of higher order nonlinear elliptic partial differential equations. Boll. UMI, 1970,3,607 -627.

Gajewski H.
[1] Analysis und Numerik des Ladungstransports Halbleitern, GAMM-Mitteilungen, 1993,16,35-57.

Gajewski H. and Gröger K.
[1] On the basic equations for carrier transport in semiconductors, Jnl. Math. Anal. Appl. 1986, 113, 12-35.

Gajewski H., Jäger W. and Koshelev A.
[1] About loss of regularity and "blow-up" of solutions for quasilinear parabolic systems. IAAS, Berlin, preprint N 70, 1993.

Giaquinta M.
[1] Sistemi ellittici nonlineari: Convegno sui sistemi ellittici e applicazioni. Ferrara: Quaderno C.N.R.,1978.

[2] Multiple integrals in the calculus of variations and nonlinear elliptic systems. Prin-

ceton (N.J.), Univ Press, 1983.

Giaquinta M., Giusti E.
[1] Partial regularity for the solutions to nonlinear parabolic systems. Annali di Matematica pura ed applicata (IV). vol. XXVII, 1973, 253-266, Bologna.

Giusti E.
[1] Regolaritá parziale delle soluzioni di sistemi ellittici quasi-lineari di ordine arbitrario. - Ann. della Sc.Norm. Super. di Pisa, 1969, 23, 1, 115-141.

Giusti E., Miranda M.
[1] Un esempio di soluzioni discontinue per un problema di minimo relativo ad un integrale di calcolo delle variazioni. Boll. Unione Mat. Ital., 1968, 12, 219-226.

Gluško V.
[1] About potential type operators and embedding theorems. Russian Doklady Acad. Nauk SSSR, 1959, 126:3, 467-470.

Gluško V., Krein S.
[1] Inequalities for norms of derivatives in weighted L_p spaces. (Russian) Sibirsk. Mat. Z., 1960, 343-382.

Gunter N.
[1] Theory of the potential and its application to the basic problems of mathematical physics. (Russian). Gosudarstv. Izdat. Tehn.-Teor. Lit. Moscow. 415 pp.

Hardy G., Littlewood J., Pólya G.
[1] Inequalities. University Press, Cambridge, 1967.

Hildebrandt S., Widman K.
[1] Sätze von Liouvillischen Typ für quasilinearen elliptische Gleichungen und systeme. Nachr. Akad. Wiss. Göttingen II, Math. Phys. Klasse, 1979, 4, 41-59.

I.I.Hirschman Jr.
[1] A note on the heat equation. Duke Mathematical Journal v.19, 1952 487-492.

Il'yushin A.
[1] Plasticity. Part I, GITTL, Moscow, 1948; French transl., Eyrolles, Paris, 1956.

Ivanov A.
[1] Two-Sided and One-Sided Liouville Theorems for Quasilinear Elliptic Equations of the Nondivergence Form. Journal of Soviet Mathematics, v.19, 6, 1982, 1659-1674.

Jäger W. and Luckhaus S.
[1] On explosions of solutions to a system of partial differential equations modelling chemotaxis, Transact. Math. Soc. 1992, 329, 819-824.

Kavraiskaya K.
[1] The weighted Korn inequality. English Translation: Soviet Math. (Iz. VUZ) 28, 1984, 1, 1-5.

Kondrat'ev V.
[1] Boundary problems for elliptic equations in domains with conical or angular points. Transactions of the Moscow Math. Soc. 16, 1967 209-292.

Korn A.
[1] Ueber Minimalflöchen deren Randkurven wenig von eben Kurven abweichen. Abhandlungen Preussische Akademie, 1909.

Koshelev A.
[1] Differentiability of solutions of certain problems of potential theory. Mat. Sbornik. N.S. 32(74), 653-664, 1953, Russian.

[2] A priori estimates in L_p and generalized solutions of elliptic equalities and systems. (Russian) Uspehi Mat. Nauk 13 (1958), no 4(82), 29-88. English Translation : Amer. Math. Soc. Transl. (2) 20 (1962), 105-171.

[3] Convergence of the method of successive approximations for quasilinear elliptic equations. English Translation : Amer. Math. Society Sov. Mathematics (Doklady), 3, 219-222, 1962.

[4] Certain questions on existence and approximate solution for quasilinear elliptic equations and systems in S.L. Sobolev spaces. (Russian). Sibirsk. Mat Ẑ. 9(1968), 1173-1181. English Translation: Siberian Mathematical Journal, 9, 1968, 875-882.

[5] The continuity of the solutions of systems of second order quasilinear equations with bounded nonlinearities. English Translation: Amer. Math. Society, Sov. Mathematics (Doklady) 14 N6, 1784-1787.

[6] The continuity of the solutions of a system of second order quasilinear equations with bounded nonlinearities. English Translation: Soviet Math. (Iz. VUZ) 18, 1974, 1, 47-56.

[7] The best estimates of the convergence of a certain iteration process. English translation: Soviet Math. (Iz. VUZ) 18, 1974, 5, 114-118.

[8] The smoothness of the solutions of second order quasilinear elliptic systems. English translation: Soviet Math. (Iz. VUZ) 20, 1976, N6, 38-52.

[9] The continuity of the solutions of quasilinear second order elliptic systems. English translation: Soviet Math. Dokl. 18, 1977, 1, 142-145.

[10] The continuity of the solutions of second order quasilinear elliptic systems. English translation: Differential equations 13, 1977, 7, 870-876.

[11] Regularity of the solutions of quasilinear elliptic systems. English translation: Russian Math. Surveys 33, 1978, 4, 1-52.

[12] Exact conditions for the smoothness of solutions of elliptic systems and Liouville's theorem. English translation: Soviet Math. Dokl. 26, 1982, 1, 252-254, 1983.

[13] The weighted Korn inequality and some iteration processes for quasilinear elliptic systems. English Translation: Soviet Math. Dokl. 28, 1983, 1, 209-219, 1983.

[14] The weighted Korn inequality and certain iteration processes for quasilinear elliptic systems. English Translation: Soviet Math. (Iz. VUZ) 28, 1984, N1, 6-18, 1983.

[15] Regularity of solutions for elliptic equations and systems. (Russian). Moscow, Nauka. 1986.

[16] Some Liouville Type Theorems for Quasilinear Parabolic Systems. Bonn University preprint N 15,1992.

[17] About some coercive inequalities for elementary elliptic and parabolic operators. IAAS, Berlin, preprint N 15, 1992.

[18] Regularity of solutions for some problems of mathematical physics. Ann. Inst. Henri Poincare'; Analyse non lineaire, vol.1, No4, 1995 p..

[19] Regularity of solutions for some problems of elliptic systems. Math. Nachr. 162, 1993, 59-88.

[20] About sharpness of some estimates for solutions of the Stokes system, Stuttgart University, Mathem. Institut A, preprint 95 N 6, 1995.

Koshelev A., Chelkak S.
[1] Regularity of solutions of quasilinear elliptic systems. Teubner-Texte, 77, Leipzig, 1985.

Ladyzhenskaya O.
[1] On the closure of the elliptic operator. Doklady Akad. Nauk SSSR (N.S.) 79, 723-725, 1951, Russian.

[2] The mathematical theory of viscous incompressible flow. Gordon and Breach. New York, London, Paris, 1969.

Ladyzhenskaya O., Ural'tseva N.
[1] Linear and quasilinear elliptic equations. Academic Press New York and London, 1968.

Ladyzhenskaya O.A., Ural'tseva N.N. and Solonnikov V.A.

[1] Linear and qusilinear equations of parabolic type. Amer. Math. Soc. Providence R.I., Translations of Mathematical Monographs 23, 1968.

Landis E.
[1] Second order equations of elliptic and parabolic type. Russian. Izdat Nauka, Moscow, 1971.

Leray J., Shauder J.
[1] Topologie et equations fonctionnelles. Ann. Es. N. Sup., 1934, 51, 45-78.

Lions J.L.
[1] Quelques methodes de resolution des problemes aux limites nonlinearies. Paris. 1969.

Mazja V.
[1] Examples of nonregular solutions of quasillinear elliptic equations with analytic coefficients. English Translation : Functional Anal. Appl. 2 (1968), pp. 230-234.

Mihlin S.
[1] On some estimates connected with Green's functions. Doklady Akad. Nauk SSSR (N.S.) 78, 443-446, 1951. Russian

[2] Multidimensional singular integrals and integral equations. Translation edited by I.N. Sneddon. Pergamon Press Oxford. New York, Paris, 1965.

Minty G.
[1] Monotone (nonlinerial) operators in Hilbert space. - Duke. Math. J., 1962, 29, 341-346.

Morrey Ch.
[1] On the solution of quasi-linear elliptic partial differential equations. - Trans. Amer.Math.Soc., 1938, 43, 126-166.

[2] Existence and differentiability theorems for the solutions of variational problems for multiple integrals. - Bull. Amer.Math. Soc., 1940, 40, 439-458.

[3] Multiple integral problems in the calculus of variations and related topics. - Univ. of Calif., Publ. in Math., new ser., 1943,1, 1-130.

[4] Multiple integrals in the calculus of variations. - N.Y., 1966.

Moser J.
[1] A new proof of De Giorgi's theorem concerning the regularity problems for elliptic differential equations. -Comm. Pure and Appl. Math., 1960, 13, 3. 457-468.

Mosolov P. and Myasnikov V.
[1] The proof of Korn's inequality, Dokl. Akad. Nauk SSSR 201 (1971) 1, 36-39;

English translation in Soviet Math. Dokl. 12(1971).

Nash T.
[1] Continuity of solutions of parabolic and elliptic equations. - Amer. Journ Math., 1958, 80., 4, 931-954.

Necas J.
[1] On the regularity of weak solutions to nonlinear elliptic systems of partial differential equations. Lectures Scuola Normale Superiore. Pisa, 1979.

[2] Introduction to the theory of nonlinear elliptic equations. Leipzig, 1983. Serie Teubner-Texte zur Mathematik, Band 52.

Neĉas J., Ŝvera'k V.
[1] On regularity of solutions of nonlinear parabolic systems. Ann Scuola Norm. Sup. Pisa, Cl.Sci.(4) 18 (1991) 1, 1-11.

Neĉas J. and Oleinik O.
[1] Liouville theorems for elliptic systems. Soviet Math. Dokl 21, 1980, N3.

Oleinik O.
[1] On Hilbert nineteenth problem. Hilbert's problems (Russian) 216-219. Izdat. "Nauka", Moscow, 1969.

Peletier L., Serrin J.
[1] Gradient bounds and Liouville theorems for quasilinear elliptic equations. Ann. Scuola Normale Superiori Pisa, ser.4,5, 1978, N1, pp65-104.

Petrovsky I.G.
[1] Sur l'ananycite des solutions des systemes d'equations differentielles. -Math.Sbornik, 1939, 5(47), 3-70.

Schauder J.
[1] Über lineare elliptische Differentialgleichungen zweiter Ordnung. -Math. Zeitschrift, 1934, 38, 257-382.

Serrin J.
[1] Liouville theorems and gradient bounds for quasi-linear elliptic systems.-Arch. Rat. Mech. Anal., 1977, 66, 295-310.

Solonnikov V.
[1] The differential properties of weak solutions of quasilinear elliptic equations (Russian). Zap. Nauĉn. Sem. Leningr. Otdel. Mat. Inst. Steklov (LOMI), 39, 1974, 110-119.

[2] Estimates for solutions of a nonstationary linearized system of Navier-Stokes equations, Amer. Math. Soc. Transl. 75, 1-116,1968.

[3] Estimates for solutions of nonstationary Navier-Stokes equations, J. Soviet Math.8, 467-529,1977.

Stara J., John O., Danécék J.
[1] Liouville type condition and the interior regularity of quasilinear parabolic systems. Comment. Math. Univ. Carolin. 28, nol, 1987.

Stara J., John O., Mali J.
[1] Contrexample of regularity of weak solution of the quasilinear parabolic system. - Comm. Math. Univ. Carolina, 27, N1, 1986,123-136.

Stein E.
[1] Singular integrals and differentiabilty properties on functions. Princeton University Press. New Jersey, 1970.

Temam R.
[1] Navier-Stokes equations, theory and numerical analysis, 3 rd.rev.ed. North. Holland, Amsterdam, 1984.

Vishik M.
[1] Quasi-linear strongly elliptic systems of differential equations in divergence form. English translation: Transactions of the Moscow Math. Society, v.12, 1963, pp. 125-184.

Vorovich I. and Krasovskij J.
[1] About the method of elastic solutions. English Translation: Soviet Phys.Dokl. 4, 1959/60.

von Wahl W.
[1] The equations of Navier-Stokes and Abstract Parabolic Equations, 1985 Friedr. Vieweg and Sohn. Braunschweig / Wiesbaden.

Widman K.
[1] Hölder continuity of solutions of elliptic systems. -Manuscr.math.,1971,5,4, 299-308.

[2] Local bound of higher order nonlinear elliptic partial differential equations. - Mathematische Zeitschrift, 1971, 21, 81-95.

Vol. 1518: H. Stichtenoth, M. A. Tsfasman (Eds.), Coding Theory and Algebraic Geometry. Proceedings, 1991. VIII, 223 pages. 1992.

Vol. 1519: M. W. Short, The Primitive Soluble Permutation Groups of Degree less than 256. IX, 145 pages. 1992.

Vol. 1520: Yu. G. Borisovich, Yu. E. Gliklikh (Eds.), Global Analysis – Studies and Applications V. VII, 284 pages. 1992.

Vol. 1521: S. Busenberg, B. Forte, H. K. Kuiken, Mathematical Modelling of Industrial Process. Bari, 1990. Editors: V. Capasso, A. Fasano. VII, 162 pages. 1992.

Vol. 1522: J.-M. Delort, F. B. I. Transformation. VII, 101 pages. 1992.

Vol. 1523: W. Xue, Rings with Morita Duality. X, 168 pages. 1992.

Vol. 1524: M. Coste, L. Mahé, M.-F. Roy (Eds.), Real Algebraic Geometry. Proceedings, 1991. VIII, 418 pages. 1992.

Vol. 1525: C. Casacuberta, M. Castellet (Eds.), Mathematical Research Today and Tomorrow. VII, 112 pages. 1992.

Vol. 1526: J. Azéma, P. A. Meyer, M. Yor (Eds.), Séminaire de Probabilités XXVI. X, 633 pages. 1992.

Vol. 1527: M. I. Freidlin, J.-F. Le Gall, Ecole d'Eté de Probabilités de Saint-Flour XX – 1990. Editor: P. L. Hennequin. VIII, 244 pages. 1992.

Vol. 1528: G. Isac, Complementarity Problems. VI, 297 pages. 1992.

Vol. 1529: J. van Neerven, The Adjoint of a Semigroup of Linear Operators. X, 195 pages. 1992.

Vol. 1530: J. G. Heywood, K. Masuda, R. Rautmann, S. A. Solonnikov (Eds.), The Navier-Stokes Equations II – Theory and Numerical Methods. IX, 322 pages. 1992.

Vol. 1531: M. Stoer, Design of Survivable Networks. IV, 206 pages. 1992.

Vol. 1532: J. F. Colombeau, Multiplication of Distributions. X, 184 pages. 1992.

Vol. 1533: P. Jipsen, H. Rose, Varieties of Lattices. X, 162 pages. 1992.

Vol. 1534: C. Greither, Cyclic Galois Extensions of Commutative Rings. X, 145 pages. 1992.

Vol. 1535: A. B. Evans, Orthomorphism Graphs of Groups. VIII, 114 pages. 1992.

Vol. 1536: M. K. Kwong, A. Zettl, Norm Inequalities for Derivatives and Differences. VII, 150 pages. 1992.

Vol. 1537: P. Fitzpatrick, M. Martelli, J. Mawhin, R. Nussbaum, Topological Methods for Ordinary Differential Equations. Montecatini Terme, 1991. Editors: M. Furi, P. Zecca. VII, 218 pages. 1993.

Vol. 1538: P.-A. Meyer, Quantum Probability for Probabilists. X, 287 pages. 1993.

Vol. 1539: M. Coornaert, A. Papadopoulos, Symbolic Dynamics and Hyperbolic Groups. VIII, 138 pages. 1993.

Vol. 1540: H. Komatsu (Ed.), Functional Analysis and Related Topics, 1991. Proceedings. XXI, 413 pages. 1993.

Vol. 1541: D. A. Dawson, B. Maisonneuve, J. Spencer, Ecole d' Eté de Probabilités de Saint-Flour XXI - 1991. Editor: P. L. Hennequin. VIII, 356 pages. 1993.

Vol. 1542: J.Fröhlich, Th.Kerler, Quantum Groups, Quantum Categories and Quantum Field Theory. VII, 431 pages. 1993.

Vol. 1543: A. L. Dontchev, T. Zolezzi, Well-Posed Optimization Problems. XII, 421 pages. 1993.

Vol. 1544: M.Schürmann, White Noise on Bialgebras. VII, 146 pages. 1993.

Vol. 1545: J. Morgan, K. O'Grady, Differential Topology of Complex Surfaces. VIII, 224 pages. 1993.

Vol. 1546: V. V. Kalashnikov, V. M. Zolotarev (Eds.), Stability Problems for Stochastic Models. Proceedings, 1991. VIII, 229 pages. 1993.

Vol. 1547: P. Harmand, D. Werner, W. Werner, M-ideals in Banach Spaces and Banach Algebras. VIII, 387 pages. 1993.

Vol. 1548: T. Urabe, Dynkin Graphs and Quadrilateral Singularities. VI, 233 pages. 1993.

Vol. 1549: G. Vainikko, Multidimensional Weakly Singular Integral Equations. XI, 159 pages. 1993.

Vol. 1550: A. A. Gonchar, E. B. Saff (Eds.), Methods of Approximation Theory in Complex Analysis and Mathematical Physics IV, 222 pages, 1993.

Vol. 1551: L. Arkeryd, P. L. Lions, P.A. Markowich, S.R. S. Varadhan. Nonequilibrium Problems in Many-Particle Systems. Montecatini, 1992. Editors: C. Cercignani, M. Pulvirenti. VII, 158 pages 1993.

Vol. 1552: J. Hilgert, K.-H. Neeb, Lie Semigroups and their Applications. XII, 315 pages. 1993.

Vol. 1553: J.-L- Colliot-Thélène, J. Kato, P. Vojta. Arithmetic Algebraic Geometry. Trento, 1991. Editor: E. Ballico. VII, 223 pages. 1993.

Vol. 1554: A. K. Lenstra, H. W. Lenstra, Jr. (Eds.), The Development of the Number Field Sieve. VIII, 131 pages. 1993.

Vol. 1555: O. Liess, Conical Refraction and Higher Microlocalization. X, 389 pages. 1993.

Vol. 1556: S. B. Kuksin, Nearly Integrable Infinite-Dimensional Hamiltonian Systems. XXVII, 101 pages. 1993.

Vol. 1557: J. Azéma, P. A. Meyer, M. Yor (Eds.), Séminaire de Probabilités XXVII. VI, 327 pages. 1993.

Vol. 1558: T. J. Bridges, J. E. Furter, Singularity Theory and Equivariant Symplectic Maps. VI, 226 pages. 1993.

Vol. 1559: V. G. Sprindžuk, Classical Diophantine Equations. XII, 228 pages. 1993.

Vol. 1560: T. Bartsch, Topological Methods for Variational Problems with Symmetries. X, 152 pages. 1993.

Vol. 1561: I. S. Molchanov, Limit Theorems for Unions of Random Closed Sets. X, 157 pages. 1993.

Vol. 1562: G. Harder, Eisensteinkohomologie und die Konstruktion gemischter Motive. XX, 184 pages. 1993.

Vol. 1563: E. Fabes, M. Fukushima, L. Gross, C. Kenig, M. Röckner, D. W. Stroock, Dirichlet Forms. Varenna, 1992. Editors: G. Dell'Antonio, U. Mosco. VII, 245 pages. 1993.

Vol. 1564: J. Jorgenson, S. Lang, Basic Analysis of Regularized Series and Products. IX, 122 pages. 1993.

Vol. 1565: L. Boutet de Monvel, C. De Concini, C. Procesi, P. Schapira, M. Vergne. D-modules, Representation Theory, and Quantum Groups. Venezia, 1992. Editors: G. Zampieri, A. D'Agnolo. VII, 217 pages. 1993.

Vol. 1566: B. Edixhoven, J.-H. Evertse (Eds.), Diophantine Approximation and Abelian Varieties. XIII, 127 pages. 1993.

Vol. 1567: R. L. Dobrushin, S. Kusuoka, Statistical Mechanics and Fractals. VII, 98 pages. 1993.

Vol. 1568: F. Weisz, Martingale Hardy Spaces and their Application in Fourier Analysis. VIII, 217 pages. 1994.

Vol. 1569: V. Totik, Weighted Approximation with Varying Weight. VI, 117 pages. 1994.

Vol. 1570: R. deLaubenfels, Existence Families, Functional Calculi and Evolution Equations. XV, 234 pages. 1994.

Vol. 1571: S. Yu. Pilyugin, The Space of Dynamical Systems with the C^0-Topology. X, 188 pages. 1994.

Vol. 1572: L. Göttsche, Hilbert Schemes of Zero-Dimensional Subschemes of Smooth Varieties. IX, 196 pages. 1994.

Vol. 1573: V. P. Havin, N. K. Nikolski (Eds.), Linear and Complex Analysis – Problem Book 3 – Part I. XXII, 489 pages. 1994.

Vol. 1574: V. P. Havin, N. K. Nikolski (Eds.), Linear and Complex Analysis – Problem Book 3 – Part II. XXII, 507 pages. 1994.

Vol. 1575: M. Mitrea, Clifford Wavelets, Singular Integrals, and Hardy Spaces. XI, 116 pages. 1994.

Vol. 1576: K. Kitahara, Spaces of Approximating Functions with Haar-Like Conditions. X, 110 pages. 1994.

Vol. 1577: N. Obata, White Noise Calculus and Fock Space. X, 183 pages. 1994.

Vol. 1578: J. Bernstein, V. Lunts, Equivariant Sheaves and Functors. V, 139 pages. 1994.

Vol. 1579: N. Kazamaki, Continuous Exponential Martingales and *BMO*. VII, 91 pages. 1994.

Vol. 1580: M. Milman, Extrapolation and Optimal Decompositions with Applications to Analysis. XI, 161 pages. 1994.

Vol. 1581: D. Bakry, R. D. Gill, S. A. Molchanov, Lectures on Probability Theory. Editor: P. Bernard. VIII, 420 pages. 1994.

Vol. 1582: W. Balser, From Divergent Power Series to Analytic Functions. X, 108 pages. 1994.

Vol. 1583: J. Azéma, P. A. Meyer, M. Yor (Eds.), Séminaire de Probabilités XXVIII. VI, 334 pages. 1994.

Vol. 1584: M. Brokate, N. Kenmochi, I. Müller, J. F. Rodriguez, C. Verdi, Phase Transitions and Hysteresis. Montecatini Terme, 1993. Editor: A. Visintin. VII. 291 pages. 1994.

Vol. 1585: G. Frey (Ed.), On Artin's Conjecture for Odd 2-dimensional Representations. VIII, 148 pages. 1994.

Vol. 1586: R. Nillsen, Difference Spaces and Invariant Linear Forms. XII, 186 pages. 1994.

Vol. 1587: N. Xi, Representations of Affine Hecke Algebras. VIII, 137 pages. 1994.

Vol. 1588: C. Scheiderer, Real and Étale Cohomology. XXIV, 273 pages. 1994.

Vol. 1589: J. Bellissard, M. Degli Esposti, G. Forni, S. Graffi, S. Isola, J. N. Mather, Transition to Chaos in Classical and Quantum Mechanics. Montecatini Terme, 1991. Editor: S. Graffi. VII, 192 pages. 1994.

Vol. 1590: P. M. Soardi, Potential Theory on Infinite Networks. VIII, 187 pages. 1994.

Vol. 1591: M. Abate, G. Patrizio, Finsler Metrics – A Global Approach. IX, 180 pages. 1994.

Vol. 1592: K. W. Breitung, Asymptotic Approximations for Probability Integrals. IX, 146 pages. 1994.

Vol. 1593: J. Jorgenson & S. Lang, D. Goldfeld, Explicit Formulas for Regularized Products and Series. VIII, 154 pages. 1994.

Vol. 1594: M. Green, J. Murre, C. Voisin, Algebraic Cycles and Hodge Theory. Torino, 1993. Editors: A. Albano, F. Bardelli. VII, 275 pages. 1994.

Vol. 1595: R.D.M. Accola, Topics in the Theory of Riemann Surfaces. IX, 105 pages. 1994.

Vol. 1596: L. Heindorf, L. B. Shapiro, Nearly Projective Boolean Algebras. X, 202 pages. 1994.

Vol. 1597: B. Herzog, Kodaira-Spencer Maps in Local Algebra. XVII, 176 pages. 1994.

Vol. 1598: J. Berndt, F. Tricerri, L. Vanhecke, Generalized Heisenberg Groups and Damek-Ricci Harmonic Spaces. VIII, 125 pages. 1995.

Vol. 1599: K. Johannson, Topology and Combinatorics of 3-Manifolds. XVIII, 446 pages. 1995.

Vol. 1600: W. Narkiewicz, Polynomial Mappings. VII, 130 pages. 1995.

Vol. 1601: A. Pott, Finite Geometry and Character Theory. VII, 181 pages. 1995.

Vol. 1602: J. Winkelmann, The Classification of Three-dimensional Homogeneous Complex Manifolds. XI, 230 pages. 1995.

Vol. 1603: V. Ene, Real Functions – Current Topics. XIII, 310 pages. 1995.

Vol. 1604: A. Huber, Mixed Motives and their Realization in Derived Categories. XV, 207 pages. 1995.

Vol. 1605: L. B. Wahlbin, Superconvergence in Galerkin Finite Element Methods. XI, 166 pages. 1995.

Vol. 1606: P.-D. Liu, M. Qian, Smooth Ergodic Theory of Random Dynamical Systems. XI, 221 pages. 1995.

Vol. 1607: G. Schwarz, Hodge Decomposition – A Method for Solving Boundary Value Problems. VII, 155 pages. 1995.

Vol. 1608: P. Biane, R. Durrett, Lectures on Probability Theory. VII, 210 pages. 1995.

Vol. 1609: L. Arnold, C. Jones, K. Mischaikow, G. Raugel, Dynamical Systems. Montecatini Terme, 1994. Editor: R. Johnson. VIII, 329 pages. 1995.

Vol. 1610: A. S. Üstünel, An Introduction to Analysis on Wiener Space. X, 95 pages. 1995.

Vol. 1611: N. Knarr, Translation Planes. VI, 112 pages. 1995.

Vol. 1612: W. Kühnel, Tight Polyhedral Submanifolds and Tight Triangulations. VII, 122 pages. 1995.

Vol. 1613: J. Azéma, M. Emery, P. A. Meyer, M. Yor (Eds.), Séminaire de Probabilités XXIX. VI, 326 pages. 1995.

Vol. 1614: A. Koshelev, Regularity Problem for Quasilinear Elliptic and Parabolic Systems. XXI, 255 pages. 1995.

Vol. 1526: J. Azéma, P. A. Meyer, M. Yor (Eds.), Séminaire de Probabilités XXVI. X, 633 pages. 1992.

Vol. 1527: M. I. Freidlin, J.-F. Le Gall, Ecole d'Eté de Probabilités de Saint-Flour XX – 1990. Editor: P. L. Hennequin. VIII, 244 pages. 1992.

Vol. 1528: G. Isac, Complementarity Problems. VI, 297 pages. 1992.

Vol. 1529: J. van Neerven, The Adjoint of a Semigroup of Linear Operators. X, 195 pages. 1992.

Vol. 1530: J. G. Heywood, K. Masuda, R. Rautmann, S. A. Solonnikov (Eds.), The Navier-Stokes Equations II – Theory and Numerical Methods. IX, 322 pages. 1992.

Vol. 1531: M. Stoer, Design of Survivable Networks. IV, 206 pages. 1992.

Vol. 1532: J. F. Colombeau, Multiplication of Distributions. X, 184 pages. 1992.

Vol. 1533: P. Jipsen, H. Rose, Varieties of Lattices. X, 162 pages. 1992.

Vol. 1534: C. Greither, Cyclic Galois Extensions of Commutative Rings. X, 145 pages. 1992.

Vol. 1535: A. B. Evans, Orthomorphism Graphs of Groups. VIII, 114 pages. 1992.

Vol. 1536: M. K. Kwong, A. Zettl, Norm Inequalities for Derivatives and Differences. VII, 150 pages. 1992.

Vol. 1537: P. Fitzpatrick, M. Martelli, J. Mawhin, R. Nussbaum, Topological Methods for Ordinary Differential Equations. Montecatini Terme, 1991. Editors: M. Furi, P. Zecca. VII, 218 pages. 1993.

Vol. 1538: P.-A. Meyer, Quantum Probability for Probabilists. X, 287 pages. 1993.

Vol. 1539: M. Coornaert, A. Papadopoulos, Symbolic Dynamics and Hyperbolic Groups. VIII, 138 pages. 1993.

Vol. 1540: H. Komatsu (Ed.), Functional Analysis and Related Topics, 1991. Proceedings. XXI, 413 pages. 1993.

Vol. 1541: D. A. Dawson, B. Maisonneuve, J. Spencer, Ecole d' Eté de Probabilités de Saint-Flour XXI - 1991. Editor: P. L. Hennequin. VIII, 356 pages. 1993.

Vol. 1542: J.Fröhlich, Th.Kerler, Quantum Groups, Quantum Categories and Quantum Field Theory. VII, 431 pages. 1993.

Vol. 1543: A. L. Dontchev, T. Zolezzi, Well-Posed Optimization Problems. XII, 421 pages. 1993.

Vol. 1544: M.Schürmann, White Noise on Bialgebras. VII, 146 pages. 1993.

Vol. 1545: J. Morgan, K. O'Grady, Differential Topology of Complex Surfaces. VIII, 224 pages. 1993.

Vol. 1546: V. V. Kalashnikov, V. M. Zolotarev (Eds.), Stability Problems for Stochastic Models. Proceedings, 1991. VIII, 229 pages. 1993.

Vol. 1547: P. Harmand, D. Werner, W. Werner, M-ideals in Banach Spaces and Banach Algebras. VIII, 387 pages. 1993.

Vol. 1548: T. Urabe, Dynkin Graphs and Quadrilateral Singularities. VI, 233 pages. 1993.

Vol. 1549: G. Vainikko, Multidimensional Weakly Singular Integral Equations. XI, 159 pages. 1993.

Vol. 1550: A. A. Gonchar, E. B. Saff (Eds.), Methods of Approximation Theory in Complex Analysis and Mathematical Physics IV, 222 pages, 1993.

Vol. 1551: L. Arkeryd, P. L. Lions, P.A. Markowich, S.R. S. Varadhan. Nonequilibrium Problems in Many-Particle Systems. Montecatini, 1992. Editors: C. Cercignani, M. Pulvirenti. VII, 158 pages 1993.

Vol. 1552: J. Hilgert, K.-H. Neeb, Lie Semigroups and their Applications. XII, 315 pages. 1993.

Vol. 1553: J.-L- Colliot-Thélène, J. Kato, P. Vojta. Arithmetic Algebraic Geometry. Trento, 1991. Editor: E. Ballico. VII, 223 pages. 1993.

Vol. 1554: A. K. Lenstra, H. W. Lenstra, Jr. (Eds.), The Development of the Number Field Sieve. VIII, 131 pages. 1993.

Vol. 1555: O. Liess, Conical Refraction and Higher Microlocalization. X, 389 pages. 1993.

Vol. 1556: S. B. Kuksin, Nearly Integrable Infinite-Dimensional Hamiltonian Systems. XXVII, 101 pages. 1993.

Vol. 1557: J. Azéma, P. A. Meyer, M. Yor (Eds.), Séminaire de Probabilités XXVII. VI, 327 pages. 1993.

Vol. 1558: T. J. Bridges, J. E. Furter, Singularity Theory and Equivariant Symplectic Maps. VI, 226 pages. 1993.

Vol. 1559: V. G. Sprindžuk, Classical Diophantine Equations. XII, 228 pages. 1993.

Vol. 1560: T. Bartsch, Topological Methods for Variational Problems with Symmetries. X, 152 pages. 1993.

Vol. 1561: I. S. Molchanov, Limit Theorems for Unions of Random Closed Sets. X, 157 pages. 1993.

Vol. 1562: G. Harder, Eisensteinkohomologie und die Konstruktion gemischter Motive. XX, 184 pages. 1993.

Vol. 1563: E. Fabes, M. Fukushima, L. Gross, C. Kenig, M. Röckner, D. W. Stroock, Dirichlet Forms. Varenna, 1992. Editors: G. Dell'Antonio, U. Mosco. VII, 245 pages. 1993.

Vol. 1564: J. Jorgenson, S. Lang, Basic Analysis of Regularized Series and Products. IX, 122 pages. 1993.

Vol. 1565: L. Boutet de Monvel, C. De Concini, C. Procesi, P. Schapira, M. Vergne. D-modules, Representation Theory, and Quantum Groups. Venezia, 1992. Editors: G. Zampieri, A. D'Agnolo. VII. 217 pages. 1993.

Vol. 1566: B. Edixhoven, J.-H. Evertse (Eds.), Diophantine Approximation and Abelian Varieties. XIII, 127 pages. 1993.

Vol. 1567: R. L. Dobrushin, S. Kusuoka, Statistical Mechanics and Fractals. VII, 98 pages. 1993.

Vol. 1568: F. Weisz, Martingale Hardy Spaces and their Application in Fourier Analysis. VIII, 217 pages. 1994.

Vol. 1569: V. Totik, Weighted Approximation with Varying Weight. VI, 117 pages. 1994.

Vol. 1570: R. deLaubenfels, Existence Families, Functional Calculi and Evolution Equations. XV, 234 pages. 1994.

Vol. 1571: S. Yu. Pilyugin, The Space of Dynamical Systems with the C^0-Topology. X, 188 pages. 1994.

Vol. 1572: L. Göttsche, Hilbert Schemes of Zero-Dimensional Subschemes of Smooth Varieties. IX, 196 pages. 1994.

Vol. 1573: V. P. Havin, N. K. Nikolski (Eds.), Linear and Complex Analysis – Problem Book 3 – Part I. XXII, 489 pages. 1994.

Vol. 1574: V. P. Havin, N. K. Nikolski (Eds.), Linear and Complex Analysis – Problem Book 3 – Part II. XXII, 507 pages. 1994.

Vol. 1575: M. Mitrea, Clifford Wavelets, Singular Integrals, and Hardy Spaces. XI, 116 pages. 1994.

Vol. 1576: K. Kitahara, Spaces of Approximating Functions with Haar-Like Conditions. X, 110 pages. 1994.

Vol. 1577: N. Obata, White Noise Calculus and Fock Space. X, 183 pages. 1994.

Vol. 1578: J. Bernstein, V. Lunts, Equivariant Sheaves and Functors. V, 139 pages. 1994.

Vol. 1579: N. Kazamaki, Continuous Exponential Martingales and *BMO*. VII, 91 pages. 1994.

Vol. 1580: M. Milman, Extrapolation and Optimal Decompositions with Applications to Analysis. XI, 161 pages. 1994.

Vol. 1581: D. Bakry, R. D. Gill, S. A. Molchanov, Lectures on Probability Theory. Editor: P. Bernard. VIII, 420 pages. 1994.

Vol. 1582: W. Balser, From Divergent Power Series to Analytic Functions. X, 108 pages. 1994.

Vol. 1583: J. Azéma, P. A. Meyer, M. Yor (Eds.), Séminaire de Probabilités XXVIII. VI, 334 pages. 1994.

Vol. 1584: M. Brokate, N. Kenmochi, I. Müller, J. F. Rodriguez, C. Verdi, Phase Transitions and Hysteresis. Montecatini Terme, 1993. Editor: A. Visintin. VII, 291 pages. 1994.

Vol. 1585: G. Frey (Ed.), On Artin's Conjecture for Odd 2-dimensional Representations. VIII, 148 pages. 1994.

Vol. 1586: R. Nillsen, Difference Spaces and Invariant Linear Forms. XII, 186 pages. 1994.

Vol. 1587: N. Xi, Representations of Affine Hecke Algebras. VIII, 137 pages. 1994.

Vol. 1588: C. Scheiderer, Real and Étale Cohomology. XXIV, 273 pages. 1994.

Vol. 1589: J. Bellissard, M. Degli Esposti, G. Forni, S. Graffi, S. Isola, J. N. Mather, Transition to Chaos in Classical and Quantum Mechanics. Montecatini Terme, 1991. Editor: S. Graffi. VII, 192 pages. 1994.

Vol. 1590: P. M. Soardi, Potential Theory on Infinite Networks. VIII, 187 pages. 1994.

Vol. 1591: M. Abate, G. Patrizio, Finsler Metrics – A Global Approach. IX, 180 pages. 1994.

Vol. 1592: K. W. Breitung, Asymptotic Approximations for Probability Integrals. IX, 146 pages. 1994.

Vol. 1593: J. Jorgenson & S. Lang, D. Goldfeld, Explicit Formulas for Regularized Products and Series. VIII, 154 pages. 1994.

Vol. 1594: M. Green, J. Murre, C. Voisin, Algebraic Cycles and Hodge Theory. Torino, 1993. Editors: A. Albano, F. Bardelli. VII, 275 pages. 1994.

Vol. 1595: R.D.M. Accola, Topics in the Theory of Riemann Surfaces. IX, 105 pages. 1994.

Vol. 1596: L. Heindorf, L. B. Shapiro, Nearly Projective Boolean Algebras. X, 202 pages. 1994.

Vol. 1597: B. Herzog, Kodaira-Spencer Maps in Local Algebra. XVII, 176 pages. 1994.

Vol. 1598: J. Berndt, F. Tricerri, L. Vanhecke, Generalized Heisenberg Groups and Damek-Ricci Harmonic Spaces. VIII, 125 pages. 1995.

Vol. 1599: K. Johannson, Topology and Combinatorics of 3-Manifolds. XVIII, 446 pages. 1995.

Vol. 1600: W. Narkiewicz, Polynomial Mappings. VII, 130 pages. 1995.

Vol. 1601: A. Pott, Finite Geometry and Character Theory. VII, 181 pages. 1995.

Vol. 1602: J. Winkelmann, The Classification of Three-dimensional Homogeneous Complex Manifolds. XI, 230 pages. 1995.

Vol. 1603: V. Ene, Real Functions – Current Topics. XIII, 310 pages. 1995.

Vol. 1604: A. Huber, Mixed Motives and their Realization in Derived Categories. XV, 207 pages. 1995.

Vol. 1605: L. B. Wahlbin, Superconvergence in Galerkin Finite Element Methods. XI, 166 pages. 1995.

Vol. 1606: P.-D. Liu, M. Qian, Smooth Ergodic Theory of Random Dynamical Systems. XI, 221 pages. 1995.

Vol. 1607: G. Schwarz, Hodge Decomposition – A Method for Solving Boundary Value Problems. VII, 155 pages. 1995.

Vol. 1608: P. Biane, R. Durrett, Lectures on Probability Theory. VII, 210 pages. 1995.

Vol. 1609: L. Arnold, C. Jones, K. Mischaikow, G. Raugel, Dynamical Systems. Montecatini Terme, 1994. Editor: R. Johnson. VIII, 329 pages. 1995.

Vol. 1610: A. S. Üstünel, An Introduction to Analysis on Wiener Space. X, 95 pages. 1995.

Vol. 1611: N. Knarr, Translation Planes. VI, 112 pages. 1995.

Vol. 1612: W. Kühnel, Tight Polyhedral Submanifolds and Tight Triangulations. VII, 122 pages. 1995.

Vol. 1613: J. Azéma, M. Emery, P. A. Meyer, M. Yor (Eds.), Séminaire de Probabilités XXIX. VI, 326 pages. 1995.

Vol. 1614: A. Koshelev, Regularity Problem for Quasilinear Elliptic and Parabolic Systems. XXI, 255 pages. 1995.

Vol. 1615: D. B. Massey, Lê Cycles and Hypersurface Singularities. XI, 131 pages. 1995.